Electronic Games
for the Evil Genius

Evil Genius Series

Electronic Games
for the Evil Genius

TOM PETRUZZELLIS

New York   Chicago   San Francisco   Lisbon
London   Madrid   Mexico City   Milan   New Delhi
San Juan   Seoul   Singapore   Sydney   Toronto

Library of Congress Cataloging-in-Publication Data

Petruzzellis, Thomas.
    Electronic games for the evil genius / Tom Petruzzellis. – 1st ed.
        p.   cm.
    Includes index.
    ISBN 0-07-147556-7 (alk. paper)
    1. Electronic games.   2. Electronics–Amateurs' manuals.   I. Title.
    TK7882.G35P49   2007
    794.8—dc22   2006021280

1   2   3   4   5   6   7   8   9   0   QPD/QPD   0   1   3   2   1   0   9   8   7   6

ISBN-13: 978-0-07-147556-3
ISBN-10: 0-07-147556-7

The sponsoring editor for this book was Judy Bass, the editing supervisor was David E. Fogarty, and the production supervisor was Pamela A. Pelton. It was set in Times New Roman by Keyword Group Ltd. The art director for the cover was Anthony Landi.

Printed and bound by Quebecor/Dubuque.

This book is printed on acid-free paper.

McGraw-Hill books are available at special quantity discounts to use as premiums and sales promotions, or for use in corporate training programs. For more information, please write to the Director of Special Sales, McGraw-Hill Professional, Two Penn Plaza, New York, NY 10121-2298. Or contact your local bookstore.

Thomas Petruzzellis is an electronics engineer currently working at the geophysical laboratory at the State University of New York, Binghamton. Also an instructor at Binghamton, with 30 years' experience in electronics, he is a veteran author who has written extensively for industry publications, including *Electronics Now*, *Modern Electronics*, *QST*, *Microcomputer Journal*, and *Nuts & Volts*. Tom wrote five previous books, including an earlier volume in this series, *Electronic Sensors for the Evil Genius*. He is also the author of *Create Your Own Electronics Workshop*; *STAMP 2 Communications and Control Projects*; *Optoelectronics, Fiber Optics, and Laser Cookbook*; *Alarm, Sensor, and Security Circuit Cookbook*, all from McGraw-Hill. He lives in Vestal, New York.

Contents

Contents

Acknowledgments

I would like to thank the following companies listed below for their help in making this book possible. I also thank these same companies for keeping the joy of electronics kit building alive, at a time when electronic "hands-on" fun has almost disappeared. Many young people can now discover the fun of learning about electronics while building and enjoying fun and useful projects and gadgets. I would also like to thank senior editor Judy Bass for her help in getting this book to print as well as to all the others in the production staff at McGraw-Hill Publications, who had a part in bringing this book to print.

Chaney Electronics (CH)
Jaycar Electronics Group (JC)
Talking Electronics (TE)
Elenco Electronics (EE)
NTE Electronics
Future Kit (FK)

Electronic Games for the Evil Genius is a fun book which was created for electronics enthusiasts of all ages from the young electronics neophyte to the seasoned engineer or technician who like to get their hands into electronics circuits. *Electronic Games for the Evil Genius* is not only fun for electronics buffs but people who enjoy games as well; electronic enthusiasts who are interested in games will really love this hands-on book.

The first chapter will introduce the readers to electronic components, what they are used for and how they work in a circuit. In Chapter 2, we will move on to identifying and reading electronic circuit diagrams as well as how to read pictorial diagrams and how to install the components onto printed circuit boards. Chapter 3 will introduce the reader to proper soldering techniques and how to check their circuits for good solder joints when completed.

Next we delve into actual hands-on Electronic Games projects. Projects are arranged from least difficult to most difficult, from skill level 1 to skill level 4. Projects are presented with all diagrams, schematics and layout or pictorial diagrams where possible as well as charts and tables to help readers understand how to successfully build and test and utilize their new project. Projects range from simple transistor-driven circuits and then on to integrated circuits, and finally microprocessor game projects. The reader will become familiar with all types of electronics circuits while actually building fun games, gadgets and gag projects.

Our first project, the **Wonky Wire Game**, is a simple fun game where you guide a wire loop through a maze without touching the maze wire; if you touch the wire, a buzzer makes a noise letting you know you don't have a steady hand. This is a skill level 1 game.

The **Lie Detector** project allows you to test friends to see if they are telling a big fib or lie. You simply ask a set of questions while your friend places his finger on a touch pad. You begin with a set of simple base-line questions and then ask the most pertinent questions for some really interesting results which may embarrass you or the questioned person! This project is a skill level 1 game.

Macho Meter is a great party game or "ice-breaker" for your macho male friends to see who is the most macho of all. This is a skill level 1 game.

Mood Meter is a simple, yet fun game to check the moods of your friends or family. This is a skill level 1 game.

Reaction Timer game seeks to finds out how fast you and your friends' reflexes are; a great party game for all ages. This project is a skill level 1 game.

Coin Tosser/Decision Maker can be used to make an "executive" decision based on an electronics circuit. For all who procrastinate, this circuit will help you decide your next big decision. This is a skill level 1 project.

The **"Tingler"** is the classic gag project which you can use to give non-hazardous shocks to your friends, siblings or enemies. You can disguise the project and fool your friends into getting a "tingle." This is a skill level 1 project.

The **Electronic Strobe** project can be used for party strobe lights to liven-up your gatherings. This is a skill 1 level project.

The **Siren** sounds just like your local police and fire sirens. You can use it for fun or with a game or put it on your bike to clear the sidewalks. This is a skill level 1 project.

The **Electronic Dice Project** can be used with any type of game of chance where you need to roll the dice—from card games to board games. This is a skill level 2 game project.

Use the **Quiz Game** project with any type of "contestant" game where players are asked questions and need to see who is the fastest to respond. This game is a skill level 2 project.

With the **Wheel of Fortune** project you can have fun designing your own Wheel of Fortune game. This project provides the spinning wheel and sound "tick" as the wheel turns. The wheel will stop at a number or letter, just like the TV game. This is a skill level 2 project.

Shoot Skeet Game will surely test your hand–eye coordination? Rifle enthusiasts test their skills by shooting "skeets," or clay targets, catapulted into the sky. The "skeet" shooting game is a much less dangerous way to test your accuracy, in this skill level 2 game.

Grab the Gold Game features a bright red LED pyramid with a clear LED on the top of the pyramid. When power is applied each of the ten LEDs lights up in a sequence pattern. If you press and hold the pushbutton, just before the top clear LED is to light, this LED will stay on illuminating its bright orange glow, indicating that you made it to the top and picked up the gold. If your timing is off, one of the red LEDs will light up indicating "you lost." This is a skill level 2 game.

Press the switch on the **Stairway to Heaven** when the bi-color LED is green and the chain of LEDs will gradually light up. But press it when the bi-color LED is red and all your hard work is gone, and then you have to start all over again! This is a skill level 3 game project.

The **Cricket** is a most annoying project. The Cricket is left in a room and it will make cricket sounds when the lights go out, and it's almost impossible to find. Build this skill level 2 project and have some devious fun.

The **Blacklight** project is great for lighting up your cool posters. The blacklight can also be used for adding a funky twist to your parties. The blacklight circuit can also be used to fluoresce certain mineral specimens, to watch them glow

pretty colors in the dark! The blacklight circuit is a skill level 2 project.

With the **Electronic Organ** you can learn to play your favorite tunes and impress your friends and parents. Have fun with this skill level 3 project.

Connect the **Color Organ** to your radio or stereo and watch the color organ divide up the sounds into three channels, which drive a three-color lamp display. Build this skill level 3 project—it will be a great addition to any party.

The **8-watt Audio Amplifier** is a project designed to be used to amplify output sounds from the Electronic Theremin, the Electronic Organ, the Voice Changer, as well as the Tetris Game projects. This is a skill level 3 project.

The **IR Target Game** consists of an IR light "gun" and an IR detector "shield" which a player wears. The game operates just like the commercial laser tag game. A skill level 3 game project, the game is a challenge for all ages.

The electronic **Hourglass** is a cool project which can be used to keep track of short time periods; it works just like a real hourglass where all the lights diminish as the "sand" runs out, only for all the LEDs on the other side of the glass to begin lighting up. This is a fascinating skill level 3 project.

The **Windmill** project is a great attention getter for displays and parties, and is a skill level 3 project which is sure to mesmerize its viewers.

The **Roulette Game** simulates a real wooden roulette wheel—place your bets now! You can use this game by itself, or it can be used with other games of chance. This is a skill level 3 game project.

The **Hangman** is an electronic variation on the old word game, but it's more graphic as you can watch the man get hanged from the tree. This is a skill level 3 game project.

The **Cliff-Hanger Game** will make you feel like you had the challenge of mountain climbing. In this game, you will have to press the pushbutton at the precise moment to advance to the next level on your way up to the top of the mountain. If the control button is pressed at the wrong time, you will fall to the bottom. You will not break any bones but you will have to start all over again. This is a skill level 3 game.

In the **Alien Attack Game** aliens are coming in their warships and are dropping their bombs on you! The only way to defend yourself is by neutralizing their bombs before they reach the ground. The bombs fall at different speeds and in different directions. The Alien Attack Game will provide hours of fun and challenges for both young and old. This is a challenging skill level 3 game project.

The electronic **Starburst** creates a hypnotic effect that everyone will love. A great project for highlighting attention to displays, posters and projects. This is a skill level 4 project.

The **Voice Changer** circuit changes your voice to Donald Duck type voice for strange effects, for your home radio station or parties. This is a skill level 4 project, which is a lot of fun to demonstrate.

The **Racquetball Game** allows you to "play" against the computer opponent. The racquet ball is represented by LEDs, and you have to "hit" the ball by pressing a button. If you do, the ball changes direction, travels up the court, bounces off the wall, then returns for you to "hit" once again, but that's not all! Suddenly the speed of the ball changes and you have to play faster or slower to keep up with the game. The project includes an adjustable speed control to allow beginner to expert play. You have to keep "on your toes" though, because if you miss the ball, the game stops and you will have to start over again. This is a skill level 4 project.

With the **Electronic Theremin** you can make strange electronic music and sound effects. This is a skill level 4 project for music and/or sound recording buffs.

The **Multi-Chip Programmer** project is used with the three microprocessor game projects. Once built, the programmer is used to "burn" a program into a microprocessor for each game. This is a skill level 4 project.

The electronic **Simon Memory Game** is a simple-to-play, but hard-to-master game. This project requires loading of a program into a microprocessor, and is a skill level 4 project.

The **Tetris** game is a variation on the classic game, a microprocessor-controlled game that you play on your television. This project requires loading a program into a microprocessor and is a skill level 4 project.

Tic-Tac-Toe is a compact electronic version of the classic game. You play against the computer, for hours of fun. This project requires loading a program into the microprocessor and is a skill level 4 project.

A number of the projects in this book are available in kit form from the respective companies listed in the Appendix.

Chapter 1

Identifying Electronic Components

Electronic circuits comprise electronic components such as resistors and capacitors, diodes, semiconductors, and LEDs, etc. Each component has a specific purpose that it accomplishes in a particular circuit. In order to understand and construct electronic circuits it is necessary to be familiar with the different types of components and how they are used. You should also know how to read resistor and capacitor color codes, and recognize physical components and their representative diagrams and pin-outs. You will also want to know the difference between a schematic and a pictorial diagram. First, we will discuss the actual components and their functions and then move on to reading schematics, then we will help you learn how to insert the components into the circuit board. In the next chapter we will discuss how to solder the components to the circuit board.

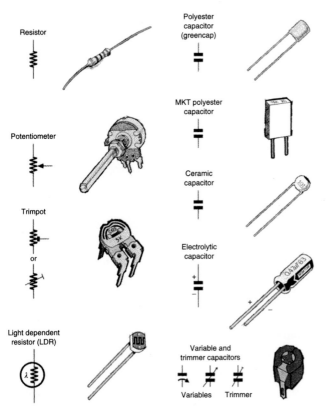

Figure 1-1 *Electronic components 1*

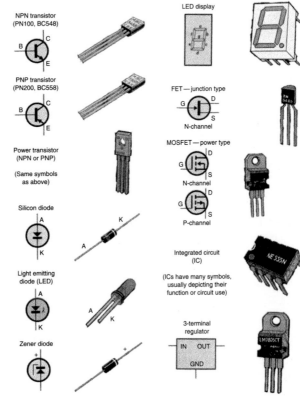

Figure 1-2 *Electronic components 2*

The diagrams shown in Figures 1-1, 1-2 and 1-3 illustrate many of the electronic components that we will be using in the projects presented in this book.

Types of Resistors

Resistors are used to regulate the amount of current flowing in a circuit. The higher the resistor's value or resistance the less current flows, and conversely a lower resistor value will permit more current to flow in a circuit. Resistors are measured in ohms (Ω) and are identified by color bands on the resistor body. The first band at one end is the resistor's first digit, the second

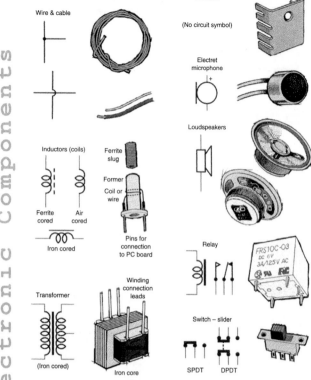

Wire & cable

Heatsink

(No circuit symbol)

Electret microphone

Loudspeakers

Inductors (coils)

Ferrite slug

Former

Coil or wire

Ferrite cored   Air cored

Iron cored

Pins for connection to PC board

Relay

FRS10C-03
DC 6V
3A/125V AC

Transformer

Winding connection leads

(Iron cored)

Iron core

Switch – slider

SPDT   DPDT

Figure 1-3 *Electronic components 3*

color band is the resistor's second digit, and the third band is the resistor's multiplier value. A fourth color band on a resistor represents the resistor's tolerance value. A silver band denotes a 10% tolerance resistor, while a gold band denotes a 5% resistor tolerance. No fourth band denotes that a resistor has a 20% tolerance. As an example, a resistor with a brown, black, and red band will represent the digit (1), the digit (0), with a multiplier value of (00), or 1000, so the resistor will have a value of 1 k or 1000 ohms. There are a number of different styles and sizes of resistor. Small resistors can be carbon, thin film, or metal. Larger resistors are made to dissipate more power and they generally have an element wound from wire.

A potentiometer or (pot) is basically a variable resistor. A pot generally has three terminals and is fitted with a rotary control shaft which varies the resistance as it is rotated. A metal wiping contact rests against a circular carbon or wire-wound resistance track. As the wiper arm moves about the circular resistance, the resistance to the output terminals changes. Potentiometers are commonly used as volume controls in amplifiers and radio receivers.

A trimpot is a special type of potentiometer which, although variable, is intended to be adjusted once or only occasionally. For this reason a control shaft is not provided, but a small slot is provided in the center of the control arm. Trimpots are generally used on printed circuit boards.

A light-dependent resistor (LDR) is a special type of resistor that varies its resistance value according to the amount of light falling on it. When it is in the dark, an LDR will typically have a very high resistance, i.e., millions of ohms. When light falls on the LDR the resistance drops to a few hundred ohms.

Types of Capacitors

Capacitors block DC current while allowing varying or AC current signals to pass. They are commonly used for coupling signals from one part of a circuit to another part of a circuit; they are also used in timing circuits. There are a number of different types of capacitor as described below.

Polyester capacitors use polyester plastic film as their insulating dielectric. Some polyester capacitors are called greencaps as they are coated with a green or brown color on the outside of the component. Their values are specified in microfarads or (µF), nanofarads (nF), or picofarads (pF) and range from 1 nF up to about 10 µF. These capacitors do not have polarity but have fixed values.

MKT capacitors are another type of capacitor, but they are rectangular or (block) in shape and are usually yellow in color. One of the major advantages of these capacitors is a more standardized lead spacing, making them more useful for PC board projects. The components can generally be substituted for polyester types.

Ceramic capacitors use a tiny disk of ceramic or porcelain material in their construction for a dielectric and they range in value from 1 pF up to 2.2 µF. Those with values above 1 nF are often made with multiple layers of metal electrodes and dielectric, to allow higher capacitance values in smaller bodies. These capacitors are usually called "multilayer monolithics" and distinguished from lower value disk ceramic types. Ceramic capacitors are often used in RF radio circuits and filter circuits.

Electrolytic capacitors use a very thin film of metal oxide as their dielectric, which allows them to provide a large amount of capacitance in a very small volume. They range in value from 100 nF up to hundreds and thousands of microfarads. They are commonly used to filter power supply circuits, coupling audio circuits, and in timing circuits. Electrolytic capacitors have polarity and must be installed with respect to these polarity markings. The capacitor will have either a white or black band denoting polarity with a plus (+) or minus (−) marking next to the color band.

Variable capacitors are used in circuits for (trimming) or adjustment, i.e, for setting a frequency. A variable capacitor has one set of fixed plates and one set of plates which can be moved by turning a knob. The dielectric between the plates is usually a thin plastic film. Most variable capacitors have low values up to a few tens of picofarads and a few hundreds of microfarads for larger variable capacitors

Diodes

A diode is a semiconductor device which can pass current in one direction only. In order for current to flow the anode (A) must be positive with respect to the cathode (K). In this condition, the diode is said to be forward biased and a voltage drop of about 0.6 volt appears across its terminals. If the anode is less than 0.6 volt positive with respect to the cathode, negligible current will flow and the diode behaves as an open circuit.

Types of Transistors

Transistors are semiconductor devices that can be used either as electronic switches or to amplify signals. They have three leads, called the collector, base, and emitter. A small current flowing between base and emitter (junction) causes a much larger current to flow between the emitter and collector (junction). There are two basic types of transistors, PNP and NPN styles.

Field effect transistors (FETs) are different types of transistors, which usually still have three terminals but work in a different way. Here the control element is the "gate" rather than the base, and it is the "gate" voltage which controls the current flowing in the "channel" between the other terminals—the "source" and the "drain." Like ordinary transistors FETs can be used as

electronic switches or as amplifiers; they also come in P-channel and N-channel types, and are available in small signal types as well as power FETs.

Power transistors are usually larger than the smaller signal type transistors. Power transistors are capable of handling larger currents and voltages. Often metal tabs and heatsinks are used to remove excess heat from the part. These devices are usually bolted to the chassis and are used for amplifying RF or audio energy.

Integrated Circuits

Integrated circuits (ICs) contain all, or most, of the components necessary for a particular circuit function in one package. Integrated circuits contain as few as ten transistors or many millions of transistors, plus many resistors, diodes and other components. There are many shapes, styles and sizes of integrated circuits: in this book we will use the dual-in-line style IC, either 8- , 14- or 16-pin devices.

Three-terminal regulators are special types of integrated circuits, which supply a regulated or constant and accurate voltage from the output regardless (within limits) of the voltage applied to the input. They are most often used in power supplies. Most regulators are designed to give specific output voltages, i.e. an "LM7805" regulator provides a 5-volt output, but some IC regulators can provide adjustable output based on an external potentiometer which can vary the output voltage.

Heatsinks

Many electronic components generate heat when they are operating. Generally heatsinks are used on semiconductors like transistors to remove heat. Overheating can damage a particular component or the entire circuit. The heatsink cools the transistor and ensures a long circuit life by removing the excess heat from the circuit area.

Light-emitting Diodes

Light-emitting diodes, or LEDs, are special diodes which have a plastic translucent body (usually clear, red, yellow, green or blue in color) and a small semiconductor element which emits light when the diode passes a small current. Unlike an incandescent lamp, an LED does not need to get hot to produce light. LEDs must always be forward biased to operate. Special LEDs can also produce infrared light.

LED displays consist of a number of LEDs together in a single package. The most common type has seven elongated LEDs arranged in an "8" pattern. By choosing which combinations of LEDs are lit, any number of digits from "0" through "9" can be displayed. Most of these "seven-segment" displays also contain another small round LED which is used as a decimal point.

Types of Inductors

Inductors or "coils" are basically a length of wire, wound into a cylindrical spiral (or layers of spirals) in order to increase their inductance. Inductance is the ability to store energy in a magnetic field. Many coils are wound on a former of insulating material, which may also have connection pins to act as the coil's terminals. The former may also be internally threaded to accept a small core or "slug" of ferrite, which can be adjusted in position relative to the coil itself to vary the coil inductance.

A transformer consists of a number of coils or windings of wire wound on a common former, which is also inside a core of iron alloy, ferrite or other magnetic material. When an alternating current is passed through one of the windings (primary), it produces an alternating magnetic field in the core and this in turn induces AC voltages in the other (secondary) windings. The voltages produced in the other winding depend on the number of turns in those windings, compared with the turns in the primary winding. If a secondary winding has fewer turns than the primary, it will produce a lower voltage, and be called a step-down transformer. If the secondary winding has more windings than the primary, then the transformer will produce a higher voltage and it will be a step-up transformer. Transformers can be used to change the voltage levels of AC power and they are available in many different sizes and power-handling capabilities.

Microphones

A microphone converts audible sound waves into electrical signals which can then be amplified. In an electret microphone, the sound waves vibrate a circular diaphragm made from very thin plastic material which has a permanent charge in it. Metal films coated on each side form a capacitor, which produces a very small AC voltage when the diaphragm vibrates. All electret microphones also contain FET which amplifies the very small AC signals. To power an FET amplifier the microphone must be supplied with a small DC voltage.

Loudspeakers

A loudspeaker converts electrical signals into sound waves that we can hear. It has two terminals which go to a voice coil, attached to a circular cone made of either cardboard or thin plastic. When electrical signals are applied to the voice coil, it creates a varying magnetic field from a permanent magnet at the back of the speaker. As a result, the cone vibrates in sympathy with the applied signal to produce sound waves.

Relays

Many electronic components are not capable of switching higher currents or voltages, so a device called a relay is used. A relay has a coil which forms an electromagnet, attracting a steel "armature" which itself pushes on one or more sets of switching contacts. When a current is passed through the coil to energize it, the moving contacts disconnect from one set of contacts to another, and when the coil is de-energized the contacts go back to their original position. In most cases a relay needs a diode across the coil to prevent damage to the semiconductor driving the coil.

Switches

A switch is a device with one or more sets of switching contacts, which are used to control the flow of current in a circuit. The switch allows the contacts to be controlled by a physical actuator of some kind—such as a pressbutton toggle lever, rotary knob, etc. As the name denotes, this type of switch has an actuator bar which slides back and forth between the various contact positions. In a single-pole, double throw or (SPDT) slider switch, a moving contact links the center contact to either of the two end contacts. In contrast, a double-pole double throw (DPDT) slider switch has two of these sets of contacts, with their moving contacts operating in tandem when the slider is actuated.

Wire

A wire is simply a length of metal conductor, usually made from copper since its conductivity is good, which means its resistance is low. When there is a risk of a wire touching another wire and causing a short, the copper wire is insulated or covered with a plastic coating which acts as an insulating material. Plain copper wire is not usually used since it will quickly oxidize or tarnish in the presence of air. A thin metal alloy coating is often applied to the copper wire; usually an alloy of tin or lead is used.

Single or multi-strand wire is covered in colored PVC plastic insulation and is used quite often in electronic applications to connect circuits or components together. This wire is often called "hookup" wire. On a circuit diagram, a solid dot indicates that the wires or PC board tracks are connected together or joined, while a "loop-over" indicates that they are not joined and must be insulated. A number of insulated wires enclosed in an outer jacket is called an electrical cable. Some electrical cables can have many insulated wires in them.

Semiconductor Substitution

Oftentimes when building an electronic circuit, it is difficult or impossible to find or locate the original transistor or integrated circuit. There are a number of circuits shown in this book which feature transistors, SCRs, UJTs, and FETs that are specified but cannot be found in some countries. Where possible many of these foreign components are converted to substitute values, either with a direct replacement or close substitution. Many foreign parts can be easily converted directly to a commonly used transistor or component. Occasionally, an outdated component has no direct common replacement, so the closest specifications of that component are attempted. In some instances we have specified replacement components with substitution components from the NTE brand or replacements. Most of the components for the project used in this book are quite common and easily located or substituted without difficulty.

When substituting components in the circuit, make sure that the pin-outs match the original components. Sometimes, for example, a transistor may have bottom view drawing, while the substituted value may have a drawing with a top view. Also be sure to check the pin-outs or the original components versus the replacement. As an example, some transistors will have EBC versus ECB pin-outs, so be sure to look closely at possible differences which may occur.

Chapter 2

Reading Schematics

Before we take a look at some electronics schematics, it is important that you understand the relationship between voltage and current and resistance. This relationship is called Ohm's Law and it's the backbone of simple formulas for electronics. In its simplest form it is shown as the equation $V = I * R$, which says voltage (V) is equal to current (I) multiplied by resistance. Voltage is represented as electromotive force where V is used instead of E. Current which is represented as (I) is in amperes or amps; in typical electronics circuits milliamps may be used, and is shown as 1/1000 or an ampere. One milliamp = 0.001 amp and is abbreviated as MA or ma. The three most important Ohm's Law equations are $V = I * R$; $I = V/R$; and $R = V/I$. The most-used Ohm's Law power equation is shown as $P = I * V$, where P is power in watts or wattage which is equal to the current of a circuit multiplied by the voltage in the circuit. Thus a voltage of 100 volts multiplied by a current consumption of two amps of current would equate to a power consumption of 200 watts.

The diagram shown in Figure 2-1 depicts a very common circuit called a voltage divider. The basic voltage divider consists of two resistors usually connected in series as shown. The total resistance is simply the sum of the two. In this example, it would be 22 k ohms plus 33 ohms = 22,033 ohms. If a volt signal is applied to the input end of R1 the 22 k ohm resistor, the current through the whole circuit would be $I = V/R = $ 1/22033 or 0.0000453864 amps or about 0.05

milliamps. Voltage dividers are used to reduce or step down voltages or signal levels to voltage converter circuits, amplifiers, etc. Voltage dividers are often set up using resistor ratios of 1000:1, 100:1, or 10:1; for example, a 10 to 1 resistor divider would have a 10 ohms and a 1 ohm resistor for R1 and R2 values.

The introduction of the operational amplifier or op-amp in the late 1960s and early 1970s transformed the electronics industry like no device before it. Op-amps began life as large integrated amplifier modules, some measuring up to three inches in length. Early op-amps were large and noisy but nonetheless a monumental development in electronics. The op-amp permitted a simple building block approach to electronic circuit design. Early op-amps were primarily used for pre-amps and instrumentation amplifiers. As op-amp quality progressed and their sizes and power consumption reduced over time, op-amps were used in all types of electronic circuit designs. Op-amps are now utilized in audio amplifiers, filter design, oscillators, comparators, regulators, etc. Since op-amps are now found in most all electronics circuits, we will spend some time identifying different op-amp configurations and applications.

Integrated circuits are used throughout much of modern electronics, so understanding how they are represented is important. Integrated circuits generally contain many individual circuits or components, and are shown schematically as functional blocks. Op-amps can be configured in many different ways, including non-inverting amplifiers.

The ubiquitous operational amplifier or op-amp as shown in the diagram of Figure 2-2 is depicted as a triangle with a plus (+) and minus (–) input and a single output at the point of the triangle. This type of op-amp configuration is called inverting input amplifier circuit, since the input is fed through the minus (–) input of the op-amp. Note that the signal input is applied through the input resistor at R1. Also notice that a second resistor R2, is shown from the minus input across to the

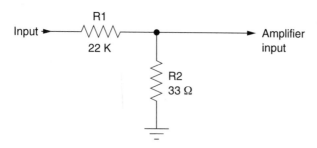

Figure 2-1   *Voltage divider*

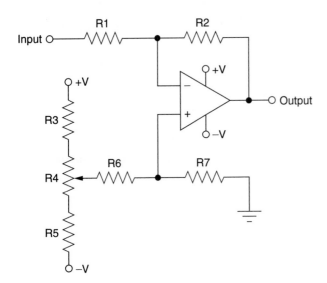

Figure 2-2 *Inverting amplifier*

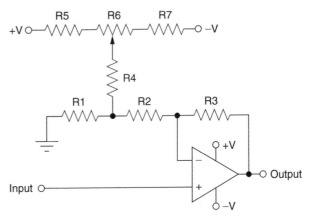

Figure 2-3 *Non-inverting amplifier*

output of the op-amp. Resistor R2 is called the feedback resistor. The plus (+) input of the op-amp is shown connected to a resistor network used to balance the input through a bias control at resistor R4. Both a plus and minus voltage source are connected to the bias network.

Simulate driving a current through the inverting input by placing a 1 volt on the input at R1 and assume that the right end has 0 volts on it. The current will be $I = V/R = 1/1\ k = 1$ ma. The voltage output will try to counter this by driving a current of the opposite polarity through the feedback resistor into the inverting input. The required voltage to do that will be $V = -(I * R) = -(1\ ma * 10\ k) = -10$ V. Thus we get a voltage to current conversion, a current to voltage conversion, a polarity inversion and, most importantly, amplification. $G = -(\text{Feedback Resistor/Input Resistor})$. In this example, it is shown as Gain or $G = -(R2/R1)$.

This inverting input op-amp configuration is commonly in modern amplifiers and filter, etc. Op-amps generally require both a plus (+) and minus (−) voltages for operation; this allows a voltage swing either side of zero.

The diagram illustrated in Figure 2-3 shows a non-inverting op-amp configuration. This configuration is used when you wish to maintain the same phase or polarity from the input, through to the output. Notice in this configuration the signal input is applied to the plus (+) input of the op-amp. The balance and trim capability is performed via the bias network formed by resistors

R4 through R7 which connect to both the plus and minus power supply.

The circuit shown in Figure 2-4 depicts a basic differential op-amp instrumentation amplifier. In this circuit, notice that there are two separate inputs and they are referenced against ground connection which is separate from the inputs. This type of circuit is used in high gain instrument amplifiers, when you wish to keep common modes noise from reaching the op-amp. This configuration is commonly used for high input impedance sensor inputs, where low leads may be used and low noise is required. Note the balance or trim for op-amp is provided through the resistor network composed of R4 through R7, which allow bias from both the plus and minus power sources.

Figure 2-5 illustrates how an op-amp can be used as a comparator to trigger an LED when a certain voltage

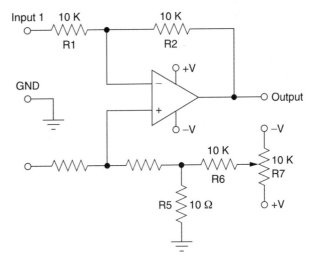

Figure 2-4 *Op-amp differential amplifier*

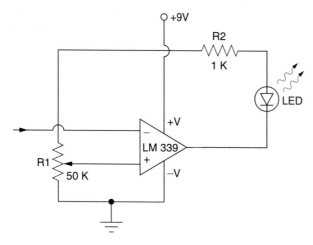

Figure 2-5   *Op-amp differential comparator*

form a timing network which is used to establish the frequency of the square-wave generator. The duty cycle is set up through resistors to produce a symmetrical square-wave output.

As you can see from these examples, the op-amp is a very powerful electronic building block tool for greatly simplifyng the design of electronic circuits. We have covered only a small number of op-amp applications. Op-amps are also used for filters, power regulators, integrators, voltage to current and current to voltage convertors. For more information on linear op-amps theory and applications look for the *IC Op-Amp Cookbook* by Walter Jung or point your browser to http://w1.859.telia.com/~u85920178/begin/opamp00.htm for a discussion on *Operational Basics* by Harry Lythall.

The above examples illustrate linear op-amps, and they represent nearly half of the spectrum of integrated circuits. On the other side of the spectrum of integrated circuits are the digital integrated circuits which generally contain multiple digital gates, switches, and memory functions all in one package. One of the more common digital integrated circuits is the Quad 2-input

threshold is reached. In this configuration, the plus (+) input of the op-amp is connected to a 50 k potentiometer which is connected between the plus 9-volt source and ground. Note in this configuration, only a single voltage is required. The minus (–) input lead of the op-amp is connected to the voltage source you wish to monitor. In operation, you would apply a voltage to the minus input pin of the op-amp and adjust the R1, so the LED is not lit. Raising the voltage at the input would now offset the comparator and its trip point and the LED would become lit.

Op-amps can also be used in oscillators and function generators as shown in Figure 2-6. In this circuit the LM339 comparator op-amp is used to generate a square-wave signal. A resistor and the capacitor at C1,

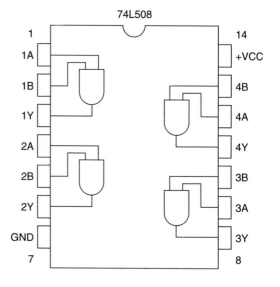

Function table

Inputs		Output
A	B	T
H	H	H
L	X	L
X	L	L

Figure 2-7   *"And" gate*

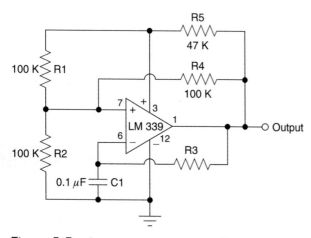

Figure 2-6   *Op-amp square-wave oscillator*

AND gate, shown in Figure 2-7. This diagram illustrates a 14 pin 74LS08 Quad 2-input AND gate, and its truth table. The 74LS08 contains four AND gates. The inputs to the first AND gate are at 1A and 1B represented by pins 1 and 2, while the output is at pin 3. The second AND gate has inputs 2A and 2B on pins 4 and 5 with its output on pin 6. The common ground connection is shown on pin 7, while 5 volt power is supplied to pin 14. The third AND gate has its inputs at 3A and 3B, on pins 9 and 10, with its output on pin 8. The fourth AND gate shown as 4A and 4B are on pins 12 and 13, with its output on pin 11. The truth table or function table for the 74LS08 describes the input vs. the output condition for each of the gates. If inputs 1A and 1B are HIGH then the resulting output at 1Y will be HIGH. If input 1A is LOW and input at 1B is either HIGH or LOW, then the output at 1Y will be LOW. If input 1B is LOW and input 1A is either HIGH or LOW the output at 1Y will be LOW once again.

In working with electronic circuits, you will come across many schematics as well as pictorial diagrams. Schematics are the electronic representation on paper of an electronic circuit. A pictorial diagram is generally a physical layout diagram of the same circuit. In electronics work you will come across both types of diagrams, they look quite different but they are really the same circuit shown differently. Experience in looking and comparing these two types of diagrams will help you enormously in your electronics projects.

The diagram shown in Figure 2-8 illustrates a schematic of a two-transistor audio amplifier circuit. The input jack at J1 is coupled to an input capacitor at

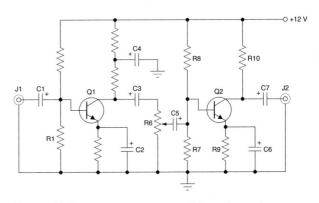

Figure 2-8 *Two-transistor amplifier schematic*

C1. Capacitor C1 couples the microphone of input source to the amplifier. It acts as a coupling device as well as a blocking device which can keep any constant DC component or voltage from reaching the input of the amplifier. The resistors R1 and R2 form the input impedance and frequency response characteristics of the input to the amplifier. The NPN transistor at Q1 forms the first stage of amplification of the amplifier. Resistors R4 and R5 form the bias or voltage supply for Q1. The output of the first amplifier stage at Q1 is connected to capacitor C3, which is used to couple the first amplifier stage to the next amplifier stage. Capacitor C3 is next fed to variable resistor R6, which acts as the systems volume control. Capacitor C5 couples the audio signal to the second amplifier stage at Q2. Resistor R10 is used to bias the second amplifier stage. Capacitor C7 is used to connect the output of transistor Q2 to the final amplifier stage or to the input of a transmitter, at J2. Note the ground bus or common connection for ground

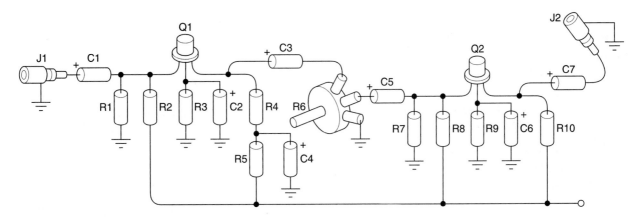

Figure 2-9 *Two-transistor amplifier pictorial diagram*

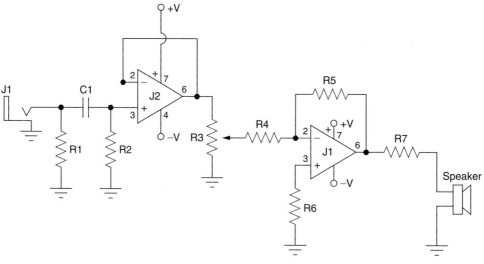

Figure 2-10 *Op-amp amplifier schematic diagram*

at the bottom of the circuit, and the power bus connection at the top of the schematic diagram.

The pictorial diagram shown in Figure 2-9 serves to illustrate the same two-transistor amplifier circuit shown in Figure 2-8. Both diagrams are electrically the same, but look quite different. The pictorial diagram shows the circuit how it would look wired and laid out in a chassis box. Compare the two diagrams so that you can see and understand how they differ from each other.

A two-op-amp guitar amplifier is shown schematically in Figure 2-10. The input to the amplifier is shown at input jack J1. Resistor R1 serves to control the input signal or current flowing into the amplifier circuit. Resistor R1 establishes the input impedance to the amplifier circuit. R1 is next fed to the 0.1-μF capacitor at C1. Capacitor C1 is used to block low frequency signals and couple the input to the next stage as well as keep any constant DC voltage from the guitar away from the input to the op-amps. The low frequencies to be blocked are dependent upon resistors R1 and R2. The triangle symbol shown at U1 represents the first op-amp amplifier in the circuit. The op-amp has two inputs, an inverting input and a non-inverting input. The input signal from resistor R2 is fed to the non-inverting input of the op-amp. The connection from the input to the output pin of the op-amp forms the feedback network which generally determines the gain

of the amplifier. In this example, there is no resistor but a direct connection between the input and the output. Also in the example the direct connection from the input to the output of the amplifier establishes a unity gain, or no change in the signal input. The purpose of the first stage in this example is to reduce the amount of current the guitar must supply to the amplifier. Resistor R3 represents a variable resistor which is as a volume control for the amplifier. The center tap on the variable resistor is fed to resistor R4, which is used to couple the first stage of the amplifier to the second stage. Resistors R4 and R5 establish the gain of the second amplification stage. This gain stage forms the real muscle of the amplifier. The output of the second op-amp at pin 6 is fed directly to a speaker through resistor R7, which is used to protect the output of the amplifier as well as protecting the speaker from too much current.

The diagram shown in Figure 2-11 depicts the same guitar amplifier circuit but it is now shown pictorially instead of schematically. The pictorial diagram shows the circuit as it might appear physically in its enclosure or circuit board. The diagram looks quite different but it's really the same circuit drawn slightly different. You will need to recognize and become familiar with the differences between these types of drawings. The best way to familiarize yourself between schematics and pictorial diagrams is to see and compare a number of them over time. Once you get some practice it will

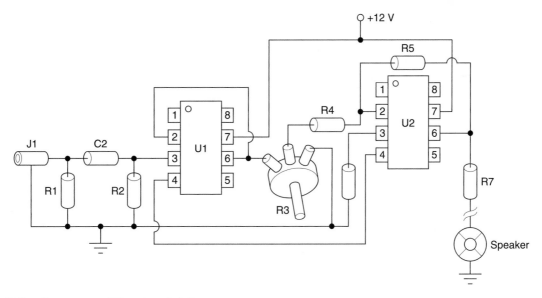

Figure 2-11 *Op-amp amplifier pictorial diagram*

become second nature to you. There are a number of good books on how to read schematics which cover the topic in more depth.

Before we discuss mounting the components to the printed circuit board we will take a few minutes to discuss the resistor and capacitor codes which are used to help identify these components. Table 2-1 lists the resistor color code information. Resistors will generally have at least three or four color bands which help identify the resistor values. The color band will start from one edge of the resistor body, this is the first color code which represents the first digit, the second color band

Table 2-1
Resistor Color Code Chart

Color Band	1st Digit	2nd Digit	Multiplier	Tolerance
Black	0	0	1	
Brown	1	1	10	1%
Red	2	2	100	2%
Orange	3	3	1000 (K)	3%
Yellow	4	4	10000	4%
Green	5	5	100000	
Blue	6	6	1000000 (M)	
Violet	7	7	10000000	
Gray	8	8	100000000	
White	9	9	1000000000	
Gold			0.1	5%
Silver			0.01	10%
No color				20%

Table 2-2
Capacitance Codebreaker Information

This table is designed to provide the value of alphanumeric coded ceramic, mylar and mica capacitors in general. They come in many sizes, shapes, values and ratings; many different manufacturers worldwide produce them and not all play by the same rules. Most capacitors actually have the numeric values stamped on them; however, some are color coded and some have alphanumeric codes. The capacitor's first and second significant number IDs are the first and second values, followed by the multiplier number code, followed by the percentage tolerance letter code. Usually the first two digits of the code represent the significant part of the value, while the third digit, called the multiplier, corresponds to the number of zeros to be added to the first two digits.

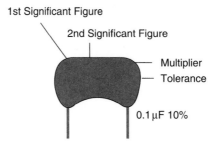

Value	Type	Code	Value	Type	Code
1.5 pF	Ceramic		1000 pF/0.001 µF	Ceramic/Mylar	102
3.3 pF	Ceramic		1500 pF/0.0015 µF	Ceramic/Mylar	152
10 pF	Ceramic		2000 pF/0.002 µF	Ceramic/Mylar	202
15 pF	Ceramic		2200 pF/0.0022 µF	Ceramic/Mylar	222
20 pF	Ceramic		4700 pF/0.0047 µF	Ceramic/Mylar	472
30 pF	Ceramic		5000 pF/0.005 µF	Ceramic/Mylar	502
33 pF	Ceramic		5600 pF/0.0056 µF	Ceramic/Mylar	562
47 pF	Ceramic		6800 pF/0.0068 µF	Ceramic/Mylar	682
56 pF	Ceramic		0.01	Ceramic/Mylar	103
68 pF	Ceramic		0.015	Mylar	
75 pF	Ceramic		0.02	Mylar	203
82 pF	Ceramic		0.022	Mylar	223
91 pF	Ceramic		0.033	Mylar	333
100 pF	Ceramic	101	0.047	Mylar	473
120 pF	Ceramic	121	0.05	Mylar	503
130 pF	Ceramic	131	0.056	Mylar	563
150 pF	Ceramic	151	0.068	Mylar	683
180 pF	Ceramic	181	0.1	Mylar	104
220 pF	Ceramic	221	0.2	Mylar	204
330 pF	Ceramic	331	0.22	Mylar	224
470 pF	Ceramic	471	0.33	Mylar	334
560 pF	Ceramic	561	0.47	Mylar	474
680 pF	Ceramic	681	0.56	Mylar	564
750 pF	Ceramic	751	1	Mylar	105
820 pF	Ceramic	821	2	Mylar	205

Table 2-2 (*Continued*)

PicoFarad (pF)	NanoFarad (nF)	MicroFarad (mF, μF or mfd)	Capacitance Code
1000	1 or 1n	0.001	102
1500	1.5 or 1n5	0.0015	152
2200	2.2 or 2n2	0.0022	222
3300	3.3 or 3n3	0.0033	332
4700	4.7 or 4n7	0.0047	472
6800	6.8 or 6n8	0.0068	682
10000	10 or 10n	0.01	103
15000	15 or 15n	0.015	153
22000	22 or 22n	0.022	223
33000	33 or 33n	0.033	333
47000	47 or 47n	0.047	473
68000	68 or 68n	0.068	683
100000	100 or 100n	0.1	104
150000	150 or 150n	0.15	154
220000	220 or 220n	0.22	224
330000	330 or 330n	0.33	334
470000	470 or 470n	0.47	474

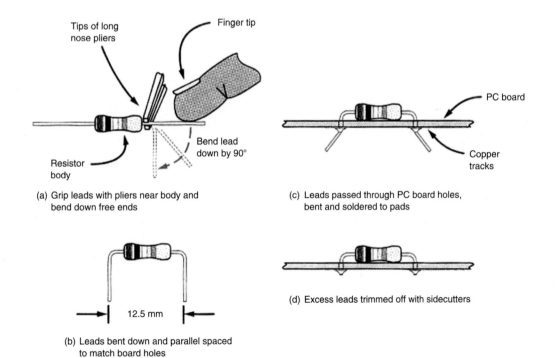

(a) Grip leads with pliers near body and bend down free ends

(b) Leads bent down and parallel spaced to match board holes

(c) Leads passed through PC board holes, bent and soldered to pads

(d) Excess leads trimmed off with sidecutters

Figure 2-12 *Preparing resistor leads for horizontal mounting*

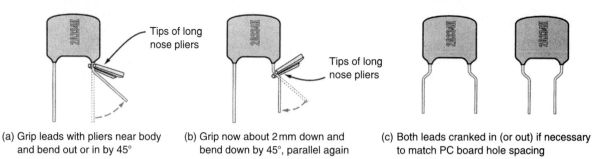

(a) Grip leads with pliers near body and bend out or in by 45°

(b) Grip now about 2 mm down and bend down by 45°, parallel again

(c) Both leads cranked in (or out) if necessary to match PC board hole spacing

Figure 2-13 *"Dressing" capacitor component leads*

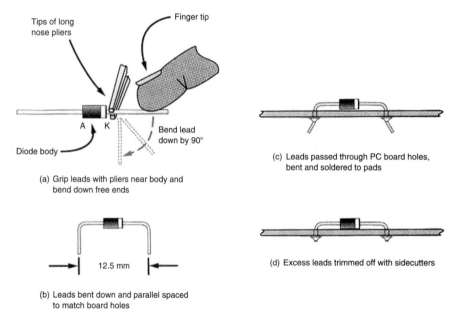

Tips of long nose pliers

Finger tip

A   K

Diode body

Bend lead down by 90°

(a) Grip leads with pliers near body and bend down free ends

12.5 mm

(b) Leads bent down and parallel spaced to match board holes

(c) Leads passed through PC board holes, bent and soldered to pads

(d) Excess leads trimmed off with sidecutters

Figure 2-14 *Preparing diode leads for horizontal mounting*

depicts the second digit, while the third band is the resistor multiplier value. The fourth color band represents the resistor tolerance value. If there is no fourth band then the resistor has a 20% tolerance, if the fourth band is silver, then the resistor has a 10% tolerance value, and if the fourth band is gold then the resistor has a 5% tolerance value. Therefore, if you have a resistor with the first band having a brown color, the second band with a black color and a third band with a red color, this resistor would have a value of 1 k or 1000 ohms.

Now let's move to installing the six capacitors; note that there are two electrolytic capacitors and four ceramic capacitors. When installing the ceramic capacitors refer to Table 2-2, which lists capacitor (code) values vs. actual capacitor values. Most large

capacitors, such as electrolytic types, will have their actual value printed on the body of the capacitor. But often with small capacitors, they are just too small to have the actual capacitor values printed on them. An abbreviated code was devised to mark the capacitor as shown in the table. A code marking of (104) would therefore denote a capacitor with a value of 0.1 µF (microfarads) or 10 nF (nanofarads). Once you understand how these component codes work, you can easily identify the resistors and capacitors and proceed to mount them on the printed circuit board.

Next you will learn how to insert the various electronic components on to a printed circuit board or PC board. The diagram shown in Figure 2-12 depicts

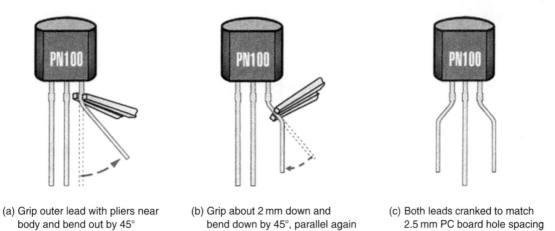

(a) Grip outer lead with pliers near body and bend out by 45°

(b) Grip about 2 mm down and bend down by 45°, parallel again

(c) Both leads cranked to match 2.5 mm PC board hole spacing

Figure 2-15 *"Dressing" TO-92 transistor leads*

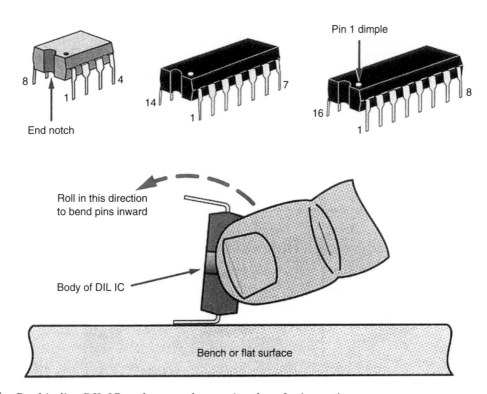

Figure 2-16 *Dual in-line DIL IC packages and preparing them for inspection*

how to "dress" or prepare the resistor leads. Resistor leads are first bent to fit the PC board component holes, then the resistor is mounted to the circuit board. Figure 2-13 depicts how capacitor leads are bent to fit the PC board holes. Once the capacitor leads are prepared, the capacitor can be installed onto the circuit board. The diagram shown in Figure 2-14 illustrates how to prepare diode leads before mounting them to the PC board. Finally, Figure 2-15 depicts transistor lead preparation, while Figure 2-16 shows how integrated circuit leads are prepared for installation.

Chapter 3

Soldering

Everyone working in electronics needs to know how to solder well. Before you begin working on a circuit, carefully read this chapter on soldering. In this section you will learn how to make good solder joints when soldering point-to-point wiring connections as well as for PC board soldering connections.

In all electronics work the wiring connections must be absolutely secure. A loose connection in a radio results in noise, scratching sounds, or no sound at all. In a TV, poor connections can disrupt the sound or picture. The safe operation of airplanes and the lives of astronauts in flight depend on secure electronic connections.

Soldering joins two pieces of metal, such as electrical wires, by melting them together with another metal to form a strong, chemical bond. Done correctly, it unites the metals so that electrically they act as one piece of metal. Soldering is not just gluing metals together. Soldering is tricky and intimidating in practice, but easy to understand in theory. Basic supplies include a soldering iron, which is a prong of metal that heats to a specific temperature through electricity, like a regular iron. The solder, or soldering wire, often an alloy of aluminum and lead, needs a lower melting point than the metal you're joining. Finally, you need a cleaning resin called flux that ensures the joining pieces are incredibly clean. Flux removes all the oxides on the surface of the metal that would interfere with the molecular bonding, allowing the solder to flow into the joint smoothly. You also need two things to solder together.

The first step in soldering is cleaning the surfaces, initially with sandpaper or steel wool and then by melting flux onto the parts. Sometimes, flux is part of the alloy of the soldering wire, in an easy-to-use mixture. Then, both the pieces are heated above the melting point of the solder (but below their own melting point) with the soldering iron. When touched to the joint, this precise heating causes the solder to "flow" to the place of highest temperature and makes a chemical

bond. The solder shouldn't drip or blob, but spread smoothly, coating the entire joint. When the solder cools, you should have a clean, sturdy connection.

Many people use soldering in their field, from electrical engineering and plumbing to jewelry and crafts. In a delicate procedure, a special material, called solder, flows over two pre-heated pieces and attaches them through a process similar to welding or brazing. Various metals can be soldered together, such as gold and sterling silver in jewelry, brass in watches and clocks, copper in water pipes, or iron in leaded glass stained windows. All these metals have different melting points, and therefore use different solder. Some "soft" solder, with a low melting point, is perfect for wiring a circuit board. Other "hard" solder, such as for making a bracelet, needs a torch rather than a soldering iron to get a hot enough temperature. Electrical engineers and hobbyists alike can benefit from learning the art and science of soldering.

Solder

The best solder for electronics work is 60/40 rosin-core solder. It is made of 60% tin and 40% lead. This mixture melts at a lower temperature than either lead or tin alone. It makes soldering easy and provides good connections. The rosin keeps the joint clean as it is being soldered. The heat of the iron often causes a tarnish or oxide to form on the surface. The rosin dissolves the tarnish to make the solder cling tightly. Solders have different melting points, depending on the ratio of tin to lead. Tin melts at 450° F and lead at 621° F. Solder made from 63% tin and 37% lead melts at 361° F, the lowest melting point for a tin and lead mixture. Called 63-37 (or eutectic), this type of solder also provides the most rapid solid-to-liquid transition and the best stress resistance. Solders made with different lead/tin ratios have a plastic state at some temperatures. If the solder is deformed while it is in the plastic state, the deformation remains when the solder freezes into the solid state. Any stress or motion applied to "plastic solder" causes a poor solder joint.

The 60-40 solder has the best wetting qualities. Wetting is the ability to spread rapidly and bond materials uniformly; 60-40 solder also has a low melting point. These factors make it the most commonly used solder in electronics.

Some connections that carry high current can't be made with ordinary tin-lead solder because the heat generated by the current would melt the solder. Automotive starter brushes and transmitter tank circuits are two examples. Silver-bearing solders have higher melting points, and so prevent this problem. High-temperature silver alloys become liquid in the 1100° F to 1200° F range, and a silver-manganese (85-15) alloy requires almost 1800° F.

Because silver dissolves easily in tin, tin bearing solders can leach silver plating from components. This problem can be greatly reduced by partially saturating the tin in the solder with silver or by eliminating the tin. Tin-silver or tin-lead-silver alloys become liquid at temperatures from 430° F for 96.5-3.5 (tin-silver), to 588° F for 1.0-97.5-1.5 (tin-lead-silver). A 15.080.0-5.0 alloy of lead-indium-silver melts at 314° F.

Never use acid-core solder for electrical work. It should be used only for plumbing or chassis work. For circuit construction, only use fluxes or solder-flux combinations that are labeled for electronic soldering.

The rosin or the acid is a flux. Flux removes oxide by suspending it in solution and floating it to the top. Flux is not a cleaning agent! Always clean the work before soldering. Flux is not part of a soldered connection—it merely aids the soldering process. After soldering, remove any remaining flux. Rosin flux can be removed with isopropyl or denatured alcohol. A cotton swab is a good tool for applying the alcohol and scrubbing the excess flux away. Commercial flux-removal sprays are available at most electronic-part distributors.

The Soldering Iron

Soldering is used in nearly every phase of electronic construction so you'll need soldering tools. A soldering tool must be hot enough to do the job and lightweight enough for agility and comfort. A temperature

controlled iron works well, although the cost is not justified for occasional projects. Get an iron with a small conical or chisel tip. Soldering is not like gluing; solder does more than bind metal together and provide an electrically conductive path between them. Soldered metals and the solder combine to form an alloy.

You may need an assortment of soldering irons to do a wide variety of soldering tasks. They range in size from a small 25-watt iron for delicate printed-circuit work to larger 100 to 300-watt sizes used to solder large surfaces. If you could only afford a single soldering tool when initially setting up your electronics workbench, then an inexpensive to moderately priced pencil-type soldering iron with between 25 and 40-watt capacity is the best for PC board electronics work. A 100-watt soldering gun is overkill for printed-circuit work, since it often gets too hot, cooking solder into a brittle mess or damaging small parts of a circuit. Soldering guns are best used for point-to-point soldering jobs, for large mass soldering joints or large components. Small "pencil" butane torches are also available, with optional soldering-iron tips. A small butane torch is available from the Solder-It Company. Butane soldering irons are ideal for field service problems and will allow you to solder where there is no 110-volt power source. This company also sells a soldering kit that contains paste solders (in syringes) for electronics, pot metal and plumbing. *See Appendix 1* for the address information.

Keep soldering tools in good condition by keeping the tips well tinned with solder. Do not run them at full temperature for long periods when not in use. After each period of use, remove the tip and clean off any scale that may have accumulated. Clean an oxidized tip by dipping the hot tip in sal ammoniac (ammonium chloride) and then wiping it clean with a rag. Sal ammoniac is somewhat corrosive, so if you don't wipe the tip thoroughly, it can contaminate electronic soldering. You can also purchase a small jar of "Tip Tinner," a soldering iron tip dresser from your local Radio Shack store. Place the tip of the soldering iron into the "Tip Tinner" after every few solder joints.

If a copper tip becomes pitted, file it smooth and bright and then tin it immediately with solder. Modern soldering iron tips are nickel or iron clad and should not be filed. The secret of good soldering is to use the right amount of heat. Many people who will have not soldered before use too little heat, dabbing at the joint to be soldered and making little solder blobs that cause unintended short circuits. Always use caution when soldering. A hot soldering iron can burn your hand badly or ruin a tabletop. It's a good idea to buy or make a soldering iron holder.

Soldering Station

Often when building or repairing a circuit, your soldering iron is kept switched "on" for unnecessarily long periods, consuming energy and allowing the soldering iron tip to burn and develop a buildup of oxide. Using this soldering-iron temperature controller, you will avoid destroying sensitive components when soldering.

Buying a lower wattage iron may solve some of the problems, but new problems arise when you want to solder some heavy-duty component, setting the stage for creating a "cold" connection. If you've ever tried to troubleshoot some instrument in which a cold solder joint was at the root of the problem, you know how difficult such defects are to locate. Therefore, the best way to satisfy all your needs is to buy a temperature controller electronics workbench.

A soldering station usually consists of a temperature controlled soldering iron with an adjustable heat or temperature control and a soldering iron holder and cleaning pad. If you are serious about your electronics hobby, or if you have been involved with electronics building and repair for any length of time, you will eventually want to invest in a soldering station at some point in time. There are real low cost soldering stations for hobbyists for under $30, but it makes more sense to purchase a moderately priced soldering station such as the quality Weller series. A typical soldering station is shown in Figure 3-1.

Soldering Gun

An electronics workbench would not be complete without a soldering gun. Soldering guns are useful for soldering large components to terminal strips, or splicing wires together, or when putting connectors on coax cable. There are many instances where more heat is needed than a soldering iron can supply. For example, a

Figure 3-1   *Temperature controlled soldering station*

large connector mass cannot be heated with a small soldering iron, so you would never be able to "tin" a connector with a small wattage soldering iron. A soldering gun is a heavy-duty soldering device which does in fact look like a gun. Numerous tips are available for a soldering gun and they are easily replaceable using two small nuts on the side arm of the soldering gun. Soldering guns are available in two main heat ranges. Most soldering guns have a two-step "trigger" switch which enables you to select two heat ranges for different soldering jobs. The most common soldering gun provides both a 100-watt setting when the "trigger" switch is pressed to its first setting, and as the "trigger" switch is advanced to the next step, the soldering gun will provide 150 watts, when more heat is needed. A larger or heavy-duty soldering gun is also available, but a little harder to locate is the 200 to 250-watt solder gun: the first "trigger" switch position provides 200 watts, while the second switch position provides the 250-watt heat setting. When splicing wires together using either the "Western Union" or parallel splice or the end splice, a soldering gun should be used especially if the wire gauge is below size 22 ga. Otherwise the solder may not melt properly and the connections may reflect a "cold" solder joint and therefore a poor or noisy splice. Soldering wires to binding post connections should be performed with a soldering gun to ensure proper heating to the connection. Most larger connectors should be soldered or pre-tinned using a soldering gun for even solder flow.

Preparing the Soldering Iron

If your iron is new, read the instructions about preparing it for use. If there are no instructions, follow this procedure:

It should be hot enough to melt solder applied to its tip quickly (half a second when dry, instantly when wet with solder). Apply a little solder directly to the tip so that the surface is shiny. This process is called "tinning" the tool. The solder coating helps conduct heat from the tip to the joint face, the tip is in contact with one side of the joint. If you can place the tip on the underside of the joint, do so. With the tool below the joint, convection helps transfer heat to the joint. Place the solder against the joint directly opposite the soldering tool. It should melt within a second for normal PC connections, within two seconds for most other connections. If it takes longer to melt, there is not enough heat for the job at hand. Keep the tool against the joint until the solder flows freely throughout the joint. When it flows freely, solder tends to form concave shapes between the conductors. With insufficient heat solder does not flow freely; it forms convex shapes or blobs. Once solder shape changes from convex to concave, remove the tool from the joint. Let the joint cool without movement at room temperature. It usually takes no more than a few seconds. If the joint is moved before it is cool, it may take on a dull, satin look that is characteristic of a "cold" solder joint. Reheat cold joints until the solder flows freely and hold them still until cool. When the iron is set aside, or if it loses its shiny appearance, wipe away any dirt with a wet cloth or sponge. If it remains dull after cleaning, tin it again.

Overheating a transistor or diode while soldering can cause permanent damage. Use a small heatsink when you solder transistors, diodes or components with plastic parts that can melt. Grip the component lead with a pair of pliers up close to the unit so that the heat is conducted away You will need to be careful, since it is easy to damage delicate component leads. A small alligator clip also makes a good heatsink to dissipate from the component.

Mechanical stress can damage components, too. Mount components so there is no appreciable mechanical strain on the leads.

Soldering to the pins of coil forms male cable plugs can be difficult. Use a suitable small twist drill to clean the inside of the pin and then tin it with resin-core solder. While it is still liquid, clear the surplus solder from each pin with a whipping motion or by blowing through the pin from the inside of the form or plug. Watch out for flying hot solder, you can get severe burns. Next, file the nickel tip, then insert the wire and solder it. After soldering, remove excess solder with a file, if necessary. When soldering to the pins of plastic coil-forms, hold the pin to be soldered with a pair of heavy pliers to form a heatsink. Do not allow the pin to overheat; it will loosen and become misaligned.

Preparing Work for Soldering

If you use old junk parts, be sure to completely clean all wires or surfaces before applying solder. Remove all enamel, dirt, scale, or oxidation by sanding or scraping the parts down to bare metal. Use fine sandpaper or emery paper to clean flat surfaces or wire. (Note, no amount of cleaning will allow you to solder to aluminum. When making a connection to a sheet of aluminum, you must connect the wire by a solder lug or a screw.)

When preparing wires, remove the insulation with wire strippers or a pocketknife. If using a knife, do not cut straight into the insulation; you might nick the wire and weaken it. Instead, hold the knife as if you were sharpening a pencil, taking care not to nick the wire as you remove the insulation. For enameled wire, use the back of the knife blade to scrape the wire until it is clean and bright. Next, tin the clean end of the wire. Now, hold the heated soldering-iron tip against the under surface of the wire and place the end of the rosin-core solder against the upper surface. As the solder melts, it flows on the clean end of the wire. Hold the hot tip of the soldering iron against the under surface of the tinned wire and remove the excess solder by letting it flow down on the tip. When properly tinned, the exposed surface of the wire should be covered with a thin, even coating of solder.

How to Solder

The two key factors in quality soldering are time and temperature. Generally, rapid heating is desired, although most unsuccessful solder jobs fail because insufficient heat has been applied. Be careful; if heat is applied too long, the components or PC board can be damaged, the flux may be used up and surface oxidation can become a problem. The soldering-iron tips should be hot enough to readily melt the solder without burning, charring or discoloring components, PC boards or wires. Usually, a tip temperature about 100° F above the solder melting point is about right for mounting components on PC boards. Also, use solder that is sized appropriately for the job. As the cross-section of the solder decreases, so does the amount of heat required to melt it. Diameters from 0.025 to 0.040 inches are good for nearly all circuit wiring.

Always use a good quality multicore solder. A standard 60% tin, 40% lead alloy solder with cores of non-corrosive flux will be found easiest to use. The flux contained in the longitudinal cores of multicore solder is a chemical designed to clean the surfaces to be joined of deposited oxides, and to exclude air during the soldering process, which would otherwise prevent these metals coming together. Consequently, don't expect to be able to complete a joint by using the application of the tip of the iron loaded with molten solder alone, as this usually will not work. Having said that, there is a process called tinning where conductors are first coated in fresh, new solder prior to joining by a hot iron. Solder comes in gauges like wire. The two most common types of solder, are 18 ga, used for general work, and the thinner 22 ga, used for fine work on printed circuit boards.

A Well-Soldered Joint Depends On

1. Soldering with a clean, well-tinned tip.

2. Cleaning the wires or parts to be soldered.

3. Making a good mechanical joint before soldering.

4. Allowing the joint to get hot enough before applying solder.

5. Allowing the solder to set before handling or moving soldered parts.

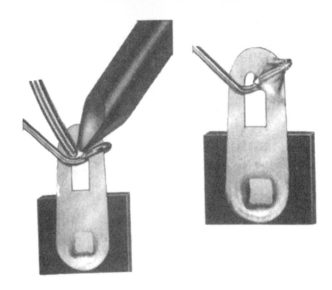

Figure 3-2   *Wire to lug soldering joint*

Making a Good Mechanical Joint

Unless you are creating a temporary joint, the next step is to make a good mechanical connection between the parts to be soldered. For instance, wrap the wire carefully and tightly around a soldering terminal or soldering lug, as shown in Figure 3-2. Bend wire and make connections with long-nosed pliers. When connecting two wires together, make a tight splice before soldering. Once you have made a good mechanical contact, you are ready for the actual soldering.

The next step is to apply the soldering iron to the connection, soldering the connection as shown. In soldering a wire splice, hold the iron below the splice and apply solder to the top of the splice. If the tip of the iron has a bit of melted solder on the side held against the splice, heat is transferred more readily to the splice and the soldering is done more easily. Don't try to solder by applying solder to the joint and then pressing down on it with the iron. Be sure not to disturb the soldered joint until the solder has set. It may take a few seconds for the solder to set, depending upon the amount of solder used in making the joint. Now take a good look at the joint. It should have a shiny, smooth appearance—not pitted or grainy. If it does have a pitted, granular appearance as seen in Figure 3-3, reheat the joint, scrape off the solder, and clean the connection.

Table 3-1

Soldering Check List

1. Prepare the joint. Clean all surfaces and conductors thoroughly with fine steel wool. First, clean the circuit traces, then clean the component leads.

2. Prepare the soldering iron or gun. The soldering device should be hot enough to melt solder applied to the tip. Apply a small amount of solder directly to the tip, so that the surface is shiny.

3. Place the tip in contact with one side of the joint; if possible, place the tip below the joint.

4. Place the solder against the joint directly opposite the soldering tool. The solder should melt within two seconds; if it takes longer use a larger iron.

5. Keep the soldering tool against the joint until the solder flows freely throughout the joint. When it flows freely the joint should form a concave shape; insufficient heat will form a convex shape.

6. Let the joint cool without any movement; the joint should cool and set-up within a few seconds. If the joint is moved before it cools the joint will look dull instead of shiny and you will likely have a cold solder joint. Re-heat the joint and begin anew.

7. Once the iron is set aside, or if it loses its shiny appearance, wipe away any dirt with a wet cloth or wet sponge. When the iron is clean the tip should look clean and shiny. After cleaning the tip apply some solder.

Then start over again. After the solder is well set, pull on the wire to see if it is a good, tight connection. If you find that you made a poor soldering job don't get upset, be thankful you found it and do it over again. A quick reference solder check list is shown in the listing in Table 3-1.

Soldering Printed Circuit Boards

Most electronic devices use one or more printed circuit (PC) boards. A PC board is a thin sheet of fiberglass or phenolic resin that has a pattern of foil conductors "printed" on it. You insert component leads into holes in the board and solder the leads to the foil pattern. This method of assembly is widely used and you will probably encounter it if you choose to build from a kit. Printed circuit boards make assembly easy. First, insert component leads through the correct holes in the circuit board. Mount parts tightly against the circuit board unless otherwise directed. After inserting a lead into the board, bend it slightly outward to hold the part in place.

When the iron is hot, apply some solder to the flattened working end at the end of the bit, and wipe it on a piece of damp cloth or sponge so that the solder forms a thin film on the bit. This is tinning the bit.

Melt a little more solder on to the tip of the soldering iron, and put the tip so it contacts both parts of the joint. It is the molten solder on the tip of the iron that allows the heat to flow quickly from the iron into both parts of the joint. If the iron has the right amount of solder on it and is positioned correctly, then the two parts to be joined will reach the solder's melting temperature in a couple of seconds. Now apply the end of the solder to

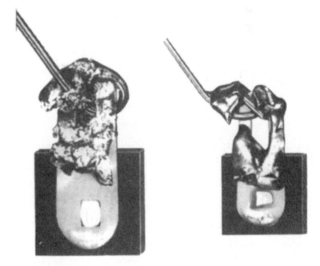

Figure 3-3   *Inferior to lug solder joint*

the point where both parts of the joint and the soldering iron are all touching one another. The solder will melt immediately and flow around all the parts that are at, or over, the melting part temperature. After a few seconds remove the iron from the joint. Make sure that no parts of the joint move after the soldering iron is removed until the solder is completely hard. This can take quite a few seconds with large joints. If the joint is disturbed during this cooling period it may become seriously weakened.

The most important point in soldering is that both parts of the joint to be made must be at the same temperature. The solder will flow evenly and make a good electrical and mechanical joint only if both parts of the joint are at an equal high temperature. Even though it appears that there is a metal-to-metal contact in a joint to be made, very often there exists a film of oxide on the surface that insulates the two parts. For this reason it is no good applying the soldering iron tip to one half of the joint only and expecting this to heat the other half of the joint as well.

It is important to use the right amount of solder, both on the iron and on the joint. Too little solder on the iron will result in poor heat transfer to the joint, too much and you will suffer from the solder forming strings as the iron is removed, causing splashes and bridges to other contacts. Too little solder applied to the joint will give the joint a half finished appearance: a good bond where the soldering iron has been, and no solder at all on the other part of the joint.

The hard cold solder on a properly made joint should have a smooth shiny appearance and if the wire is pulled it should not pull out of the joint. In a properly made joint the solder will bond the components very strongly indeed, since the process of soldering is similar to brazing, and to a lesser degree welding, in that the solder actually forms a molecular bond with the surfaces of the joint. Remember it is much more difficult to correct a poorly made joint than it is to make the joint properly in the first place. Anyone can learn to solder, it just takes practice.

The diagram in Figure 3-4 shows how to solder a component lead to a PC board pad. The tip of the soldering iron heats both the lead and the copper pad, so the end of the solder wire melts when it's pushed into the contact. The diagram illustrated in Figure 3-5 shows how a good solder joint is obtained. Notice that it has a

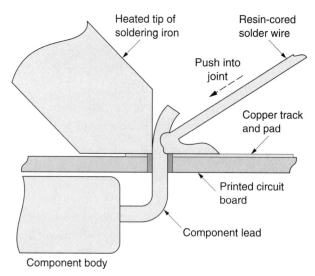

Figure 3-4 *Proper technique for soldering components on PC board*

smooth and shiny "fillet" of solder metal, bonding all around to both the component lead and the copper pad of the PC board. This joint provides a reliable electrical connection.

Try to make the solder joint as quickly as possible because the longer you take, the higher the risk that the component itself and the printed circuit board pad and track will overheat and be damaged. But don't work so quickly that you cannot make a good solder joint. Having to solder the joint over again always increases the risk of applying too much heat to the PCB.

As the solder solidifies, take a careful look at the joint you have made, to make sure there's a smooth and fairly shiny metal "fillet" around it. This should be broadly concave in shape, showing that the solder has formed a good bond to both metal surfaces. If it has a rough and dull surface or just forms a "ball" on the

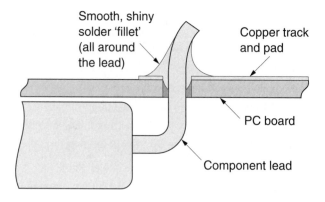

Figure 3-5 *Excellent solder joint*

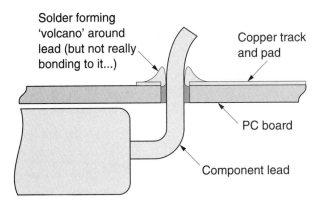

Solder forming 'volcano' around lead (but not really bonding to it...)

Copper track and pad

PC board

Component lead

Figure 3-6 *Inferior solder joint. Example 1*

component lead, or a "volcano" on the PCB pad with the lead emerging from the crater, you have a "dry joint." If your solder joint looks like the picture shown in Figure 3-6, you will have to re-solder the joint over again. Figure 3-7 shows another type of dry solder joint which would have to be re-soldered. These types of "dry" solder joints if now redone will cause the circuit to be unreliable and intermittent.

For projects that use one or more integrated circuits, with their leads closely-spaced pins, you may find it easier to use a finer gauge solder, i.e. less than 1 mm in diameter. This reduces the risk of applying too much solder to each joint, and accidently forming "bridges" between pads to PC "tracks."

The finished connection should be smooth and bright. Re-heat any cloudy or grainy-looking connections. Finally, clip off the excess wire length, as shown in Figure 3-8.

Occasionally a solder "bridge" will form between two adjacent foil conductors. You must remove these bridges, otherwise a short circuit will exist between the

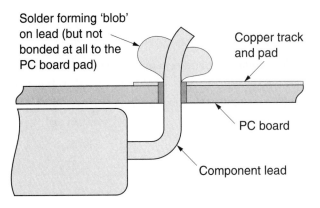

Solder forming 'blob' on lead (but not bonded at all to the PC board pad)

Copper track and pad

PC board

Component lead

Figure 3-7 *Inferior solder joint. Example 2*

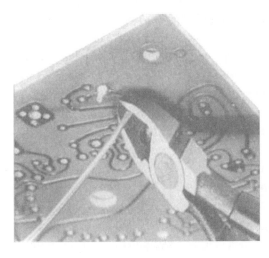

Figure 3-8 *Trimming excess component leads*

two conductors. Remove a solder bridge by heating the bridge and quickly wiping away the melted solder with a soft cloth. Often you will find a hole on the board plugged by solder from a previous connection. Clear the hole by heating the solder while pushing a component lead through the hole from the other side of the board. Good soldering is a skill that is learnt by practice.

How to Unsolder

In order to remove components, you need to learn the art of de-soldering. You might accidentally make a wrong connection or have to move a component that you put in an incorrect location. Take great care while unsoldering to avoid breaking or destroying good parts. The leads on components such as resistors or transistors and the lugs on other parts may sometimes break off when you are unsoldering a good, tight joint. To avoid heat damage, you must use as much care in unsoldering delicate parts as you do in soldering them. There are three basic ways of unsoldering. The first method is to heat the joint and "flick" the wet solder off. The second method is to use a metal wick or braid to remove the melted solder. This braid is available at most electronics parts stores; use commercially made wicking material (braid). Place the braid against the joint that you want to unsolder. Use the heated soldering iron to gently press the braid against the joint. As the solder melts, it is pulled into the braid. By repeating this process, you can remove virtually all the solder from the joint. Then reheat the joint and lift off the component leads. Another useful tool is an air-suction solder remover.

Most electronics parts stores have these devices. Before using a desolder squeeze bulb, use your soldering iron to heat the joint you want to unsolder until the solder melts. Then squeeze the bulb to create a vacuum inside. Touch the tip of the bulb against the melted solder. Release the bulb to suck up the molten solder. Repeat the process until you've removed most of the solder from the joint. Then reheat the joint and gently pry off the wires. This third method is easy, and is the preferred method, since it is fast and clean. You can use a vacuum device to suck up molten solder. There are many new styles of solder vacuum devices on the market that are much better than the older squeeze bulb types. The new vacuum de-soldering tools are about 8 to 12 inches long with a hollow Teflon tip. You draw the vacuum with a push handle and set it. As you reheat the solder around the component to be removed, you push a button on the device to suck the solder into the chamber of the de-soldering tool.

De-soldering Station

A de-soldering station is a very useful addition to an electronics workshop or workbench, but in many cases they just cost too much for most hobbyists. De-soldering stations are often used in production environments or as re-work stations, when production changes warrant changes to many circuit boards in production. Some repair shops used de-soldering stations to quickly and efficiently remove components.

Another useful de-soldering tool is one made specifically for removing integrated circuits. The specially designed de-soldering tip is made the same size as the integrated circuit, so that all IC pins can be de-soldered at once. This tool is often combined with a vacuum suction device to remove the solder as all the IC pins are heated. The IC de-soldering tips are made in various sizes; there are 8 pin, 14 pin and 16 pin versions, which are used to uniformly de-solder all IC pins quickly and evenly, so as not to destroy the circuit board. The specialized soldering tips are often used in conjunction with vacuum systems to remove the solder at the same time.

Remember These Things When Un-soldering:

1. Be sure there is a little melted solder on the tip of your iron so that the joint will heat quickly.

2. Work quickly and carefully to avoid heat damage to parts. Use long-nosed pliers to hold the leads of components just as you did while soldering.

3. When loosening a wire lead, be careful not to bend the lug or tie point. Use pieces of wire or some old radio parts and wire. Practice until you can solder joints that are smooth, shiny, and tight. Then practice unsoldering connections until you are satisfied that you can do them quickly and without breaking wires or lugs.

Caring for Your Soldering iron

To get the best service from your soldering iron, keep it cleaned and well tinned. Keep a damp cloth on the bench as you work. Before soldering a connection, wipe the tip of the iron across the cloth, then touch some fresh solder to the tip. The tip will eventually become worn or pitted. You can repair minor wear by filing the tip back into shape. Be sure to tin the tip immediately after filing it. If the tip is badly worn or pitted, replace it. Replacement tips can be found at most electronics parts stores. Remember that oxidation develops more rapidly when the iron is hot. Therefore, do not keep the iron heated for long periods unless you are using it. Do not try to cool an iron rapidly with ice or water. If you do, the heating element may be damaged and need to be replaced or water might get into the barrel and cause rust. Take care of your soldering iron and it will give you many years of useful service.

Remember, soldering equipment gets hot! Be careful. Treat a soldering burn as you would any other. Handling lead or breathing soldering fumes is also hazardous. Observe these precautions to protect yourself and others!

Ventilation

Properly ventilate the work area where you will be soldering. If you can smell fumes, you are breathing them. Often when building a new circuit or repairing a "vintage" circuit you may be soldering continuously for a few hours at a time. This can mean you will be breathing solder fumes for many hours and the fumes can cause you to get dizzy or lightheaded. This is dangerous because you could fall down and possibly hurt yourself in the process. Many people highly allergic are also allergic to the smell of solder fumes. Solder fumes can cause sensitive people to get sinus infections. So ventilating solder fumes is an important subject. There a few different ways to handle this problem. One method is to purchase a small fan unit housed with a carbon filter which sucks the solder fumes into the carbon filter to eliminate them. This is the most simple method of reducing or eliminating solder fumes from the immediate area. If there is a window near your soldering area, be sure to open the window to reduce the exposure to solder fumes.

Another method of reducing or eliminating solder fumes is to buy or build a solder smoke removal system. You can purchase one of these systems but they tend to be quite expensive. You can create you own solder smoke removal system by locating or purchasing an 8 to 10 foot piece of 2 inch diameter flexible hose, similar to your vacuum cleaner hose. At the solder station end of the hose you can affix the hose to a wooden stand in front of your work area. The other end of the hose is funneled into a small square "muffin" type fan placed near a window. Be sure to wash your hands after soldering, especially before handling food, since solder contains lead; also try to minimize direct contact with flux and flux solvents.

Chapter 4

Wonky Wire Game

Parts List

Parts Bin

R1,R3 1k ohm, 5%,
¼-watt resistor

R2 47k ohm
potentiometer

R4,R5 47k ohm, 5%,
¼-watt resistor

C1,C2 4.7nF, 35-volt
capacitor

Q1,Q2 2N3904 NPN
transistor or NTE123AP

BZ piezo buzzer

B1 9-volt transistor
radio battery

J1,J2 binding post
jacks

(2) 3 foot lengths of
#22 gauge stranded
insulated wire

(1) old ball point pen
housing

(1) #16 gauge un-
insulated solid wire

(1) #12-16 g solid un-
insulated wire Wonky
"Maze" wire

Misc PC board, wire,
chassis box, hardware,
etc.

The Wonky Wire Game is an old game idea that has lasted the "test of time"; it is a fun game for all ages. The Wonky Wire Game tests your hand–eye coordination as well as the stability of your hands. The Wonky Wire Game consists of an audio oscillator circuit shown in Figure 4-1 and a wire "maze" and handle as shown in

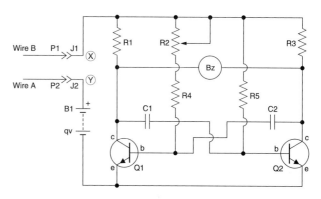

Figure 4-1 *Wonky Wire Game*

Figures 4-2 and 4-3. The object of the game is to guide the wire loop through the wire "maze," from one end of the wire to the opposite end, without touching the center loop to the "maze wire." If the wire loop touches the "maze wire" the buzzer will sound, telling you that your hand is unsteady, and maybe you should try again for a perfect score. The wire "maze" can be any length of stiff un-insulated solid gauge #18-12 ga. You can shape the "maze" wire in any configuration you desire with large or small loops or squares or bends.

The Wonky Wire circuit consists of an audio oscillator or astable multivibrator whose timing components are formed by R5/C1 and R4/C2. The oscillator produces an audio tone of about 3kHz. The oscillator is built around a piezo electric speaker which sounds when the handheld loop touches the "maze" wire. Note that the piezo speaker is connected between the collector leads of the two transistors Q1 and Q2. The collector voltage of Q1 oscillates between 0 and 9 volts, as does the collector voltage of Q2, but in an opposite way. This has the desirable effect of producing a voltage variation of 18 volts across the piezo speaker making the volume quite loud. The potentiometer at R2 is used to vary the pitch of the oscillator tone, so you can adjust the sound to your liking. The Wonky Wire Game is powered by a 9-volt transistor radio battery, which will last for a long time. The circuit requires no on-off switch since the Wonky wires complete the circuit when the loop touches the "maze" wire.

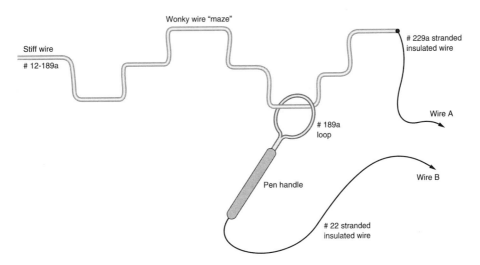

Figure 4-2   *Wonky Wire Game Pattern A*

The Wonky Wire Game circuit can be built in a number of different ways. You could elect to build the circuit using perf-board construction or point-to-point wiring, or you could choose to build the circuit on a small circuit board. The prototype circuit was built on a small circuit board as shown. Building time for the Wonky Wire circuit should only take about an hour or two, since there are no critical parts, integrated circuits or inductors.

Before we start the construction of the Wonky Wire Game, we are going to take a few moments to find a large table or workbench in a well-lit area in which to build the Wonky Wire project. First, locate a 27 to 33-watt pencil soldering iron with either a small

flat-blade tip or a small pointed tip. Also you will want to find a spool of #22 ga 60/40 rosin core solder along with a wet sponge, and some small tools, such as a pair of small needle-nose pliers, a pair of end-cutters or diagonals, a magnifying glass and a pair of tweezers. Round out your tools with a small flat-blade screwdriver and a Phillips screwdriver. Procure a small jar of "Tip Tinner," a cleaner/dressing compound and an anti-static wrist strap from your local Radio Shack store. The anti-static wrist band is a good idea as it will help prevent damage to the integrated circuits when handling them. Often just moving about near your workbench or getting up and down from your chair will be enough to cause large static voltages, which can damage the integrated

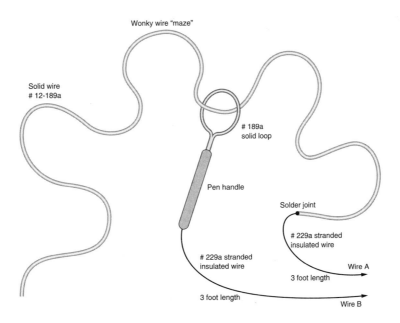

Figure 4-3   *Wonky Wire Game Pattern B*

Table 4-1
Resistor Color Code Chart

Color Band	1st Digit	2nd Digit	Multiplier	Tolerance
Black	0	0	1	
Brown	1	1	10	1%
Red	2	2	100	2%
Orange	3	3	1000 (K)	3%
Yellow	4	4	10000	4%
Green	5	5	100000	
Blue	6	6	1000000 (M)	
Violet	7	7	10000000	
Gray	8	8	100000000	
White	9	9	1000000000	
Gold			0.1	5%
Silver			0.01	10%
No color				20%

circuits. With all your tools in place, we can heat up the solder iron and begin constructing the project.

Now, place all the project parts in front of you along with the schematic, layout diagrams and charts and we will begin building the circuit. Refer to the chart shown in Table 4-1, which illustrates the resistor color chart and values. Notice that the first column in the table shows the resistor's first digit, from 0 to 9. The second column represents the resistor's second digit, from 0 to 9. The third color band represents the resistor's multiplier value. If the resistor has a fourth color band which is silver, then the resistor will have a 10% tolerance value, but if the resistor has a gold band then the resistor will have a 5% tolerance value. If the resistor has no fourth band then the resistor will have a 20% tolerance value. Let's try and find resistor R1. Look for a resistor which has a black color band, a brown color band and a red third band, this resistor will have a value of 1000 ohms or 1k ohms, and this resistor will be R1. Install resistor R1 on the circuit board as you refer to the schematic and pictorial or layout diagram. Next, identify the remaining resistors and install them on the circuit board. Now, solder all of the resistors to the circuit board, then follow-up by cutting the excess resistor leads with your end-cutters. Cut the resistor leads flush to the edge of the PC board.

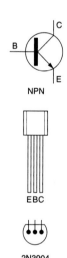

Figure 4-4   *Wonky Wire transistors*

Now, locate the potentiometer at P1. The potentiometer is a 47 k ohm trimmer type soldered to the circuit board. Potentiometers will have all three leads in a straight row, or it will have two outside leads with a center offset lead which is the adjustable wiper lead. Referring to the schematic and layout diagram, install the potentiometer to the PC board, then solder it in place. Remember to cut and remove the component leads.

Locate the chart at Table 4-2, which illustrates the code used for identifying small capacitors. In many instances,

Table 4-2

Capacitance Codebreaker Information

This table is designed to provide the value of alphanumeric coded ceramic, mylar and mica capacitors in general. They come in many sizes, shapes, values and ratings; many different manufacturers worldwide produce them and not all play by the same rules. Most capacitors actually have the numeric values stamped on them; however, some are color coded and some have alphanumeric codes. The capacitor's first and second significant number IDs are the first and second values, followed by the multiplier number code, followed by the percentage tolerance letter code. Usually the first two digits of the code represent the significant part of the value, while the third digit, called the multiplier, corresponds to the number of zeros to be added to the first two digits.

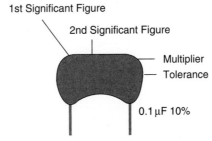

1st Significant Figure
2nd Significant Figure
Multiplier
Tolerance
0.1 µF 10%

Value	Type	Code	Value	Type	Code
1.5 pF	Ceramic		1000 pF/0.001 µF	Ceramic/Mylar	102
3.3 pF	Ceramic		1500 pF/0.0015 µF	Ceramic/Mylar	152
10 pF	Ceramic		2000 pF/0.002 µF	Ceramic/Mylar	202
15 pF	Ceramic		2200 pF/0.0022 µF	Ceramic/Mylar	222
20 pF	Ceramic		4700 pF/0.0047 µF	Ceramic/Mylar	472
30 pF	Ceramic		5000 pF/0.005 µF	Ceramic/Mylar	502
33 pF	Ceramic		5600 pF/0.0056 µF	Ceramic/Mylar	562
47 pF	Ceramic		6800 pF/0.0068 µF	Ceramic/Mylar	682
56 pF	Ceramic		0.01	Ceramic/Mylar	103
68 pF	Ceramic		0.015	Mylar	
75 pF	Ceramic		0.02	Mylar	203
82 pF	Ceramic		0.022	Mylar	223
91 pF	Ceramic		0.033	Mylar	333
100 pF	Ceramic	101	0.047	Mylar	473
120 pF	Ceramic	121	0.05	Mylar	503
130 pF	Ceramic	131	0.056	Mylar	563
150 pF	Ceramic	151	0.068	Mylar	683
180 pF	Ceramic	181	0.1	Mylar	104
220 pF	Ceramic	221	0.2	Mylar	204
330 pF	Ceramic	331	0.22	Mylar	224
470 pF	Ceramic	471	0.33	Mylar	334
560 pF	Ceramic	561	0.47	Mylar	474
680 pF	Ceramic	681	0.56	Mylar	564
750 pF	Ceramic	751	1	Mylar	105
820 pF	Ceramic	821	2	Mylar	205

(Continued)

Table 4-2 (*Continued*)

PicoFarad (pF)	NanoFarad (nF)	MicroFarad (mF, μF or mfd)	Capacitance Code
1000	1 or 1n	0.001	102
1500	1.5 or 1n5	0.0015	152
2200	2.2 or 2n2	0.0022	222
3300	3.3 or 3n3	0.0033	332
4700	4.7 or 4n7	0.0047	472
6800	6.8 or 6n8	0.0068	682
10000	10 or 10n	0.01	103
15000	15 or 15n	0.015	153
22000	22 or 22n	0.022	223
33000	33 or 33n	0.033	333
47000	47 or 47n	0.047	473
68000	68 or 68n	0.068	683
100000	100 or 100n	0.1	104
150000	150 or 150n	0.15	154
220000	220 or 220n	0.22	224
330000	330 or 330n	0.33	334
470000	470 or 470n	0.47	474

capacitors will have their values printed on the side of the capacitor body, but often smaller capacitors will not have room to print the full value on the capacitor body, so a three-digit code is used. This project uses two capacitors which will be small disk type capacitors which will most likely use the three-digit code. Look for two capacitors which will be marked with (472), which represents a 4.7 nF capacitor. Locate these capacitors and install them on the circuit board. Solder the capacitors to the PC board, and then remember to cut the excess component leads from the board.

Next we will install the two NPN transistors at Q1 and Q2. Look at the diagram depicted in Figure 4-4, which shows the transistor pin-outs. Note that the transistor's TO-92 plastic package is viewed from the bottom. When constructing the circuit be sure to install the transistors in the correct way to avoid damage to them when power is applied to the circuit. The BASE lead of the transistor is the one that enters perpendicular into the vertical line in the transistor symbol. The lower diagonal line with the arrow on it is the emitter lead and

it connects to the ground or minus (–) lead of the battery. The collector lead is the remaining transistor lead which connects to resistors R1 and R3. The collector Q1 connects to resistor R1, while the collector lead of Q2 connects to the lower lead of resistor R3. Install the two transistors and then solder them in place on the PC board. Follow-up by removing the excess component leads.

Solder in a 9-volt battery clip lead. Locate a 9-volt transistor radio battery and connect the battery to the circuit. Take a short length of wire and connect one end of the wire to point X on the circuit and the remaining free end of the wire to the point marked Y on the circuit board. When the two ends of the wire connect between points (X) and (Y) on the circuit board, the piezo speaker should sound. If all is working correctly the oscillator should begin to sound. If you don't hear any sounds from the speaker, re-check the circuit to make sure you installed the components correctly and that there are no shorts. You may also try to reverse the leads of the piezo speaker. Finally, make sure that the polarity of the battery is correct, the minus (–) lead of the battery connects to the common emitter leads of the transistors. The red or plus lead (+) of the battery should connect to one of the Wonky Wires.

Once the circuit board has been populated, you should inspect the circuit board for 'cold' solder joints. Place the circuit board in front of you with the foil side of the PC board facing upwards toward you. You will want to look over the solder joints to make sure that all the solder joints look smooth, shiny and clean. If you find any of the solder joints look dull or "blobby," then you need to unsolder the connection, remove the solder and solder the joint over again, until the joint looks smooth, clean and shiny. You will also want to inspect the circuit board for any "stray" component leads which may still be left from cutting component leads. Make sure that there are no "shorts" between the circuit board pads that don't belong there. Shorted circuit pads can damage the circuit when power is applied. Once the circuit board has been inspected, you can move on to powering and testing the circuit.

With the circuit complete you can move on to installing the circuit into a small plastic enclosure. As there are no switches or controls to mount on the chassis the layout and mounting should be quite simple. And as the potentiometer for sound pitch adjustment is a PC board mount, you will not have to worry about mounting this on the chassis. We elected to mount two binding post jacks on the top of the plastic box to allow connection to the Wonky Wires. The binding post jacks will allow you to quickly connect a new Wonky Wire of a different design challenge. The diagrams in Figures 4-2 and 4-3 illustrate two possible configurations for different Wonky Wires. Once you tire of one type of Wonky Wire you can design another more challenging Wonky Wire to test your friends and family members. You could even have a design contest between your friends to see who can design the most intricate Wonky Wire set-up.

You will need to fabricate a small wire loop to hold in your hand which is moved along the "maze wire." You could take an old ball point pen housing and form a wire loop at the ball point end, once the internal ink-pen is removed. You can make a small 1 inch to 1½ inch loop made from un-insulated solid #16 gauge wire. Solder a two to three foot length of #22 gauge insulated wire to the end of the loop and then pass the wire through the opposite end of the pen housing. In effect you have made a handle from the pen housing with a small loop at one end of the pen housing—see the diagram. The #22 gauge wire from the pen housing loop is connected to the binding post at point (Y) on the circuit board. Once you have constructed a Wonky Wire "Maze" you can solder one end of the "maze" wire to a second 3 foot length of solid insulated #22 gauge wire. Take the #22 gauge wire you soldered to the Wonky Wire "maze" and connect the opposite end of the wire to the binding post connected to the point marked (X) on the circuit board. The steadiest hand and good hand–eye coordination and practice will make you the winner in the Wonky Wire Game. Challenge your sisters and brothers or your mother and father or your friends to some great fun!

Chapter 5

Lie Detector

Parts List

Parts Bin

R1 4.7k ohm, 5%,
     ¼-watt resistor

R2 82k ohm, 5%,
     ¼-watt resistor

C1 .01µF, 35-volt disk
     capacitor

Q1 2N3904 NPN
     transistor

Q2 2N3906 PNP
     transistor

SPRK 8 ohm 2 or 3 inch
       speaker

B1 9-volt transistor
     radio battery

Misc PC circuit board,
      wire, touch plates,
      battery clip,
      enclosure, etc.

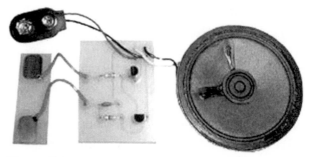

Figure 5-1 *Lie Detector*

The Lie Detector Test is a fun project that produces a tone, which varies in pitch depending upon a person's skin resistance at any given moment, see Figure 5-1. The circuit works on a principle of galvanic skin response (GSR), which is a change in the ability of the skin to conduct electricity, caused by an emotional stimulus, such as fright. A measure of electrical resistance as a reflection of changes in emotional arousal, taken by attaching electrodes to any part of the skin and recording changes in moment-to-moment perspiration and related activity of the autonomic nervous system.

   It is popularly known as a lie detector, but is also used in biofeedback conditioning. The theory is that the more relaxed you are the dryer your skin is and so the higher the skin's electrical resistance. When you are

under stress your hand sweats and then the resistance goes down. The overall range of resistance varies from 5 k to about 25 k ohms.

   The Galvanic Skin Response (or GSR) has been used by psychologists and psychiatrists for more serious research and study for many years. The simple psycho-galvanometer was one of the earliest tools of psychological research. A psycho-galvanometer measures the resistance of the skin to the passage of a very small electric current. It has been known for decades that the magnitude of this electrical resistance is affected, not only by the subject's general mood, but also by immediate emotional reactions. Although these facts have been known for over a hundred years and the first paper to be presented on the subject of the psycho-galvanometer was written by Tarchanoff in 1890, it has only been within the last 25 to 30 years that the underlying causes of this change in skin resistance have been discovered.

   The Tarchanoff Response is a change in DC potential across neurons of the autonomic nervous system connected to the sensory-motor strip of the cortex. This change was found to be related to the level of cortical arousal. The emotional charge on a word, heard by a subject, would have an immediate effect on the subject's level of arousal, and cause this physiological response. Because the hands have a particularly large representation of nerve endings on the sensory-motor strip of the cortex, hand-held electrodes are ideal. As arousal increases, the "fight or flight" stress response of the autonomic nervous system comes into action, and adrenaline causes increased sweating amongst many other phenomena, but the speed of sweating response is nowhere near as instantaneous or accurate as the Tarchanoff Response.

By virtue of the Galvanic Skin Response, autonomic nervous system activity causes a change in the skin's conductivity. The overall degree of arousal of the hemispheres, and indeed the whole brain, is shown by the readings of the GSR psychometer, which does not differentiate between the hemispheres, or between cortical and primitive brain responses. Higher arousal (such as occurs with increased involvement) will almost instantaneously (.2–.5 s) cause a fall in skin resistance; reduced arousal (such as occurs with withdrawal) will cause a rise in skin resistance.

Thus a rise or fall relates directly to reactive arousal, due to re-stimulation of repressed mental conflict. Initially this may cause a rise in resistance as this emerging, previously repressed, material is fought against. When the conflict is resolved, by the viewing of objective reality—the truth of exact time, place, form and event—there is catharsis and the emotional charge dissipates, the release of energy giving a fall in resistance.

Our low cost Lie Detector project is illustrated in the schematic shown in Figure 5-2. The project uses a two-transistor direct coupled oscillator that has a frequency determined by C1 and R2 and the skin resistance across the touch pads. As C1 and R2 are fixed values, only the skin resistance across the touch pads can vary. Transistor Q1 is an NPN type and transistor Q2 is a PNP type, which form a complementary pair oscillator. The output of Q2 is fed into a small speaker. The project relies on the fact that human skin conducts electricity. Actually, human skin is not a great conductor, but it conducts well enough, that at high voltage, we can receive dangerous shocks. The Lie Detector project operates from a low voltage source and is completely safe for adults and children. When a person gets nervous, you perspire and it causes salt to be added to your skin. This addition of salt lowers the skin resistance, causing the detector to produce a higher frequency sound from the speaker.

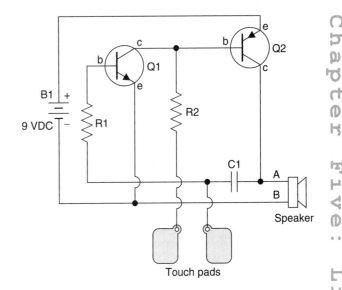

Figure 5-2 *Lie Detector schematic*

Using the Lie Detector can be great fun at a party. It does not truly tell if a person told a lie but it does indicate stress. Professional Lie Detection machines monitor up to six or more simultaneous physiological body functions at the same time and correlates all the parameters at once, thus trying to eliminate any false indications. However, professional Lie Detectors and their interpreters are not foolproof.

The "trick" to getting a person to respond to the Lie Detector is to ask a number of baseline questions which are not too stressful, like What is your name?; What is your address? After a few non-stress questions are asked, then the operator will ask a stressful or highly emotional question. If you know a person has stolen something, or if they have had relations with someone they think you do not know about, these are the type of questions that make people get stressed.

The circuit is quite simple, so you can build this fun Lie Detector project in about an hour or so from start to finish. You can build the circuit on a perf-board or on a small circuit board as desired. The touch pads for the prototype Lie Detector were made from a single piece of circuit board material. Two ¾-inch square copper pads were left on an etched PC board to form the touch pads. You could make the pads with other materials if desired; you just want to ensure clean solid connections between skin and the touch pads. So silver or gold connections would be even better than copper.

Before we begin the actual construction of the Lie Detector, let's take a moment to secure a table or workbench in a well-ventilated and well-lit area. You will need to locate a pencil tip soldering iron and a small roll of #22 ga 60/40 rosin core solder for this project. You will also want to collect a few small tools, such as a pair or needle-nose pliers, a pair of end-cutters or diagonals, a small flat-blade and a Phillips screw driver, as well as a pair of tweezers and a magnifying glass. Head out to your local Radio Shack store and grab a small jar of "Tip Tinner" which is used to clean and dress the soldering iron tip. Finally, find a small wet sponge for cleaning residue from the soldering iron tip. Place the schematic and the pictorial or layout diagram, shown in Figure 5-3, in front of you along with the project components and we will begin constructing the circuit. Heat-up your soldering iron and we will start the project.

Refer to the chart shown in Table 5-1, which represents the resistor color codes and their values. The first color band closest to the edge of the resistor end is the first digit. The second color band represents the second digit, while the third band is the resistor multiplier number. Often there is a fourth band which

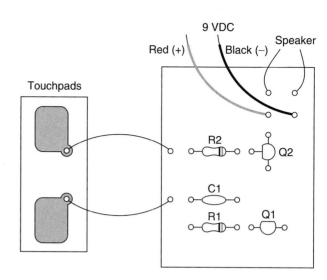

Figure 5-3   *Lie Detector pictorial diagram*

represents the resistor tolerance. If there is no band then the resistor has a 20% tolerance value. If the fourth band is silver then the resistor has a tolerance of 10%, while a gold band represents a 5% tolerance resistor. For example, if you have a resistor with its first band (yellow), the second band (violet) and the third band (red) then the resistor's value is 4.7 k or 4700 ohms. This will be resistor be R1. Now install the resistor R1

Table 5-1
Resistor Color Code Chart

Color Band	1st Digit	2nd Digit	Multiplier	Tolerance
Black	0	0	1	
Brown	1	1	10	1%
Red	2	2	100	2%
Orange	3	3	1000 (K)	3%
Yellow	4	4	10000	4%
Green	5	5	100000	
Blue	6	6	1000000 (M)	
Violet	7	7	10000000	
Gray	8	8	100000000	
White	9	9	1000000000	
Gold			0.1	5%
Silver			0.01	10%
No color				20%

Table 5-2

Capacitance Codebreaker Information

This table is designed to provide the value of alphanumeric coded ceramic, mylar and mica capacitors in general. They come in many sizes, shapes, values and ratings; many different manufacturers worldwide produce them and not all play by the same rules. Most capacitors actually have the numeric values stamped on them; however, some are color coded and some have alphanumeric codes. The capacitor's first and second significant number IDs are the first and second values, followed by the multiplier number code, followed by the percentage tolerance letter code. Usually the first two digits of the code represent the significant part of the value, while the third digit, called the multiplier, corresponds to the number of zeros to be added to the first two digits.

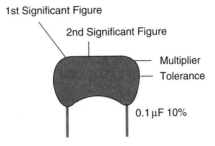

Value	Type	Code	Value	Type	Code
1.5 pF	Ceramic		1000 pF/0.001 µF	Ceramic/Mylar	102
3.3 pF	Ceramic		1500 pF/0.0015 µF	Ceramic/Mylar	152
10 pF	Ceramic		2000 pF/0.002 µF	Ceramic/Mylar	202
15 pF	Ceramic		2200 pF/0.0022 µF	Ceramic/Mylar	222
20 pF	Ceramic		4700 pF/0.0047 µF	Ceramic/Mylar	472
30 pF	Ceramic		5000 pF/0.005 µF	Ceramic/Mylar	502
33 pF	Ceramic		5600 pF/0.0056 µF	Ceramic/Mylar	562
47 pF	Ceramic		6800 pF/0.0068 µF	Ceramic/Mylar	682
56 pF	Ceramic		0.01	Ceramic/Mylar	103
68 pF	Ceramic		0.015	Mylar	
75 pF	Ceramic		0.02	Mylar	203
82 pF	Ceramic		0.022	Mylar	223
91 pF	Ceramic		0.033	Mylar	333
100 pF	Ceramic	101	0.047	Mylar	473
120 pF	Ceramic	121	0.05	Mylar	503
130 pF	Ceramic	131	0.056	Mylar	563
150 pF	Ceramic	151	0.068	Mylar	683
180 pF	Ceramic	181	0.1	Mylar	104
220 pF	Ceramic	221	0.2	Mylar	204
330 pF	Ceramic	331	0.22	Mylar	224
470 pF	Ceramic	471	0.33	Mylar	334
560 pF	Ceramic	561	0.47	Mylar	474
680 pF	Ceramic	681	0.56	Mylar	564
750 pF	Ceramic	751	1	Mylar	105
820 pF	Ceramic	821	2	Mylar	205

(Continued)

Table 5-2 (*Continued*)

PicoFarad (pF)	NanoFarad (nF)	MicroFarad (mF, μF or mfd)	Capacitance Code
1000	1 or 1n	0.001	102
1500	1.5 or 1n5	0.0015	152
2200	2.2 or 2n2	0.0022	222
3300	3.3 or 3n3	0.0033	332
4700	4.7 or 4n7	0.0047	472
6800	6.8 or 6n8	0.0068	682
10000	10 or 10n	0.01	103
15000	15 or 15n	0.015	153
22000	22 or 22n	0.022	223
33000	33 or 33n	0.033	333
47000	47 or 47n	0.047	473
68000	68 or 68n	0.068	683
100000	100 or 100n	0.1	104
150000	150 or 150n	0.15	154
220000	220 or 220n	0.22	224
330000	330 or 330n	0.33	334
470000	470 or 470n	0.47	474

on the PC circuit board, and solder it in place. Use your end-cutters to cut the excess resistor lead off the circuit board. Cut the leads flush to the edge of the PC board. Next locate resistor R2, an 82 k ohm resistor and install it on the circuit board. Remember to trim the excess resistor leads from the board.

Now refer to the chart depicted in Table 5-2; it illustrates the capacitor code often used to identify small capacitors. Large capacitors will often have their value printed directly on the body of the component. But in many instances, small capacitors will not have adequate room to print the actual value of the part on the body, so a three-digit code is used. The Lie Detector circuit has a single capacitor and it is a small disk capacitor which will be marked with (103). From the chart you will see that this capacitor has a value of 0.01 μF. Next install capacitor C1, then solder it in place, and follow-up by removing the excess component leads. Now locate the two transistors and refer to the transistor pin-out diagram shown in Figure 5-4 for installation details. These transistors will have three leads: a Base, a

Collector and an Emitter. Often the pin-outs can be different between transistors so pay careful attention to the pin-outs before installing them into the circuit.

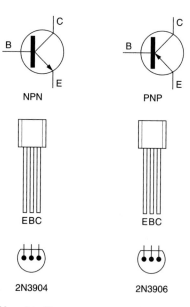

Figure 5-4 *Lie Detector semiconductor pin-out diagram*

Many times the part will be shown with either a top or bottom view, so make sure that you note the difference. The two transistors are set opposed to each other, note that Q1 is an NPN type transistor, while transistor Q2 is a PNP type. Note that the Emitter of Q1 is connected to the minus (–) of the power supply while the Emitter of Q2 is connected to the plus (+) power supply lead. The Emitter of Q1 is connected to one speaker lead while the collector of Q2 is connected to the opposite speaker lead.

Prepare two 6-inch #22 ga stranded-insulated wires for connecting the speaker to the circuit board; solder one end of each wire to the speaker. Mark one speaker wire (A) and it will go to the junction of C1 and the Collector of Q2. The other speaker lead can be marked (B) and it will be connected to the Emitter of Q1 and the minus side of the power source. Next, prepare two more 6-inch # 22 ga wires which will be used to connect the touch pads to the circuit board. One touch pad lead will connect between the Collector of Q1 and the Base of Q2 and the second touch pad lead will connect between R1 and C1. Next, solder the 9-volt battery clip leads to the circuit. The black or minus lead (–) will connect between the Emitter of Q1 and speaker lead (B). The red or positive (+) battery clip lead will be connected to the Emitter of transistor Q2.

Before applying power to the circuit, it is a good time to recheck your wiring and solder joints. Pick up the circuit board and place the PC board in front of you with the foil side facing upwards toward you. First we will inspect the circuit to make sure that all the solder joints look clean, smooth and shiny. If they don't look good, you can clean the joints and re-solder them so they look clean and shiny. Next, look over the board and be sure that there are no "stray" component leads attached to the circuit board from when you were constructing the circuit. Sometimes "cut" lead will adhere to the board and these leads could cause the circuit to "short out" thus destroying the circuit. Often the cause of component leads "sticking" on the board is the residue left from the rosin solder. It is a good idea to inspect the board and remove any foreign objects before applying power to the circuit.

Once inspection has been completed you can attach the battery to the circuit. When the battery has been attached, you can place two fingers from the same hand on the touch pads. You should hear some sound from the speaker. If you moisten your fingers and place them back on the pads, the sound should change. If all is well, you can begin using your new Lie Detector. If the circuit doesn't work, then you will need to remove the battery and reinspect the circuit for errors. First make sure that the resistors are all in the right place and that the color codes are correct for each resistor at its particular location. Next, check the transistor leads against the diagram and make sure you have connected the right pins of the transistor to the right locations on the circuit. Once you have made a final inspection, you can re-attach the battery and try the circuit again. Hopefully this time the circuit works correctly and you can begin testing your sister or brother or your friends. This circuit is great fun at parties and get togethers. Have fun!

Chapter 6

Macho Meter

Parts List

R1   4.7 k ohm, 5%, ¼-watt
resistor

R2   100 ohm, 5%, ¼-watt
resistor

C1   0.1 µF, 35-volt
capacitor (104)

U1   LM555 IC timer
(National Semiconductor)

U2   CD4017 CMOS counter
(National Semiconductor)

L1, L2, L3, L4, L5 red
LEDs

B1   9-volt transistor
radio battery

Sx   SPST optional on/off
switch

Misc PC board, battery
clip, IC socket, wire,
decal, enclosure, etc.

Are you "macho" or a "nerd," or are you a "jock" or a "wimp." What are you? This project will bring you back to the days of the "Penny Arcade" on-the-boardwalk, but with a modern twist. Turning on the Macho Meter displays five LEDs flashing in a random fashion. When you place your fingers on the "touch pad" and remove them, "your rating" will be indicated by the LED which remains "on." The Macho Meter shown in Figure 6-1 has five indicator LEDs marked MACHO, JOCK, SMART, WIMP, and NERD. This game is silly but lots of fun at parties, it's perfectly safe and operates from a single 9-volt battery.

The Macho Meter circuit shown in Figure 6-2 uses two integrated circuits to make a fun project. When you place your fingers across the "touch pads," you are

Figure 6-1   *Macho Meter*

completing the electrical path between pins 7 and 2 and 6 of U1. The purpose of U1 is to provide clock pulses to the input of U2. An LM55 timer IC is used at U1, and is configured as an astable multivibrator or oscillator. Its frequency is determined by capacitor C1, and the resistance of your fingers connected from pin 7 and pins 2 and 6. Since C1 is a fixed value, you control the frequency of the pulses by the resistance of your fingers. Once the clock pulses are generated, they are direct coupled to the input of U2. This IC is a CD4017, which is set up as a Johnson Counter. With each incoming clock pulse, the Johnson Counter shifts its

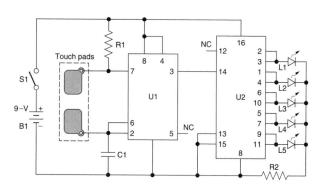

Figure 6-2   *Macho Meter schematic. Courtesy Chaney (CH)*

output to the next output pin. We have connected the 10 output pins into an arrangement of five sets of two connected in parallel. This allows us to make a random-looking display. The Macho Meter continues its sequence through the five outputs lighting each LED until the clock pulse stops, i.e. when you remove your fingers from the "touch pads." At this point, i.e. when the clock signal stops, one LED remains "on" thus indicating your "rating."

The Macho Meter was constructed on a 2 inch × 5 inch circuit board shown in the pictorial diagram Figure 6-3, where the display template lies over the LED display at the top of the circuit board. Let's first begin by preparing a clean, well-lit work area, where you can spread out all the diagrams and the circuit in front of you. Locate a pencil type 27 to 33-watt soldering, a small jar or "Tip Tinner," i.e. a tip cleaner and dresser compound, from your local Radio Shack store. Locate some #22 ga 60/40 rosin core solder, and a small wet sponge to clean the soldering tip. Next, try to find an anti-static wrist band, which is used to handle integrated circuits thus preventing damage from static build-up when assembling the circuit. High voltage static charges can readily mount up as you are moving around building a circuit, getting up from your chair or walking around during construction and can build-up a significant charge which can damage integrated circuits. Heat up the solder iron and clean the solder iron tip and we will begin.

Place the schematic and pictorial diagram in front of you, along with all the project components. Next, locate the project resistors; there are only two resistors in this project. You will need first to identify the color codes on each of the resistors to determine their values before installing them on the circuit board. Table 6-1 lists the color code information for resistors. Most resistors will usually have three or four color bands on them. The color bands usually start at one edge of the resistor body. The first color band is the first digit, and the second color band is the second digit of the resistor's value. The third color band is the tolerance value of the resistor. If there is no fourth color band then the resistor has a 20% tolerance, if the fourth band is silver then the resistor has a 10% tolerance, and if there is a gold color band then the resistor has a 5% tolerance value. Try to locate resistor R1; its first color band will be yellow, its second color band will be violet and the third color

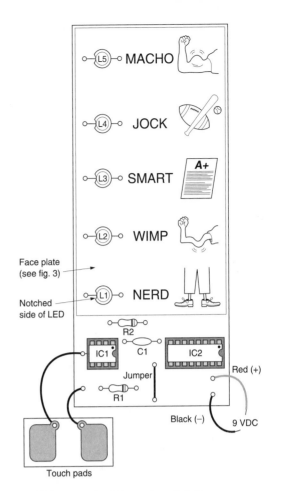

Figure 6.3 *Macho Meter pictorial diagram*

band will be red, thus resistor value will 4,700 ohms. Now go ahead and install the two resistors and solder them in place: take your end-cutters and cut the excess resistor leads flush to the edge of the PC board.

Next, refer to Table 6-2, which lists the capacitor codes which you may find on project capacitors. Try to locate capacitor C1; it should be a red or orange disk capacitor with a code marking of (104). You can look up the code for (104) and you will see that the capacitor will have a value of 0.1 μF. Go ahead and install the capacitor and solder it in place on the circuit board. Remember to remove the excess leads.

Next, you will want to locate two integrated circuit sockets for Macho Meter project. IC sockets will make your life a lot easier in the event of a circuit failure, should this ever occur. Locate an 8 pin and a 14 pin IC socket and install them on the circuit board. On the socket, you should see a small notch or cutout. Just to the left of the notch is pin 1. Next, identify the

Table 6-1

Resistor Color Code Chart

Color Band	1st Digit	2nd Digit	Multiplier	Tolerance
Black	0	0	1	
Brown	1	1	10	1%
Red	2	2	100	2%
Orange	3	3	1000 (K)	3%
Yellow	4	4	10000	4%
Green	5	5	100000	
Blue	6	6	1000000 (M)	
Violet	7	7	10000000	
Gray	8	8	100000000	
White	9	9	1000000000	
Gold			0.1	5%
Silver			0.01	10%
No color				20%

integrated circuits; you will notice that the ICs will have a notch, cutout or small indented circle at one end of the IC package. Just to the left of the notch or cutout will be pin 1. Locate both integrated circuits from the parts pile. Refer to the semiconductor pin-out diagram shown in Figure 6-4. Integrated circuit U1 is an LM555 timer chip, while U2 is a CD4017 CMOS counter chip. Make sure that you can identify pin 1 on the IC socket and that it corresponds to pin 1 on the IC itself before inserting the IC into its socket. Installing the IC backwards can often damage the component permanently. If you are having difficulty identifying the IC pin-outs, ask a knowledgeable electronics friend help you.

Now, let's locate the red LEDs and identify their leads before installing them. Refer to the diagram shown in Figure 6-5, which illustrates the LED leads. You will note that LEDs usually have a flat edge on one side of the IC package, this flat edge corresponds to the shorter lead which is the ground or minus (–) lead of the LED. The minus (–) lead also corresponds to the dark half moon on the diagram; this should help you to identify the LED leads. You can begin installing the LEDs onto the circuit board. Note the ground or minus leads will all be connected together and join the free

end of resistor R2, as shown. As discussed earlier, the output pins on U2 are tied together and connected to an LED. For example, note that output pins 2 and 3 are connected together to LED L1, and pins 1 and 4, which connect to LED L2, and so forth. Install the LEDs onto the circuit board, then solder them in place. Be sure to cut the excess leads with your diagonals or end-cutters.

The "touch pads" were constructed from two small scrap pieces of circuit board with copper pads on each of them. You could also use a single piece of circuit board with two copper pads on the board. You could also elect to use two large screw heads as possible "touch pads." Each screw would have to be mounted on the main circuit board and a lead wire from one screw "touch pad" would have to be connected to pins 2 and 6 of U1, while the second "touch pad" would be connected to R1 and pin 7 of U1. If your "touch pads" are remote to the main circuit board then you will have to prepare two six inch lengths of number #22 ga stranded-insulated wires to connect the "touch pads" to the main circuit board.

The Macho Meter is powered from an ordinary 9-volt transistor radio battery. In the schematic, there is an optional SPST switch that was placed in series with the red or plus (+) battery clip lead to allow the circuit to be

Table 6-2

Capacitance Codebreaker Information

This table is designed to provide the value of alphanumeric coded ceramic, mylar and mica capacitors in general. They come in many sizes, shapes, values and ratings; many different manufacturers worldwide produce them and not all play by the same rules. Most capacitors actually have the numeric values stamped on them; however, some are color coded and some have alphanumeric codes. The capacitor's first and second significant number IDs are the first and second values, followed by the multiplier number code, followed by the percentage tolerance letter code. Usually the first two digits of the code represent the significant part of the value, while the third digit, called the multiplier, corresponds to the number of zeros to be added to the first two digits.

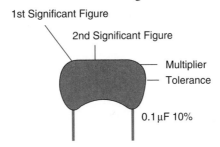

Value	Type	Code	Value	Type	Code
1.5 pF	Ceramic		1000 pF/0.001 µF	Ceramic/Mylar	102
3.3 pF	Ceramic		1500 pF/0.0015 µF	Ceramic/Mylar	152
10 pF	Ceramic		2000 pF/0.002 µF	Ceramic/Mylar	202
15 pF	Ceramic		2200 pF/0.0022 µF	Ceramic/Mylar	222
20 pF	Ceramic		4700 pF/0.0047 µF	Ceramic/Mylar	472
30 pF	Ceramic		5000 pF/0.005 µF	Ceramic/Mylar	502
33 pF	Ceramic		5600 pF/0.0056 µF	Ceramic/Mylar	562
47 pF	Ceramic		6800 pF/0.0068 µF	Ceramic/Mylar	682
56 pF	Ceramic		0.01	Ceramic/Mylar	103
68 pF	Ceramic		0.015	Mylar	
75 pF	Ceramic		0.02	Mylar	203
82 pF	Ceramic		0.022	Mylar	223
91 pF	Ceramic		0.033	Mylar	333
100 pF	Ceramic	101	0.047	Mylar	473
120 pF	Ceramic	121	0.05	Mylar	503
130 pF	Ceramic	131	0.056	Mylar	563
150 pF	Ceramic	151	0.068	Mylar	683
180 pF	Ceramic	181	0.1	Mylar	104
220 pF	Ceramic	221	0.2	Mylar	204
330 pF	Ceramic	331	0.22	Mylar	224
470 pF	Ceramic	471	0.33	Mylar	334
560 pF	Ceramic	561	0.47	Mylar	474
680 pF	Ceramic	681	0.56	Mylar	564
750 pF	Ceramic	751	1	Mylar	105
820 pF	Ceramic	821	2	Mylar	205

(Continued)

Table 6-2 (Continued)

PicoFarad (pF)	NanoFarad (nF)	MicroFarad (mF, µF or mfd)	Capacitance Code
1000	1 or 1n	0.001	102
1500	1.5 or 1n5	0.0015	152
2200	2.2 or 2n2	0.0022	222
3300	3.3 or 3n3	0.0033	332
4700	4.7 or 4n7	0.0047	472
6800	6.8 or 6n8	0.0068	682
10000	10 or 10n	0.01	103
15000	15 or 15n	0.015	153
22000	22 or 22n	0.022	223
33000	33 or 33n	0.033	333
47000	47 or 47n	0.047	473
68000	68 or 68n	0.068	683
100000	100 or 100n	0.1	104
150000	150 or 150n	0.15	154
220000	220 or 220n	0.22	224
330000	330 or 330n	0.33	334
470000	470 or 470n	0.47	474

turned "on" and "off" easily. The red or plus (+) battery wire is connected to R1 and pins 4 and 8 of U1, while the black or minus (–) battery lead is connected to pin 1 of U1 and pin 8 of U2 as shown.

Let's take a short break and when we return we will inspect the printed circuit board for any errors. First turn the circuit board over, so that the copper foil is facing upwards towards you. Take a look at all the solder joints to make sure that they all look smooth, clean and shiny. If you spot any solder joints that look dull or "blobby" then you will want to de-solder the connection and re-solder the joint to make sure it looks good. "Cold" solder joints can cause an electronic circuit to fail prematurely, as the connection oxidizes. Next, take a second look at the PC board; this time you will be looking for any "stray" component leads which may have attached themselves after the component leads were cut. These "stray" leads can cause the circuit to "short-out" causing the circuit to fail upon power-up.

Let's take another short break and when we return we are going to inspect the circuit board for any possible "cold" solder joints or "short" circuits. Pick up the circuit board, so that the foil side of the PC board is facing upwards toward you. First we are going to look for cold solder joints. It is important to locate "cold" solder joints early, since they can cause circuit failure later if not found: the joint will oxidize and form a semiconductor where you do not want a semiconductor. You will want to make sure that all the solder joints look clean, smooth and shiny. If you find a solder joint that looks dull or "blobby," you will want to un-solder the joint and clean it up and then re-solder it again, so the joint looks clean, smooth and shiny like the others. Next, we are going to inspect the circuit board for any possible "short" circuits which could be caused by "stray" wires or excess component leads that got "stuck" to the circuit board. Oftentimes a small amount of residue from the rosin inside the solder which is sticky will grab and hold foreign object to the board. You will want to remove any metal pieces, flecks, wire or leads from the bottom of the circuit board to avoid any shorts which could destroy the circuit when power is first applied.

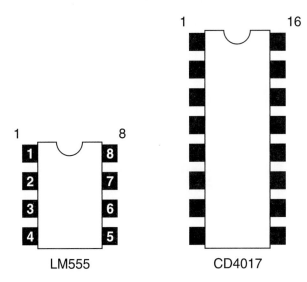

Figure 6-4   *Macho Meter semiconductor diagram*

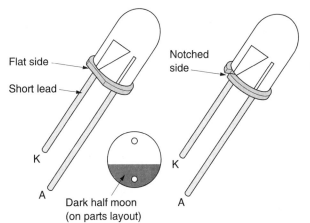

Figure 6-5   *LED identification*

At this point, if everything looks good then you are ready to connect up a 9-volt battery and test the circuit. Insert a battery into the battery clip and turn "on" the optional switch that is installed. One of the LEDs should light. Now place your fingers across the pads on the "touch plate." The LEDs L1-L5 should begin to randomly flash. When you remove your finger from the "touch pads," your Macho "rating" is shown by which LED remains "on." If your Macho Meter is working, punch holes in the decal with a paper punch and mount on the circuit board or case, if desired. If you elect to use a plastic enclosure to house the Macho Meter, then you can simply glue the "touch pads" to the top of the outside of the case. Remember to remove the battery or turn the unit off when finished, to ensure that the battery will not be drained by accidently leaving the unit in the "on" position.

In the event that your Macho Meter does not work properly, you will need to remove the battery and inspect the circuit board for any wiring mistakes or parts placement errors. It is very to easy to make a mistake when inserting components into the circuit board.

There are at least three possible causes for circuit failure. First, you may have inserted the integrated circuits backwards in their respective sockets. Make sure you have inserted the IC into its socket with the correct orientation, and that pin 1 of the IC is aligned with pin 1 of the socket and that pin 1 of the IC is connected to the correct surrounding components. Another possible cause for error is installing the LEDs backwards. Recheck the installation of the LEDs, and refer to the LED mounting diagram to make sure that you have installed the LEDs correctly. Re-check that you have connected the battery clips to right locations on the circuit board. Once you have re-checked for all of these possible causes for error, then you can reconnect the battery and try the circuit out once again. Hopefully you have discovered your error and corrected it without damaging the circuit. With the circuit now working correctly, you and your friends can have the Most Macho Contest! Have fun!

Chapter 7

Mood Meter Project

Part List

Parts Bin

R1, R2, R3 470 ohm, 5%, ¼-watt resistor

R4, R5, R6 470 ohm, 5%, ¼-watt resistor

R7 390 k ohm, 5%, ¼-watt-resistor

R8, R9 12 ohm, 5%, ¼-watt resistor

P1 5 k trimmer potentiometer

U1 LM339 comparator (National Semiconductor)

L1 Blue LED

L2, L3 Green LEDs

L4, L5, L6 Yellow LEDs

L7, L8, L9, L10 Red LEDs

B1 9-volt battery

S1 SPST on/off switch (optional)

Misc PC board, IC socket, wire, label, touch pads, battery clip, etc.

With the Mood Meter you can "see" who is the most "cool" or the most calm of all your friends and family! The Mood Meter is an easy-to-build and fun project, see Figure 7-1. You simply set two fingers on the two "sensor pads" and the sophisticated comparator circuit will determine your skin resistance and light up the appropriate rating. The blue band indicates "Cool as a

Figure 7-1 *Mood Meter Game*

Cucumber," the green band indicates "Easy Going," while the yellow band says "Warm Hearted," and finally the red band claims you're "Smokin' Hot". The LEDs light up the rainbow in color and more color "bands" if your skin resistance is lower. So if the red band lights, all the other colors will be on also. This project operates on a 9-volt battery and is safe for children.

The Mood Meter project uses a quad comparator integrated circuit, an LKM339 that is configured to measure skin resistance, see Figure 7-2. The four comparators inside U1 each drive an LED or a group of LEDs. As the resistance between the "touch pads" decreases more current flows into the input pins of the comparator at pins 6, 4, 8 and 10, causing more groups of LEDs to light up. As soon as the resistance of your fingers between the "touch pads" is removed the LEDs go out. Trimmer potentiometer P1 sets the "turn on" point of the comparator circuit and is used to set the triggering level. Resistors R8, R9, R1 and R2 limit the drive current to the LEDs. Resistors R7, R6, R5, R3 and R4 set the level of the "turn on" current for the comparators. Power for the circuit is supplied by an ordinary 9-volt transistor radio battery.

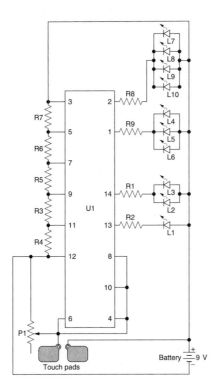

Figure 7-2 *Mood Meter schematic. Courtesy Chaney (CH)*

Are you ready to build the Mood Meter? We built the Mood Meter on a 3½ inch × 4 inch circuit board, which hosted the entire circuit with LEDs and "touch pads."

First let's prepare a suitable clean and well-lit work area with enough room to spread out the project parts and all the diagrams. Locate a small pencil tip 27 to 33-watt soldering iron. Next, find some "Tip Tinner" from your local Radio Shack store. "Tip Tinner" is a soldering iron tip cleaner and dresser. Now locate some # 22 ga 60/40 rosin core solder, as well as a small wet sponge, used to clean the soldering iron tip. You will also want to locate a pair of needle-nose pliers, a pair of diagonals or wire-cutters, a pair of tweezers and a magnifying glass. Try to locate a small flat-head and Phillips screwdrivers. Finally, you will want to find an anti-static wrist strap, usually available from Radio Shack. The anti-static will prevent high voltage static charges from damaging the integrated circuits as they are being handled. Often just by getting up from your chair or moving or walking around your worktable, you can build up static charges. The anti-static wrist strap will plug into your household outlet and ground the strap. Locate the schematic shown in Figure 7-2 and the pictorial or layout diagram shown in Figure 7-3, then place all the project parts in front of you. Plug in your soldering iron, and clean the tip.

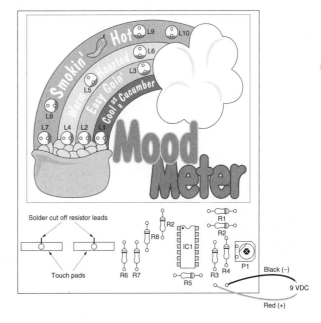

Figure 7-3 *Mood Meter pictorial diagram*

Refer to the resistor color code chart in Table 7-1. Locate and identify all the project resistors. Each resistor will have three or four color codes which identify the resistor value. The first color band is the first digit, the second color band is the second digit, while the third color band is the multiplier value; the fourth band, if there is one, is the tolerance value for the resistor. No fourth color band indicates a 20% tolerance value, while a silver band denotes a 10% tolerance and a gold band represents a 5% tolerance value. Locate resistor R1; it should have a 470 ohm value, so its first band would be yellow, its second band would be violet and its third band would be brown. Make sure you can identify all of the resistors. Note that there are six 470 ohm resistors, two 12 ohm resistors and one 390 k ohm resistor in this project. When you have identified the resistor color codes, you can locate the circuit board placement and solder the resistors into the circuit. After soldering the resistors to the circuit board, you can use your end-cutters to cut the excess resistor leads flush from the edge of the PC board.

Next, locate the trimmer potentiometer P1. Trimmer potentiometers will have three leads in a row, with the center lead denoting the wiper arm of the potentiometer. But some trimmers will have two end leads with an offset center wiper arm lead. Install the potentiometer now. Solder it in place and remember to remove the excess components leads.

Table 7-1

Resistor Color Code Chart

Color Band	1st Digit	2nd Digit	Multiplier	Tolerance
Black	0	0	1	
Brown	1	1	10	1%
Red	2	2	100	2%
Orange	3	3	1000 (K)	3%
Yellow	4	4	10000	4%
Green	5	5	100000	
Blue	6	6	1000000 (M)	
Violet	7	7	10000000	
Gray	8	8	100000000	
White	9	9	1000000000	
Gold			0.1	5%
Silver			0.01	10%
No color				20%

There are no capacitors in this project, so we will not have to worry about them: there is just a single integrated circuit. Locate a 14 pin IC socket for the LM339 comparator IC. The use of IC sockets is a good idea, since no one can predict if the circuit will ever fail, and if it does you can easily replace the IC without having to un-solder the part from the circuit board. Most people cannot un-solder integrated circuits without damaging the PC board, so an IC socket is a really good idea. The Mood Meter has only a single integrated circuit, as can be seen in the semiconductor pin-out diagram shown in Figure 7-4. Find the LM339 comparator IC, and take a look at the IC package. Carefully handle the IC using your anti-static wrist band, so that you do not damage the IC with static electrical potential. When you handle an IC make sure that you are either standing or seated but not moving around. If you have an anti-static wrist band, ground it at an outlet and place the wrist band on the hand that is handling the IC. Most integrated circuits will have some sort of identification or marking which indicates its orientation. Look at the IC and you will find either a small indented circle at one end of the IC package or a small notch or cutout at the center of one edge of the package. Note that pin 1 of the IC will be to the left of

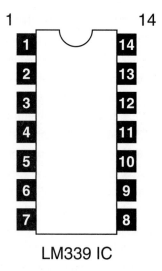

Figure 7-4 *Mood Meter semiconductor diagram*

either of these identifiers. Make sure that you can also identify which socket hole on the IC socket is pin 1 before inserting the IC into the socket. If the IC is not oriented correctly upon power-up, the circuit could be damaged. Refer to the schematic and layout diagrams to help you. If you find you are still too confused, contact a knowledgeable electronic friend who can help you.

Now let's move on to installing the LEDs into the circuit. Refer to the diagram in Figure 7-5, which

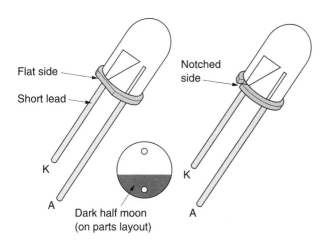

Flat side

Short lead

Notched side

K

A

Dark half moon
(on parts layout)

K

A

Figure 7-5 *LED identification*

illustrates the LED packaging and pin-outs. Note that each LED should have a flat edge on one side of the LED package. The shorter LED lead corresponds to the flat edge and the LEDs cathode. Look at the diagram and you will also see that the negative lead or cathode is depicted as the dark half moon. When placing the LEDs into the circuit board refer to the pictorial diagram in Figure 7-3, which shows the LED layout scheme. Identify the LEDs and note that the blue LED is L1 and it is mounted on the bottom left in the "Cool as a Cucumber" blue color band. LEDs L2 and L3 are green LEDs mounted in the green color band. LEDs marked L4, L5 and L6 are yellow LEDs which are mounted on the yellow color band. Finally red LEDs L7, L8, L9 and L10 are all mounted in the red color band marked "Smokin' Hot" at the far left.

The "touch pads" were integrated on the PC board, but you could choose to fabricate a small separate circuit board for them if desired. One "touch pad" is connected to the junction of pins 4, 6, 8 and 10 of U1 and the potentiometer P1. The other "touch pad" is connected to the plus (+) battery connection.

Once the circuit has been completed you can solder in the 9-volt battery clip. The red or positive (+) lead connects to the LED common point and one of the "touch pads" as mentioned. The black or minus (–) battery lead is connected to pin 12 of U1 and the junction of P1.

Take a short break and then when we return we will inspect the PC board for bad solder joints and possible "short" circuits. Turn the circuit board over, so that the foil or copper side faces up toward you. Make sure that

all the solder joints look clean, smooth and shiny. If a solder joint looks dull or "blobby" then you will have to remove the solder with a "wick" or solder "sucker" and re-solder the joint so that it looks clean, smooth and shiny.

Now let's move on to inspecting the PC board for possible "shorts." Look over the board carefully and see if there are any pieces of bare copper wire "stuck" to the copper foil side of the board. Often "cut" component leads stick to the bottom side of the PC board. If these leads are left on the PC board they can cause a "short" circuit or "bridge" and possibly damage the circuit when power is first applied.

Your Mood Meter is now complete and ready to test, and the "moment of truth" is at hand! Snap in a 9-volt battery. Note that you may want to install an optional on/off switch in series with the red or plus (+) battery in order to prevent the battery running down when the circuit is not in use, and if you do not remove the battery promptly when you are finished using the circuit.

With power first applied, adjust potentiometer P1, so that all the LEDS are lit up. To use the Mood Meter, simply have a person rest two fingers on the "touch plate." If the person's skin resistance is low (from exercise), all the LEDS will light up. If the person's skin resistance is higher, only the blue LED will light or possibly also the green LEDs and yellow LEDs. Note, the lower the skin resistance, the more LEDS will light up. Re-adjust P1 if necessary. Also note that pressing too firmly on the "touch pads" with your fingers will almost always cause all the lights to light up. This is not how the project should be operated. The person to be "tested" should just "rest" his or her fingers on the "touch pads."

If the Mood Meter circuit did not "come to life" when power was applied, then you will have to remove the battery and re-inspect the circuit to make sure that the integrated circuit and the LEDs have been installed correctly. If a particular LED does not light-up when it should, you will know that the LED is probably installed backwards. Just unsolder it, remove the solder from the joint with a "wick" or "solder sucker" and then re-solder the joint. If the circuit does not appear to work at all then you will have to troubleshoot farther into the circuit. Next, you will need to make sure that you installed the correct value of resistors at the right

Table 7-2

Capacitance Codebreaker Information

This table is designed to provide the value of alphanumeric coded ceramic, mylar and mica capacitors in general. They come in many sizes, shapes, values and ratings; many different manufacturers worldwide produce them and not all play by the same rules. Most capacitors actually have the numeric values stamped on them; however, some are color coded and some have alphanumeric codes. The capacitor's first and second significant number IDs are the first and second values, followed by the multiplier number code, followed by the percentage tolerance letter code. Usually the first two digits of the code represent the significant part of the value, while the third digit, called the multiplier, corresponds to the number of zeros to be added to the first two digits.

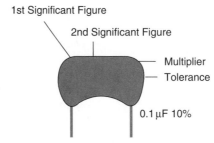

Value	Type	Code	Value	Type	Code
1.5 pF	Ceramic		1000 pF/0.001 µF	Ceramic/Mylar	102
3.3 pF	Ceramic		1500 pF/0.0015 µF	Ceramic/Mylar	152
10 pF	Ceramic		2000 pF/0.002 µF	Ceramic/Mylar	202
15 pF	Ceramic		2200 pF/0.0022 µF	Ceramic/Mylar	222
20 pF	Ceramic		4700 pF/0.0047 µF	Ceramic/Mylar	472
30 pF	Ceramic		5000 pF/0.005 µF	Ceramic/Mylar	502
33 pF	Ceramic		5600 pF/0.0056 µF	Ceramic/Mylar	562
47 pF	Ceramic		6800 pF/0.0068 µF	Ceramic/Mylar	682
56 pF	Ceramic		0.01	Ceramic/Mylar	103
68 pF	Ceramic		0.015	Mylar	
75 pF	Ceramic		0.02	Mylar	203
82 pF	Ceramic		0.022	Mylar	223
91 pF	Ceramic		0.033	Mylar	333
100 pF	Ceramic	101	0.047	Mylar	473
120 pF	Ceramic	121	0.05	Mylar	503
130 pF	Ceramic	131	0.056	Mylar	563
150 pF	Ceramic	151	0.068	Mylar	683
180 pF	Ceramic	181	0.1	Mylar	104
220 pF	Ceramic	221	0.2	Mylar	204
330 pF	Ceramic	331	0.22	Mylar	224
470 pF	Ceramic	471	0.33	Mylar	334
560 pF	Ceramic	561	0.47	Mylar	474
680 pF	Ceramic	681	0.56	Mylar	564
750 pF	Ceramic	751	1	Mylar	105
820 pF	Ceramic	821	2	Mylar	205

(Continued)

Table 7-2 (*Continued*)

PicoFarad (pF)	NanoFarad (nF)	MicroFarad (mF, μF or mfd)	Capacitance Code
1000	1 or 1n	0.001	102
1500	1.5 or 1n5	0.0015	152
2200	2.2 or 2n2	0.0022	222
3300	3.3 or 3n3	0.0033	332
4700	4.7 or 4n7	0.0047	472
6800	6.8 or 6n8	0.0068	682
10000	10 or 10n	0.01	103
15000	15 or 15n	0.015	153
22000	22 or 22n	0.022	223
33000	33 or 33n	0.033	333
47000	47 or 47n	0.047	473
68000	68 or 68n	0.068	683
100000	100 or 100n	0.1	104
150000	150 or 150n	0.15	154
220000	220 or 220n	0.22	224
330000	330 or 330n	0.33	334
470000	470 or 470n	0.47	474

location. It is easy to make a mistake and install the wrong resistor in the wrong place. Re-check the color code chart against the resistor values and locations. Next, check the orientation of the integrated circuits to make sure that they are in the correct location and orientated properly. Re-check the LED placement and orientation. When you are satisfied that the re-inspection has been completed, you can connect up the battery once again and test the Mood Meter circuit. Hopefully you found the problem and corrected it and the circuit works properly this time around. Have fun with your new Mood Meter! The Mood Meter is a great "ice breaker" or party game for all to enjoy!

Chapter 8

Super-Reaction Time Tester

Parts List

Parts Bin

R1   100 ohm, 5%,
¼-watt resistor

R2   1k ohm, 5%,
¼-watt resistor

R3,R5   20k ohm, 5%,
¼-watt resistor

R4   12 ohm, 5%,
¼-watt resistor

C1, C2, C3   0.33 µF,
35-volt, mono capacitor

P1   5 megohm trimmer
potentiometer

U1   LM555 Timer IC

U2   CD4017 CMOS Johnson
Counter

S1,S3   DPDT push on/off
switch

S2   SPST momentary
pushbutton switch

B1   9-volt transistor
radio battery

Misc   PC board, IC
sockets, wire, battery
clip

The Super-Reaction Time Tester, illustrated in
Figure 8-1, is a fun project designed to test your
reaction time. It is easy to "test" a friend or family
member, by simply depressing the "start" button with
your finger. Eight LEDs begin to flash in sequence.
Then as fast as possible, press and release the "Stop"
button with the same finger. You reaction time "rating"

Figure 8-1   *Super-reaction Timer*

is indicated by the label next to the "bank" of LEDs.
This project can be used to train yourself for faster
hand–eye coordination. You can challenge your friends
or family to a "test" to see who is the fastest on the
"draw."

The Super-Reaction Time Tester schematic is shown in
Figure 8-2. The Super-Reaction Time Tester is centered
around an LM555 timer/oscillator integrated circuit. The
LM555 at U1 is configured as an astable or free running
oscillator. The LM555 is used to provide the clock pluses
necessary to allow the CD4017 at U2 to begin counting.

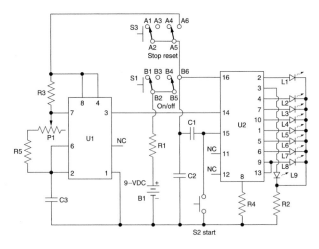

Figure 8-2   *Super-reaction Timer Game schematic*

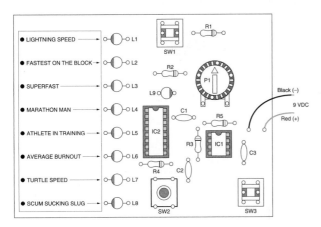

Figure 8-3 *Super-reaction Timer pictorial diagram*

The integrated circuit at U2 is a CMOS decade counter IC. The clock pulses from U1 at pin 3 are sent to pin 14 of the decade counter. However, U2 does not start sequencing its output until S2 is pushed or "started." Pushing S2 momentarily connects pin 15 of U2 to ground. In order to stop the sequencing of U2, the "gamer" must press S3 as fast as possible. This action removes the positive voltage from U1, which in turn causes U1 to stop producing clock pulses. Without any clock pulses, U2 stops sequencing and displays your speed "rating." The reaction time "rating" is a template, shown in Figure 8-3 which is placed next to the series of LEDs.

Before we begin assembly of the Super-Reaction Time Tester, let's take a moment to locate a large, clean, well-lit worktable or workbench. You will want to gather some small tools, such as a wire-cutter, needle-nose pliers, a magnifying glass, and a few small screwdrivers. Also you will want to locate a 27 to 33-watt pencil tip soldering iron with a small pointed tip or a small flat-edge wedge type tip. Also grab a small roll of #22 ga 60/40 rosin core solder, a small wet sponge and a small can of "Tip Tinner," which is used to dress and clean your soldering iron tip. A magnifying glass and a pair of tweezers would also be suggested. Try to locate an anti-static wrist strap, which will be helpful in order to prevent static discharge from damaging the integrated circuits as they are being handled. Locate all the project parts and place them in front of you; locate the schematic diagram, shown in Figure 8-2, the pictorial diagram in Figure 8-3 and all the pin-out diagrams and place them in front of you

as well. It is also recommended that you obtain an anti-static wrist band. An anti-static wrist band prevents damage to the integrated circuits when you are handling them. Often moving around while you assemble a circuit, such as getting up from a chair and moving across a carpet, can give static damage to integrated circuits.

The prototype Reaction Time Tester was built on a 2½ inch × 1½ inch printed circuit board for reliable operation. Integrated circuits were used for the two integrated circuits, in the event of a circuit failure, if it ever occurs. Both the integrated circuit and IC socket will have some sort of orientation markings. Look for a small notch or cutout on both the IC and socket. Look just to the left of the notch or cutout and you will find pin 1 on both the socket and the IC package itself. Pin 1 of U1 is connected to the minus (–) terminal of the battery, while pin 8 of U1 is connected to pin 4 of U1 as well as R3 as shown. Pin 1 of U2 is connected to LED L5, while pin 16 is connected to B6 of S1. Refer to the schematic and pictorial diagrams when installing the IC sockets. If you are confused or puzzled, ask a knowledgeable electronics friend for help in identifying the pin-outs.

After installing the IC sockets, you can locate the resistors and compare the resistors with the resistor color code chart shown in Table 8-1. The first color band is the first digit and the second color is the second digit. The third color band is the multiplier. The fourth color band, if there is one, will note the resistor tolerance value. A silver band notes a 10% tolerance value for the resistor, while a gold band denotes a 5% tolerance value. The color bands begin at one end of the resistor body. So, let's try to locate resistor R1. Look through the resistors for one which has a (brown) first band, followed by a (black) band and then a (brown) band, this resistor will have a 100 ohm value. Identify the remaining resistors and install them all on the circuit board and then solder them in place on the circuit board in their respective locations. After soldering the resistors in place, cut the excess component leads flush to the circuit board.

Next, we will install the potentiometer at P1. Trimmer potentiometers will have three leads, these three leads may be all in a row or the center lead may be offset from the two end leads. The center lead of the

Table 8-1
Resistor Color Code Chart

Color Band	1st Digit	2nd Digit	Multiplier	Tolerance
Black	0	0	1	
Brown	1	1	10	1%
Red	2	2	100	2%
Orange	3	3	1000 (K)	3%
Yellow	4	4	10000	4%
Green	5	5	100000	
Blue	6	6	1000000 (M)	
Violet	7	7	10000000	
Gray	8	8	100000000	
White	9	9	1000000000	
Gold			0.1	5%
Silver			0.01	10%
No color				20%

potentiometer is the wiper arm or adjustable lead pin. Insert the trimmer pot on the board and solder it in place, then remove the excess leads.

After installing the resistors, you can move on to installing the capacitors. The capacitors in this circuit have no polarity consideration, since there are no electrolytic capacitors. Capacitors will often have their values printed on the body, but often small capacitors will have a three-digit code marked on them. Refer to the chart in Table 8-2 to see the code system notes. Look in the table and see if you can find a code marked (334), this will denote the value of 0.33 µF which is the value of capacitors in this project. Once all three capacitors have been placed on the PC board you can solder them in place. Remember to cut the excess leads from the capacitors.

Locate the nine LEDs. Red LEDs L1 through L8 are connected to the CD4017 decade counter chip's output pins, as shown. The final LED, L9 is connected to output pin 3 of U2. The eight LEDs, L1 through L8, are lined up in ascending order to correspond to the "Reaction Rating" template diagram shown in

Table 8-3. L1 would therefore correspond to the lowest LED designation "SLUG," and L2 would correspond to "TURTLE," etc. When installing the LEDs, ensure that the cathode lead or the "shorter" LED lead corresponds to the half moon on the LED pin-out diagram shown in Figure 8-4. The cathode leads of all the LEDS are connected to a common point, at R2 as shown. Install the LEDs on the circuit board and solder them in their respective locations. Remove the excess component leads. Paste the template alongside the LEDs on the circuit board.

The Super-Reaction Time Tester uses three switches which can be mounted at this time; see the schematic diagram. Switches S1 and S3 are both DPDT push on/off switches. You will need to solder a jumper between the center poles of each of the switches as shown in the schematic. Solder a short jumper wire from A2 to A5 and then another jumper between B2 to B5. Next solder a #22 ga 6 to 8 inch-long piece of stranded-insulated wire from A6 to point marked (X). Now connect a 2 to 3 inch wire from A5 to B6. Next, connect a 6 to 8 inch piece of #22 ga wire between A6

Table 8-2

Capacitance Codebreaker Information

This table is designed to provide the value of alphanumeric coded ceramic, mylar and mica capacitors in general. They come in many sizes, shapes, values and ratings; many different manufacturers worldwide produce them and not all play by the same rules. Most capacitors actually have the numeric values stamped on them; however, some are color coded and some have alphanumeric codes. The capacitor's first and second significant number IDs are the first and second values, followed by the multiplier number code, followed by the percentage tolerance letter code. Usually the first two digits of the code represent the significant part of the value, while the third digit, called the multiplier, corresponds to the number of zeros to be added to the first two digits.

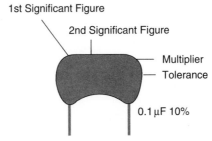

Value	Type	Code	Value	Type	Code
1.5 pF	Ceramic		1000 pF/0.001 μF	Ceramic/Mylar	102
3.3 pF	Ceramic		1500 pF/0.0015 μF	Ceramic/Mylar	152
10 pF	Ceramic		2000 pF/0.002 μF	Ceramic/Mylar	202
15 pF	Ceramic		2200 pF/0.0022 μF	Ceramic/Mylar	222
20 pF	Ceramic		4700 pF/0.0047 μF	Ceramic/Mylar	472
30 pF	Ceramic		5000 pF/0.005 μF	Ceramic/Mylar	502
33 pF	Ceramic		5600 pF/0.0056 μF	Ceramic/Mylar	562
47 pF	Ceramic		6800 pF/0.0068 μF	Ceramic/Mylar	682
56 pF	Ceramic		0.01	Ceramic/Mylar	103
68 pF	Ceramic		0.015	Mylar	
75 pF	Ceramic		0.02	Mylar	203
82 pF	Ceramic		0.022	Mylar	223
91 pF	Ceramic		0.033	Mylar	333
100 pF	Ceramic	101	0.047	Mylar	473
120 pF	Ceramic	121	0.05	Mylar	503
130 pF	Ceramic	131	0.056	Mylar	563
150 pF	Ceramic	151	0.068	Mylar	683
180 pF	Ceramic	181	0.1	Mylar	104
220 pF	Ceramic	221	0.2	Mylar	204
330 pF	Ceramic	331	0.22	Mylar	224
470 pF	Ceramic	471	0.33	Mylar	334
560 pF	Ceramic	561	0.47	Mylar	474
680 pF	Ceramic	681	0.56	Mylar	564
750 pF	Ceramic	751	1	Mylar	105
820 pF	Ceramic	821	2	Mylar	205

(Continued)

Table 8-2 (Continued)

PicoFarad (pF)	NanoFarad (nF)	MicroFarad (mF, μF or mfd)	Capacitance Code
1000	1 or 1n	0.001	102
1500	1.5 or 1n5	0.0015	152
2200	2.2 or 2n2	0.0022	222
3300	3.3 or 3n3	0.0033	332
4700	4.7 or 4n7	0.0047	472
6800	6.8 or 6n8	0.0068	682
10000	10 or 10n	0.01	103
15000	15 or 15n	0.015	153
22000	22 or 22n	0.022	223
33000	33 or 33n	0.033	333
47000	47 or 47n	0.047	473
68000	68 or 68n	0.068	683
100000	100 or 100n	0.1	104
150000	150 or 150n	0.15	154
220000	220 or 220n	0.22	224
330000	330 or 330n	0.33	334
470000	470 or 470n	0.47	474

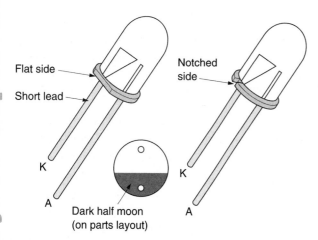

Flat side —
Short lead —
Notched side —
Dark half moon (on parts layout)

K
A
K
A

Figure 8-4 *LED identification*

and pin 16 of U2 as shown. Now connect a 6 to 8 inch #22 ga wire from B2 to the point marked (Y) or to the free end of resistor R1. Finally, you can install the momentary pushbutton switch S2. This switch is mounted just below U2 on the circuit board.

Refer to the semiconductor pin-out diagram, shown in Figure 8-5. Now, you can insert the integrated circuits into their respective sockets, but be sure to use your anti-static wrist strap when installing the IC. Orientating is very important when installing integrated circuits. Pay careful attention to the IC package and you will see that it will either have a small indented circle or a small cutout at one end of the IC package. Be sure to line-up pin 1 on the IC with pin on the IC socket before inserting the IC into the socket. Installing the IC correctly is very important to the proper operation of the circuit. Installing the IC backwards can cause damage to the circuit, when power is applied.

Why don't we take a short well-deserved break now and when we return we will check your circuit board for "cold" solder joints as well as for any "short" circuits. Pick up the circuit board and place the PC board in front of you with the foil side facing upwards toward you. Look the circuit board over carefully for possible

Table 8-3
Reaction Rating Template

Reaction Time Rating

- Lightning speed
- Fastest on the block
- Super fast
- Marathon man
- Athlete in training
- Average burnout
- Turtle spped
- Scum sucking slug

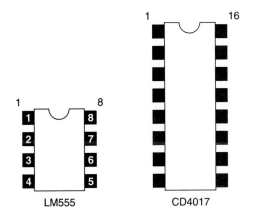

Figure 8-5 *Super-reaction Timer IC pin-outs*

"cold" solder joints. Good solder joints will look clean, smooth and shiny. If you find any solder joint that looks dull or "blobby," remove the solder from the joint, clean the joint and re-solder the joint so that it looks clean, shiny and smooth.

Now, look the circuit board over once again for any "cut" component lead which may have stuck to the board when you cut the components before installing them on the board. These "cut" leads could "bridge" the solder pads on the board and possibly short out the circuit when power is applied.

Finally, solder the battery clip leads to the circuit board. The minus (–) lead of the battery clip is connected

to the circuit common, while the (+) plus lead of the clip is connected to the free end of resistor R1.

To operate your reaction Time Tester, first place the switches S1 and S3 in the depressed position. Rotate the potentiometer control P1 to the center position and connect up your 9-volt battery. Note that the "READY" light or L9 is already lit up, as soon as you attach the battery. Now press the "START" button at S2. You will notice that LEDs L1 though L8 will light up in sequence starting with L1. If you do nothing else, the final LED at L8 will light up and the LEDs will stop sequencing. However, the goal is to test your reaction time by quickly removing your finger from switch S2 and pressing S3. When you do this, you will receive your "rating" indicated by the LED that remains "on." Now that you know how to use the Reaction Time tester, it is time to make the test "more difficult." First turn the control at P1 slightly counterclockwise. Now reset the circuit by releasing the "STOP" button S3, and then pressing S1, the reset switch, twice, leaving it in the depressed position. Now LED 9 should remain "on," but no other LEDs should be "on." You are now ready to test your reaction time more realistically. Press the "START" button at S2, and as quickly as possible, with the same finger, try to press the "STOP" button at S3. Follow the reset procedure above each time you cycle the circuit as a test "run." Note, that once you adjust P1 to setting that is realistic yet challenging, it should not need to be readjusted. You can have contests with your friends or siblings to see who has the fastest reaction speed.

In the event that the circuit does not "spring-to-life" or it does not seem to work properly, you will have to disconnect the battery and re-check the circuit board for any possible errors. If one or more LEDs do not light up, you may have installed them backwards. You can simply unsolder the LED from the board, remove the solder from the joint, rotate the LED 180 degrees, install it back into the circuit and solder it back on to the PC board. If the circuit appears "dead" and nothing lights up, then you will have to troubleshoot the circuit more carefully to find the problem. First, inspect the circuit board to make sure that all the resistors are in the correct locations and that they are the correct value for that location. Now, make sure that the integrated circuits have been installed correctly and that pin 1 of the IC matches pin 1 on the socket. Refer back to the schematic diagram. Note the locations of the components with respect to the IC pins. Finally, check all the LEDs to make sure that they have been installed correctly and not backwards. Refer to the parts layout diagram and the schematic as needed. If you are confused with the IC or LED mounting details, check with a knowledgeable electronic enthusiast for help or clarification.

Once you have re-inspected the circuit board for any errors, you can re-connect the battery and re-test the circuit; hopefully it will now work correctly. Pass your Super-Reaction Timer around to your friends and have a contest to see who is the "fastest-gun-in-the-west"! Have fun!

Chapter 9

Touch Sensitive Coin Tosser/Decision Maker

Parts Bin

R1, R3 33k ohm, 5%, ¼-watt resistor

R2, R4 510 ohm, 5%, ¼-watt resistor

P1   50k ohm trimmer potentiometer

C1, C2 0.01µF, 35-volt disk capacitor

Q1, Q2, Q3 2N3906 PNP transistor or NTE159 replacement

L1   Green LED

L2   Red LED

B1   9-volt transistor radio battery

Misc   PC board, battery clip, wire, enclosure, etc.

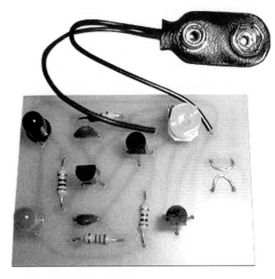

Figure 9-1   *Coin Tosser/Decision Maker*

The Coin Tosser/Decision Maker is a three-transistor, two-LED circuit which can be used to make decisions or replace coin tossing, see Figure 9-1. The project features one red and one green LED that flash at a high frequency when you touch the two sensor wires. When you quit touching the sensor your decision is registered by the LED that is remaining "on." The circuit has an "odds" control adjustment, is fun to operate and operates from a 9-volt battery for safety.

The Touch Sensitive Coin Tosser/Decision Maker, shown in the schematic in Figure 9-2, consists of a simple transistor multivibrator or oscillator circuit made up of transistors Q2, Q3 and the switching circuit made up of Q1. When a finger is placed on the "touch wires,"

the skin resistance causes a forward bias to be applied to transistor Q1, turning it "on." With Q1 biased "on," a path is created for transistor Q2 and Q3 to be biased "on." The transistors alternately turn on their respective LEDs when they are "on." This causes both LEDs to appear to light at the same time when you place your finger on the "touch wires." However, they are alternately turning "on" and "off" when you remove your finger; transistor Q1 turns "off," causing the multivibrator action to stop. Whichever transistor, Q2 or

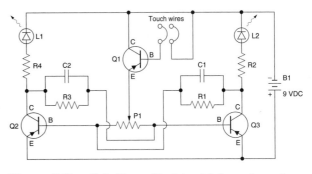

Figure 9-2   *Coin Tosser/Decision Maker schematic*

Q3, was "on" when you released your finger, stays "on" to indicate your decision.

The Touch Sensitive Coin Tosser/Decision Maker was constructed on a small printed circuit board which measured 1½ inches × 2 inches. The circuit is quite simple to construct and should only take about an hour or hour and half to build. First let's begin by locating a large table or workbench suitable for spreading out all the project parts and diagrams. Locate all the components and place them before you on a clean, well-lit table. Locate a 27 to 33-watt pencil tip soldering iron and small jar of "Tip Tinner" from your local Radio Shack store. The "Tip Tinner" is used to clean and dress the soldering iron tip. Locate some #22 ga 60/40 rosin core solder and a small wet sponge. Locate an anti-static wrist band or strap, which is used to protect integrated circuits from static charge buildup when constructing the circuit. You should also try to locate some small hand tools, such as a pair of end-cutters, a pair of small needle-nose pliers, a magnifying glass, a pair of tweezers and a small flat-blade and Phillips screwdriver.

Now refer to the parts layout diagram shown in Figure 9-3. Locate the four resistors and check the resistor color codes against the resistor color code chart in Table 9-1. Make sure you can identify each of the

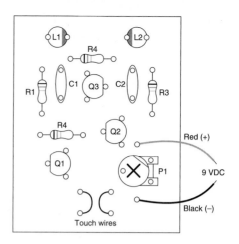

Figure 9-3 *Touch Sensitive Coin Tosser pictorial diagram*

resistors before you install them. Each resistor will have three or four color bands on it. The first color band is the first digit, the second color band is the second digit value, while the third band is the multiplier value. So look at resistor R1. The first color band is orange, so the first digit is 3; the second color band is orange, so the second digit is 3. Look at the third color band, it is orange, so the multiplier value is 1000 or 1k, so the resistor value is 33000 ohms. Resistors often have a fourth band and this represents the tolerance value of the resistor. No fourth band indicates a 20% tolerance

Table 9-1
Resistor Color Code Chart

Color Band	1st Digit	2nd Digit	Multiplier	Tolerance
Black	0	0	1	
Brown	1	1	10	1%
Red	2	2	100	2%
Orange	3	3	1000 (K)	3%
Yellow	4	4	10000	4%
Green	5	5	100000	
Blue	6	6	1000000 (M)	
Violet	7	7	10000000	
Gray	8	8	100000000	
White	9	9	1000000000	
Gold			0.1	5%
Silver			0.01	10%
No color				20%

Table 9-2

Capacitance Codebreaker Information

This table is designed to provide the value of alphanumeric coded ceramic, mylar and mica capacitors in general. They come in many sizes, shapes, values and ratings; many different manufacturers worldwide produce them and not all play by the same rules. Most capacitors actually have the numeric values stamped on them; however, some are color coded and some have alphanumeric codes. The capacitor's first and second significant number IDs are the first and second values, followed by the multiplier number code, followed by the percentage tolerance letter code. Usually the first two digits of the code represent the significant part of the value, while the third digit, called the multiplier, corresponds to the number of zeros to be added to the first two digits.

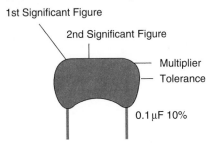

1st Significant Figure
2nd Significant Figure
Multiplier
Tolerance
0.1 µF 10%

Value	Type	Code	Value	Type	Code
1.5 pF	Ceramic		1000 pF/0.001 µF	Ceramic/Mylar	102
3.3 pF	Ceramic		1500 pF/0.0015 µF	Ceramic/Mylar	152
10 pF	Ceramic		2000 pF/0.002 µF	Ceramic/Mylar	202
15 pF	Ceramic		2200 pF/0.0022 µF	Ceramic/Mylar	222
20 pF	Ceramic		4700 pF/0.0047 µF	Ceramic/Mylar	472
30 pF	Ceramic		5000 pF/0.005 µF	Ceramic/Mylar	502
33 pF	Ceramic		5600 pF/0.0056 µF	Ceramic/Mylar	562
47 pF	Ceramic		6800 pF/0.0068 µF	Ceramic/Mylar	682
56 pF	Ceramic		0.01	Ceramic/Mylar	103
68 pF	Ceramic		0.015	Mylar	
75 pF	Ceramic		0.02	Mylar	203
82 pF	Ceramic		0.022	Mylar	223
91 pF	Ceramic		0.033	Mylar	333
100 pF	Ceramic	101	0.047	Mylar	473
120 pF	Ceramic	121	0.05	Mylar	503
130 pF	Ceramic	131	0.056	Mylar	563
150 pF	Ceramic	151	0.068	Mylar	683
180 pF	Ceramic	181	0.1	Mylar	104
220 pF	Ceramic	221	0.2	Mylar	204
330 pF	Ceramic	331	0.22	Mylar	224
470 pF	Ceramic	471	0.33	Mylar	334
560 pF	Ceramic	561	0.47	Mylar	474
680 pF	Ceramic	681	0.56	Mylar	564
750 pF	Ceramic	751	1	Mylar	105
820 pF	Ceramic	821	2	Mylar	205

(Continued)

Table 9-2 (*Continued*)

PicoFarad (pF)	NanoFarad (nF)	MicroFarad (mF, µF or mfd)	Capacitance Code
1000	1 or 1n	0.001	102
1500	1.5 or 1n5	0.0015	152
2200	2.2 or 2n2	0.0022	222
3300	3.3 or 3n3	0.0033	332
4700	4.7 or 4n7	0.0047	472
6800	6.8 or 6n8	0.0068	682
10000	10 or 10n	0.01	103
15000	15 or 15n	0.015	153
22000	22 or 22n	0.022	223
33000	33 or 33n	0.033	333
47000	47 or 47n	0.047	473
68000	68 or 68n	0.068	683
100000	100 or 100n	0.1	104
150000	150 or 150n	0.15	154
220000	220 or 220n	0.22	224
330000	330 or 330n	0.33	334
470000	470 or 470n	0.47	474

value, while a silver color band is a 10% value and a gold color band represents a 5% tolerance value. Now, go ahead and install all of the resistors, and solder them in place. Finally, with a pair of small wire-cutters remove the excess component leads from the board.

Next, refer to the capacitor code chart in Table 9-2 before installing the capacitors. Often capacitors are marked with their (coded) values and sometimes they are marked with their actual values. The 0.01 µF capacitors in this project will likely be marked with three digits (103). The value 0.01 µF is equivalent to the value of (103). You can now go ahead installing the two capacitors onto the circuit board, then solder them in place. Cut the excess component leads, flush to the edge of the PC board.

Locate the trimmer potentiometer; it will have three pins, either three pins in a row or straight line or the potentiometer may have two pins on either side of the package and a center offset pin, which represents the wiper or movable arm. Make sure when you mount the potentiometer that the center wiper pin gets mounted in the correct PC hole; refer to the schematic if needed. Solder the potentiometer in its proper location.

Locate the three transistors, refer to the diagram shown in Figure 9-4. All the transistors are PNP types, but these transistors are available in two package styles, the TO92 plastic type and the TO5 metal can type transistor. Identify which package style you have and note that the transistors have three leads. In the TO92 package the Emitter of (E) pin will be at one end of the package. Generally the transistor pin-outs are viewed from the bottom of the transistor. The transistor Base (B) pin will be in the center, while the Collector (C) will be at the opposite end of the transistor. In the TO5 metal can style transistor package the center Base pin will be offset from the Emitter and Collector pin. Note that the Emitters of both transistors are connected together and sent to the plus (+) battery terminals. Observe that the Bases of the two transistors are tied together through potentiometer P1. You will see that the

Figure 9-4 *Coin Tosser semiconductors*

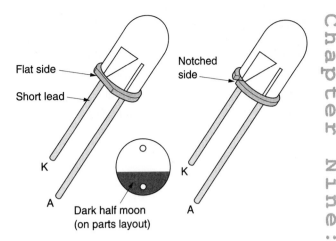

Figure 9-5 *LED identification*

Collector of Q2 feeds resistor R4 and L1, while the Collector of Q3 feeds resistor R2 and L2. Observe that the Emitter of Q1 is sent to the wiper terminal of P1. The "touch plates" are connected between the Base of Q1 and the minus (–) or ground terminal of the battery. Now you can install each of the transistors into their respective PC holes. Solder the transistors in place, when you are certain that you know exactly where they go. Cut the excess leads flush to the circuit board. If you are confused by the pin-outs or the placement of the transistors, ask a knowledgeable electronics friend for help.

Locate the red and green LEDs. Look at the diagram shown in Figure 9-5; this diagram illustrates the mounting configuration for the LEDs. The flat edge of the LED is identified with the short lead. In the diagram the dark "half moon" is associated with the flat edge or ground or minus (–) lead of the LED. Note that the L2 is the red LED connected to resistor R4, and the green LED is L1, connected to resistor R2. You can now go ahead and mount the LEDs.

Next, you can install two bare jumper wires on the PC board, these are the "touch wires." The "touch wires" essentially "bridge" or connect the circuit ground to the base of Q1 when touched by your finger. Essentially your finger completes the circuit and turns "on" the oscillator, and starts the LED alternately blinking.

You are now almost finished building the Coin Tosser/Decision Maker. Locate the 9-volt battery clip and solder it to the PC circuit. The red or plus (+) lead is connected to the Emitter of transistor Q3, while the black or minus (–) lead is connected to L2.

Take a short break and when we return you will inspect the circuit board before applying power to the circuit. Turn the PC board over so that you can look at the copper foil side of the board. Inspect the solder connections and be sure that all the solder joints look clean, smooth and shiny. If any solder joint looks dull or "blobby", then you will need to remove the solder and re-solder the connection, so that the solder joint looks clean, smooth and shiny. Now let's move to the second inspection, and we are going to look for any "short" circuits caused by "stray" component leads that may have stuck to the PC board after you cut the component. Component leads often stick to the PC board, due to the residue left from the rosin core solder. These "stray" component leads can form "bridges" and short out the PC board, possibly damaging the circuit when power is applied. If the PC board looks clean with no "bridges" or "shorts" you can go ahead and snap in a 9-volt battery and test the circuit.

With the battery installed, put your finger across the "touch wire" and the oscillator should begin alternately blinking the LEDs back and forth. When you release your finger from the "touch wire" the oscillator should stop, but one LED should remain "on."

If all goes well, your Coin Tosser/Decision Maker is ready to use; if not, you will have to carefully inspect the circuit board one more time. The most critical components for making mistakes during installation are the transistors and the LEDs. If the LEDs do not light up, you may have installed the LEDs backwards. It is a simple task to unsolder the LED from the board, remove the solder from the joint and then rotate the LED

180 degrees and re-solder the LED back on to the PC board. Now look to see if the resistors have the right color code for each particular location; use an ohmmeter to check the resistance if in doubt. Make sure you can positively identify each of the transistor leads, and when they are inserted into the circuit board. Once you have re-checked these components and made any necessary changes, you can once again apply power and re-check to see if the circuit is operating correctly. If all went well, your Decision Maker should now be operating correctly. Place your finger on the "touch wires" and the circuit will help you make your important decision. Go ahead and have fun!!!

Chapter 10

The Tingler Project

Parts List

Parts Bin

R1 15 k ohm, 5%, ¼-watt resistor

R2 7.5 k ohm, 5%, ¼-watt resistor

R3 10 ohm, 5%, ¼-watt resistor

R4 4.7 k ohm, 5%, ¼-watt resistor

P1 10 k linear slider potentiometer control

C1 1 µF, 35-volt electrolytic capacitor

Q1 2N3904 NPN transistor

U1 LM555 Timer IC

L1 Neon lamp

T1 Inverter transformer (10 k–2 k ohm)—Mouser Electronics 42TU002

B1 9-volt transistor radio battery

Misc PC board, IC socket, wire, touch plate, battery clip, etc.

The "Tingler" Project, shown in Figure 10-1, will create a shocking sensation among all your friends. Are you a "wimp" or a "strong" man? This project will give you a chance to find out! Just set the slider control to the "wimp" position and place your fingers on the touch pads. You'll then receive a "tingling" sensation from a mild electrical voltage. Now move the slider control towards "Strong Man" position and you will find out

Figure 10-1 *Tingler Project*

that it is much harder to leave your finger on the touch pads—Wow!

The "Tingler" Project uses a simple oscillator circuit made up of an LM555 timer IC which is used to convert a low voltage DC to higher voltage AC, as shown in Figure 10-2. The LM555 timer IC is configured to provide a steady stream of pulses from pin 3. From pin 3, resistor R4 and trim potentiometer P1 couple the low voltage AC to the Base of transistor Q1, where the signal is amplified and applied to the primary of transformer T1. The purpose of T1 is to amplify the 9-volt VAC signal to a large enough voltage that can light up a neon lamp; of course this "high" voltage can also be felt! Slider potentiometer P1 sets the intensity of the signal. Resistors R1, R2 and capacitor C1 control the frequency of oscillation. Finally, resistor R3 limits the current to the circuit.

The "Tingler" Project was constructed on a small circuit board, but perf-board construction could also be used. Before we begin constructing the "Tingler" Project, you will want to prepare a clean, well-lit worktable or workbench, for building the project. Next, you will want to locate a 27 to 33-watt pencil tip soldering iron with a pointed or small wedge-shaped tip.

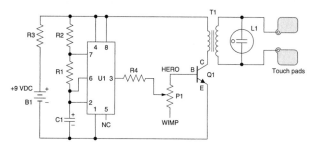

Figure 10-2 *Tingler Project schematic. Courtesy Chaney (CH)*

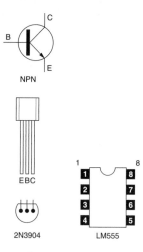

Figure 10-4 *Tingler Project semiconductors*

Now, find some 60/40 rosin type solder. Locate some "Tip Tinner," a soldering tip cleaner/dresser compound, along with a small wet sponge. Also try to locate an anti-static wrist band to place around your wrist while handling integrated circuits to prevent static damage. Often when building a circuit the builder will move around across a carpeted area, and up and down in a chair; all these movements can cause static charge buildup which can be reduced or eliminated using an anti-static wrist band. Gather a few tools, such as a pair of needle-nose pliers, a pair of end-cutters, a pair of tweezers and a magnifying glass. Round out your tools with a small flat-blade and a Phillips screwdriver.

Now, you will need to refer to the diagram shown in Figure 10-3, which illustrates the parts layout, called a pictorial diagram. The semiconductor pin-out diagram shown in Figure 10-4 will assist you when installing the transistor and the integrated circuit. Assembly will begin by referring to Table 10-1 which depicts the resistor color codes. You will notice that there are three columns denoting the 1st and 2nd digit and a multiplier. The first color band near the outside edge is the first digit, while the second represents the second digit of the

resistor value. The third color band notes the multiplier value assigned to the resistor. A fourth color, if on the resistor, illustrates the tolerance value of the resistor. No color band denotes a 20% tolerance value, while a silver color band represents a 10% tolerance value, and a gold color band notes a 5% tolerance value. If you identified a resistor with the first band of a green color and the second color band a blue color with the third color band of an orange color then the resistor will have 56000 ohms, and since there is no fourth color band the resistor has a 20% tolerance value. With this knowledge in mind you can now install all the resistors in the "Tingler" circuit board, and then solder them in place in their designated locations. With a small pair of wire-cutters, you can cut the excess component leads from the circuit board. Locate the 10 k slide potentiometer; note that most potentiometers will either have three in-line leads or two outer leads with an offset center lead which represents the movable wiper arm. Go ahead and solder the potentiometer in place and then, using a small wire-cutter, remove the excess component leads.

Next, install the capacitor labeled C1: this capacitor is an electrolytic, so you will note that there is a light or dark band with a either a plus or minus designation. Be sure to observe the band and the polarity marking on the capacitor before installing it. The plus (+) lead of C1 is connected to pins 2 and 6 of U1. Solder the capacitor in place and then remove the excess component leads. At this point you can go ahead and install the integrated circuit socket on the PC board. The IC socket will greatly aid you in the event of a circuit failure somewhere "down the road" or at a later date if a fault

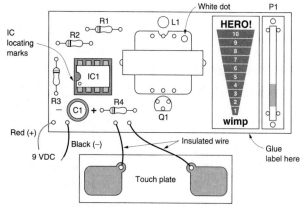

Figure 10-3 *Tingler Project pictorial diagram*

Table 10-1

Resistor Color Code Chart

Color Band	1st Digit	2nd Digit	Multiplier	Tolerance
Black	0	0	1	
Brown	1	1	10	1%
Red	2	2	100	2%
Orange	3	3	1000 (K)	3%
Yellow	4	4	10000	4%
Green	5	5	100000	
Blue	6	6	1000000 (M)	
Violet	7	7	10000000	
Gray	8	8	100000000	
White	9	9	1000000000	
Gold			0.1	5%
Silver			0.01	10%
No color				20%

develops. Because most people cannot unsolder an integrated circuit without damaging the circuit board, it is cheap insurance to install an IC socket from the beginning. Note that the IC socket will have a notch or cutout at one end of the socket, and just to the left of the notch, you will see pin 1. Pin 1 of the LM555 timer IC connects to the junction of transistor Q1 and the minus (–) supply of the battery, while pins 4 and 8 of U1 will connect to resistors R2 and R3. Solder the socket to the PC board.

Now, you can install the linear slider trimmer potentiometer onto the circuit board. Note that the potentiometer will have three leads all in a row or two outside leads with a staggered center pin. The center pin will be the wiper pin and in this project the wiper is connected to resistor R4. Observe that the pin-outs for the transistor are shown from the bottom view. Note that the transistors used in this project will likely be marked with the E,B,C which notes the Emitter lead, the Base lead in the center of the pin-outs, and the Collector lead. The Base lead of the transistor is connected to one end of the potentiometer P1, while the Collector lead is connected directly to the transformer at T1. The Emitter lead is connected to the minus (–) of the power supply or ground and pin of the LM555 IC. Install and solder

the transistor in place and follow-up by cutting the excess component leads.

Now you can move on to mounting the mini transformer. The transformer is an 8 ohm to 1000 ohm transformer. The transformer is likely to have three leads on either side of the transformer case. The outer leads on the 8 ohm side of the transformer are connected to the Collector of Q1 and to pins 4 and 8 of the LM555. The 1000 ohm secondary side of the transformer will likely also have three leads. The outer two leads are connected to the neon lamp and the touch plates, as shown in the schematic. If you are having difficulties identifying the output windings on the transformer, use an ohmmeter to measure the resistance of the winding pairs. The 1000 ohm pair will have the higher resistance. Finally, you can install the neon lamp. The neon lamp itself is somewhat fragile where the two leads exit the glass lamp envelope, so handle it with care while trying to install it onto the circuit board. You may want to wrap the neon lamp in tape or place it in a small grommet to protect it from damage.

Now is a good time to install the integrated circuits into its respective socket. Look at the IC and you will see either a small indented circle or a small notch or cutout at one end of the IC package. Pin of the IC will always be to the left of the small circuit or cutout. Before installing the IC make sure that you know which pin on the socket is pin 1, so that you can line-up pin 1 of the IC to pin 1 on the socket, turn the circuit board over and identify which socket pin is number 1. Note that pin 1 of the IC should be connected to the ground bus of the circuit and pins 4 and 8 will be connected to the plus voltage bus on the circuit as shown in the schematic diagram.

You may wish to install an SPST toggle or slide switch in series with the 9-volt battery clip leads, instead of having to remove the battery every time you are finished using the "Tingler." The black or minus (–) battery clip wire connects to the ground of the circuit at the transistors' Emitter pins and pin 1 of the U1. The red lead or plus (+) lead of the battery clip is connected in series with an SPST switch or directly to resistor R3 as shown.

Prepare two 6 to 8 inch #22 ga stranded-insulated wires which will be used to connect the touch pads to the main PC circuit board. Our touch plate was made from a small piece of circuit board material, with two copper pads left on the circuit board which act as the touch plate. You could also elect to use two small bolts on a piece of plastic as the touch plate if desired. Solder one end of each of the two wires to the touch plates. The two free wire ends are now soldered to the main "Tingler" circuit board, across the neon lamp. Note the neon lamp is across the transformer secondary which is connected in parallel with the touch plates.

Let's take a short break and when we return we will take a few moments to inspect the circuit board for "cold" solder joints and for any "shorts." First turn over the circuit board, so that the foil side of the board is facing upwards toward you and inspect the solder joints carefully. Look at the solder joints to make sure that all the joints look clean, smooth and shiny. If any of the solder joints do not look clean and shiny, then clean the joint and re-solder the connection until it looks clean and smooth and shiny. Next you will want to inspect the circuit board for any "stray" component leads which might have stuck to the board while you were cutting the excess component leads. Any "stray" leads could cause the circuit to "short" out and damage the circuit when power is first applied. Component leads can often stick to the PC board, due to the residue from the rosin core solder.

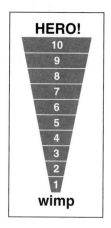

Figure 10-5   *Tingler Project template*

Well, now is the "moment of truth." Locate a 9-volt battery and snap into the battery clip. If you installed an on/off switch "turn" the switch to the "on" position. Touch two fingers of the same hand to the touch pads, while varying the slide potentiometer. Note which end of the slider produces the most tingling sensation and note it on the board with (H) for high or (L) for low. Turn off the circuit and locate the decal and paste it to the circuit board next to the slider potentiometer based on the high and low settings. The number 1 corresponds to "Wimp" or the lowest tingling, and the 10 notes "Hero" or "Champ" (see Figure 10-5).

If the circuit didn't work when the battery was installed, then you will have to do some troubleshooting. First disconnect the battery from the circuit. You can begin by re-checking the color codes on the resistors and making sure they are at the correct position. Next make sure that the transistors were installed correctly, with each transistor lead in the right location. Then check to make sure that the capacitor at C1 is installed correctly with respect to the polarity marking. Finally, make sure you have the transformer installed correctly; if it was installed backwards, you will not feel any tingling from the touch pads, since the voltage would not step up. Once you have checked and repaired the problem, you can go ahead and install the battery once again and retry the circuit again. You can disguise the circuit in a plastic box with two aluminum foil plates on the outside of the box, then have a brother, sister or friend hold the box in their hands and watch their reaction. Wow!

Chapter 11

Xenon Flasher Project

Parts List

Parts Bin

R1   33k ohm, 5%, 1-watt resistor

R2   10 megohm potentiometer (fine-rate)

R3   10 megohm, 5%, ½-watt resistor

R4   24k ohm, 5%, 1-watt resistor

C1   1000 µF, 25-volt electrolytic capacitor

C2   270 pF, 1000-volt capacitor (code)

C3a   0.15 µF, 400-volt capacitor

C3b   0.22 µF, 400-volt capacitor

C3c   0.47 uF, 400-volt capacitor (474)

C4   2.2 µF, 600-volt capacitor

D1, D2   1N4004 silicon diode

Q1   H1061 NPN transistor or NTE2515

Q2   SCR (CR02AM-8A) or NTE5645 replacement

T1   transformer 8 ohm to 1200 ohm type—Mouser 42TU003

T2   4-kV trigger transformer—Mouser 422-1304

N1   neon lamp—NE-2— Mouser 36NE007

N2   Xenon flash tube— Mouser 361-8538

S1   SPST power switch

S2   3-position rotary switch (course-rate)

F1   1-ampere fast-blow fuse

Misc   PC circuit board, wire, hardware, chassis box, etc.

So, you ask, what can I do with a Xenon flash tube circuit? I'm glad you asked!

The low voltage Xenon flash circuit can be used in a variety of applications, one of which is pure FUN! You can use the Xenon flash circuit to "spice-up" parties or to liven up the dancing, or you could use the flash circuit to draw attention to a display or project. You can even use the Xenon flash along with a simple alarm circuit to scare a thief away from a protected item or object. You may be able to think up a number of possible applications for your new Xenon flash or strobe circuit.

The low cost Xenon flash circuit is simple to construct and provides many hours of fun for the builder or your friends. The adjustable frequency strobe flasher is powered from a 6-volt battery to allow portable operation at any location, see Figure 11-1.

The Xenon flash circuit, shown in Figure 11-2, consists of three sections: an oscillator, an RC or timing network, and a flash circuit. The oscillator portion of the circuit is a self-oscillating circuit centered around transformer T1. Applying power to the circuit turns "on" transistor Q1 via current flow through resistor R1, a 33 k ohm, 1-watt resistor. This causes current to flow in the primary winding of transformer T1. The resulting magnetic field causes a voltage to be induced into the

Figure 11-1   *Strobe/Flasher Project*

secondary winding of transformer T1. The polarity of this voltage at pin 3 of the transformer is such that it turns off transistor Q1. Current stops flowing in the primary winding, the magnetic field collapses and the induced secondary voltage reverses polarity. This voltage now causes Q1 to turn off and the whole process repeats itself.

The turns ratio between the primary winding, i.e. pins 1 and 4, and the secondary windings, i.e. pins 2 and 5, is 25 turns to 1500 turns respectively. So the voltage induced at pin 2 is high. This alternating voltage is half-wave rectified by diode D3, which then charges capacitor C4. This produces a DC voltage across C4 of about 375 volts DC. This negative voltage is due to the orientation of diode D3.

The RC timing network consists of capacitor C3 and the resistor pair R2 and R3. Capacitor C3 is charged at

a rate determined by the resistors R3 and R4; this charging rate determines the flash rate.

The flash circuit—the voltage across capacitor C3—is also applied across the neon tube N1 via pins 1 and 2 of the trigger transformer T2 and the SCR at Q2. This voltage increases as C3 charges. When it reaches about 70 volts, the firing voltage of the neon tube, N1, fires and a voltage pulse is put onto the gate of the SCR at Q2. The SCR conducts thus discharging capacitor C3. This puts a voltage pulse into the trigger transformer T2, which is stepped up to hundreds of volts. When the high voltage pulse from the trigger transformer appears on the surface of the flash tube N2, the electric field inside the flash tube initiates the break-over and the tube flashes. The cycle then begins once again repeating itself.

The flash rate is determined by the RC network of C3 and resistors R2 and R3. The time constant of the RC network is given by the equation $T = R \times C$ or Time equals the resistance of R2/R3 multiplied by capacitance of C3. Reducing the value of T will reduce the charge time and increase the flash frequency rate. Conversely, increasing T will increase the charge time and reduce the flash rate. For the values supplied $R = 20$ megohms; and $C = 0.1$ μF, thus the circuit flashes about 2 flashes per second. If we lessen the value of one of the 10 megohm resistors, i.e. by varying the value of the adjustable potentiometer at R2, the R is now halved and flash rate will be about double the rate or 4 flashes per second. Slowing the flash rate is harder to do since

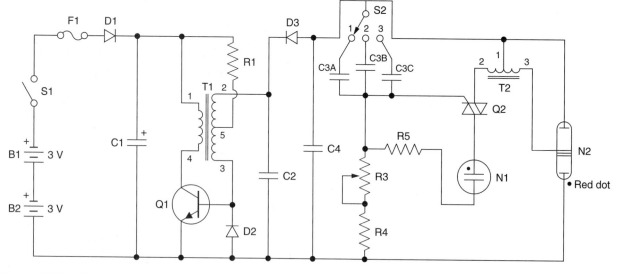

Figure 11-2   *Xenon flasher*

resistor values above 10 megohm are more difficult to obtain. The easiest thing to do is to increase the value of C3. So we have included a rotary selector switch to reduce the flash rate, by substituting different values of capacitance for C3, as shown in the schematic. Remember, the voltage values of these capacitors must rate at 400 volts since they are in the high voltage secondary portion of the circuit. To increase the brightness of the flash you can adjust the value of the capacitor at C4 from 2.2 µF to 4.4 µF.

Before we actually begin construction of the Xenon Flasher, let's prepare a large, clean, table top surface for our project building. Locate a 27 to 33-watt pencil tip soldering iron, a soldering holder, a wet sponge, some 60/40 rosin core solder and a small can of "Tip Tinner" soldering tip cleaner. Locate a small screwdriver, a wire-cutter, and needle-nose pliers. If you have an anti-static wrist band for handling it would be a good idea to prepare that by plugging in the ground tip into a well grounded outlet. The anti-static wrist band is a great tool for safely handling integrated circuits and preventing damage. Next, locate the parts for the project and place the parts in front of you as well as all of the diagrams, charts and schematics. Plug in your solder iron and let's begin.

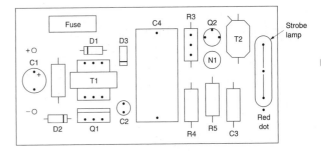

Figure 11-3 *Strobe pictorial diagram*

The Xenon flasher circuit can be built using perf-board style wiring or the circuit can be built on a printed circuit board. A 3½ inch × 4 inch circuit board was used for the prototype flasher circuit. The pictorial diagram for the Xenon Flasher circuit is shown in Figure 11-3; it will help you "see" the parts layout on the PC board.

First, refer to the chart in Table 11-1, which shows the resistor codes. Find the project resistors and let's identify them. Referring to the resistor color code, you will notice that there are three columns, the first column represents the resistor's first digit, the second column denotes the second resistor digit, and the third column shows the resistor multiplier value. The

Table 11-1
Resistor Color Code Chart

Color Band	1st Digit	2nd Digit	Multiplier	Tolerance
Black	0	0	1	
Brown	1	1	10	1%
Red	2	2	100	2%
Orange	3	3	1000 (K)	3%
Yellow	4	4	10000	4%
Green	5	5	100000	
Blue	6	6	1000000 (M)	
Violet	7	7	10000000	
Gray	8	8	100000000	
White	9	9	1000000000	
Gold			0.1	5%
Silver			0.01	10%
No color				20%

color bands on a resistor will begin from one end, that will be the first number or digit. If there is a fourth color band, that represents the resistor tolerance value. No color band is a 20% tolerance, while a silver color band is a 20% resistor and a gold band is a 5% tolerance resistor. Take a look at the project resistors; you will find one with three orange color bands (3) (3) (000) or 33 k ohms. This resistor will be resistor R1. Go ahead and mount this resistor onto the printed circuit board and solder it in place. Trim the excess leads after soldering. Now identify the rest of the resistors and place them on the circuit board in their respective locations and solder them in place.

Next, locate the capacitors for the project. Now refer to the capacitor chart in Table 11-2, which identifies the codes that are often used on capacitor bodies. Look at the project capacitors and see if you can find any codes marked on them. A code marking of (104) denotes a 0.1 µF capacitor and so forth. Locate and identify the capacitors using their codes; once you identify the capacitors, go ahead and install them on the board. Capacitors C1 and C4 are electrolytic types; these capacitors have polarity which must be observed when installing them. You will find either a white band or black band on the body of an electrolytic capacitor. This is the polarity marking. The band will have either a plus or minus symbol marked on it. Look at the capacitor to identify the polarity marking and then refer to the schematic to make sure that you can identify where it goes in the circuit and observe the polarity when installing it, in order to avoid damage to the circuit when power is first applied. Once you have identified all of the electrolytic capacitors and know where they should be placed on the PC board, then go ahead and install them into the PC board and solder them in place.

Next locate the silicon diodes at D1 and D2; note they will have a white or black band at one end of the diode body. Look at the diode in the schematic and you will see that the anode is depicted as an arrow or triangle and the cathode is shown as a single line. The colored band at one of the diodes denotes the diode's cathode end. Make sure you can identify the anode and cathode on the schematic and install the diodes so that the colored band on the diode faces the cathode symbol on the diagram, since diodes are polarity sensitive.

Note that the Xenon Flasher circuit uses both a transistor and an SCR, which are shown in the pin-out diagram in Figure 11-4. Note that both the transistor and the SCR have three leads coming out from the bottom of the devices. The transistor will have a Base lead, a Collector lead and an Emitter lead. The Base is shown in the schematic as a straight line in the transistor circle. The Emitter will have an arrow pointing toward the Base lead, and the remaining lead is the Collector lead. Be sure that you can identify each of the transistor leads using the schematic and then compare the schematic to the actual transistor when mounting it. If you are confused, ask an adult or knowledgeable friend for help. The SCR also has three leads but the pin-outs are a bit different. The SCR will have an anode and a cathode but will have an additional lead called a Gate, which goes into the cathode on the SCR's symbol. Once you can identify both the pin-outs on the transistor and SCR, you can go ahead and place them on the PC board and solder them in place. Correct orientation is essential for proper operation of the circuit and to avoid circuit failure.

The diagram in Figure 11-5 illustrates the parts pin-out, both transformers T1 and T2, and the Xenon lamp. Transformer T1 is voltage up-converter, which creates a larger voltage from the 6-volt power source. Transformer T2 is the small "trigger" transformer. Pay particular attention to the pin-outs of the transformers to ensure that the flasher circuit will work when power is first applied. Install the transformers and solder them in place. Finally, you can go ahead and install the neon lamp.

After the circuit has been built be sure to carefully re-check the circuit board and look for "cold" solder joints which may cause the circuit to fail after a short period of time. A good solder joint should look clean, smooth and shiny. If you encounter a solder joint which looks dull or "blobby" then remove the solder and solder the connection again. Now you will want to inspect the circuit board again, to ensure that there are no "stray" leads from the trimmed components which may still be left on the circuit board. You would not want to apply power to the circuit only to find that the circuit was "shorted-out" by leftover "stray" component leads. Again re-check the circuit for proper orientation of the critical components such as the semiconductors, capacitors and the transformer.

Once you are satisfied that the circuit is complete and correct you can solder the fuse leads and the switch and

Table 11-2

Capacitance Codebreaker Information

This table is designed to provide the value of alphanumeric coded ceramic, mylar and mica capacitors in general. They come in many sizes, shapes, values and ratings; many different manufacturers worldwide produce them and not all play by the same rules. Most capacitors actually have the numeric values stamped on them; however, some are color coded and some have alphanumeric codes. The capacitor's first and second significant number IDs are the first and second values, followed by the multiplier number code, followed by the percentage tolerance letter code. Usually the first two digits of the code represent the significant part of the value, while the third digit, called the multiplier, corresponds to the number of zeros to be added to the first two digits.

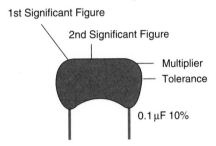

1st Significant Figure
2nd Significant Figure
Multiplier
Tolerance
0.1 µF 10%

Value	Type	Code	Value	Type	Code
1.5 pF	Ceramic		1000 pF/0.001 µF	Ceramic/Mylar	102
3.3 pF	Ceramic		1500 pF/0.0015 µF	Ceramic/Mylar	152
10 pF	Ceramic		2000 pF/0.002 µF	Ceramic/Mylar	202
15 pF	Ceramic		2200 pF/0.0022 µF	Ceramic/Mylar	222
20 pF	Ceramic		4700 pF/0.0047 µF	Ceramic/Mylar	472
30 pF	Ceramic		5000 pF/0.005 µF	Ceramic/Mylar	502
33 pF	Ceramic		5600 pF/0.0056 µF	Ceramic/Mylar	562
47 pF	Ceramic		6800 pF/0.0068 µF	Ceramic/Mylar	682
56 pF	Ceramic		0.01	Ceramic/Mylar	103
68 pF	Ceramic		0.015	Mylar	
75 pF	Ceramic		0.02	Mylar	203
82 pF	Ceramic		0.022	Mylar	223
91 pF	Ceramic		0.033	Mylar	333
100 pF	Ceramic	101	0.047	Mylar	473
120 pF	Ceramic	121	0.05	Mylar	503
130 pF	Ceramic	131	0.056	Mylar	563
150 pF	Ceramic	151	0.068	Mylar	683
180 pF	Ceramic	181	0.1	Mylar	104
220 pF	Ceramic	221	0.2	Mylar	204
330 pF	Ceramic	331	0.22	Mylar	224
470 pF	Ceramic	471	0.33	Mylar	334
560 pF	Ceramic	561	0.47	Mylar	474
680 pF	Ceramic	681	0.56	Mylar	564
750 pF	Ceramic	751	1	Mylar	105
820 pF	Ceramic	821	2	Mylar	205

(Continued)

Table 11-2 (Continued)

PicoFarad (pF)	NanoFarad (nF)	MicroFarad (mF, μF or mfd)	Capacitance Code
1000	1 or 1n	0.001	102
1500	1.5 or 1n5	0.0015	152
2200	2.2 or 2n2	0.0022	222
3300	3.3 or 3n3	0.0033	332
4700	4.7 or 4n7	0.0047	472
6800	6.8 or 6n8	0.0068	682
10000	10 or 10n	0.01	103
15000	15 or 15n	0.015	153
22000	22 or 22n	0.022	223
33000	33 or 33n	0.033	333
47000	47 or 47n	0.047	473
68000	68 or 68n	0.068	683
100000	100 or 100n	0.1	104
150000	150 or 150n	0.15	154
220000	220 or 220n	0.22	224
330000	330 or 330n	0.33	334
470000	470 or 470n	0.47	474

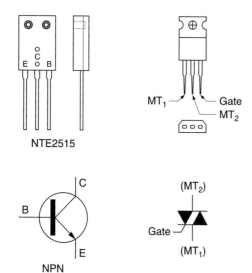

Figure 11-4 *Strobe light semiconductor pin-out diagram*

battery leads to the circuit board. The Xenon flash circuit uses a 1 ampere fast blow fuse in series with the power leads to the battery. The Xenon strobe flasher circuit is powered from a 6-volt power source, which can be made up of four "AA" cells or four "C" cells. Two dual-cell holders were wired in series to provide 6 volts to power the flasher circuit.

The Xenon flasher circuit can be mounted in a plastic enclosure to ensure that there are no exposed parts, which might "shock" an unsuspecting onlooker. Remember that this circuit is capable of producing over 375 volts in the secondary part of the circuit, so be sure to handle the circuit carefully with the power off.

If you applied power to the flasher circuit and nothing happened, then you will have to remove the battery and re-check the circuit once again to make sure that you have installed all of the components correctly. It's easy to make a mistake when installing parts on the circuit board. First check to see if the resistors have been installed in their proper location. Look at the color codes again to make sure that you have not installed a resistor in the wrong location. Next check the polarity of the electrolytic capacitors; you may have installed

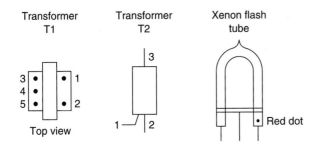

Figure 11-5 *Transformer and flash tube pin-outs*

one of them backwards. Next check the two diodes to make sure that you observed the cathode markings and installed them correctly. Now, check the transistors and SCR. Make sure that you can tell the difference between the transistor and the SCR. Have a friend help you if you are confused with these semiconductor mountings. Finally re-check the transformers, make sure that you can identify which is T1 and which is T2 and make sure that you have oriented them correctly on the circuit board. If the transformer T1 was installed incorrectly then it will not generate a high voltage. Note that the primary of the transformer has only 6 volts while the secondary transformer T1 will have about 100 volts. Once you have re-inspected the circuit for any errors, you can re-install the battery and see if the circuit works. You should see the neon lamp glow and then the flashing should start. Check the position of the power switch at S1 and also check the position of the rotary switch at S2 which controls the flash rate of the circuit.

Our Xenon flasher was housed in a 4 inch × 5 inch × 2½ inch plastic box. We elected to mount the power switch, S1 and the flash rate adjustment potentiometer R2 on the front panel of the box to allow easy access to controlling the circuit. The rotary selector switch, which also controls the "coarse" flash rate, was also mounted on the front of the chassis box for easy adjustment. The battery holders and the circuit board were mounted in a bottom portion of the chassis box. You could also make provisions for powering the Xenon flash circuit from an external power supply if desired.

Your Xenon Flasher circuit is now ready to dazzle your friends. Why not try to find an interesting application for your new flashing strobe light. Have fun!

Chapter 12

Siren Project

Parts List

Parts Bin

R1   47k ohm, 5%,
  ¼-watt resistor

R2, R3   1 megohm, 5%,
  ¼-watt resistor

R5   180 ohm, 5%,
  ¼-watt resistor

R4, R7   100k ohm, 5%,
  ¼-watt resistor

R6   10 megohm, 5%,
  ¼-watt resistor

C1   10 µF, 35-volt
  electrolytic capacitor

C2   1 µF, 35-volt
  capacitor

C3   10 nF, 35-volt
  capacitor

D1, D2   1N4148 silicon
  diode

Q1   NPN transistor
  (BC639)—NTE382
  replacement

U1   LM358 dual op-amp
  (National)

SPK   2 inch or 3 inch 8
  ohm speaker

S1   SPST toggle or slide
  switch (power)

S2   momentary pushbutton
  switch

B1   9-volt transistor
  radio battery

Misc   PC board, IC
  socket, wire, chassis
  box, battery clip,
  hardware, etc.

The Siren Project shown in Figure 12-1 is a fun little project which can be used in a number of ways. Serious applications include a siren for a bicycle, which can be used to warn walkers of your presence. The Siren Project can also be used in conjunction with a simple alarm circuit which could be used to protect something valuable. For you jokers out there, the siren can be used for gags to scare your friends or relatives and even your parents, when they least expect it. This project is great for sound effects at your next party, etc. The Siren Project is a simple straightforward circuit which utilizes a commonly available op-amp or operation amplifier integrated circuit. The siren circuit is powered by a 9-volt transistor radio battery. The siren uses a 2 or 3 inch speaker for producing the sound output.

The siren circuit is shown in the schematic at Figure 12-2. The heart of the siren circuit is based upon a National Semiconductor dual op-amp. The op-amp shown at U1:a and U1:b is a dual op-amp IC which features two op-amps in a single package which can be powered from a 9-volt battery. A positive voltage is applied to pin 8 of the op-amp while ground is applied to pin 4. The op-amp has two minus (–) inputs on pin 2 and pin 6 and two plus inputs (+) at pin 3 and pin 5.

Figure 12-1   *Siren Project*

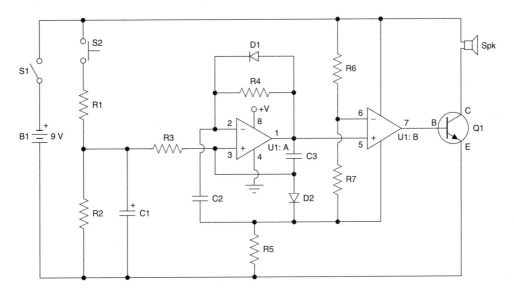

Figure 12-2 *Siren circuit*

The op-amp outputs are presented at pin 1 and pin 7, as seen.

The circuit operation begins when power is applied to the circuit at the on/off switch at S1. Switch S2 is normally an open pushbutton switch. When S2 is pressed, capacitor C1 is charged up through resistor R1, with a time constant of .47 seconds. When S2 is released capacitor C1 begins a slower discharge through resistors R2 and R3, with a time constant of 5 seconds. The op-amp is set up as a voltage controlled oscillator. The control voltage in this project is the exponential rise and fall in voltage of capacitor C1 and it charges and discharges.

When the output of the voltage controlled oscillator on pin 7 switches low, this current discharges C3 causing the voltage on pin 5 of the IC to rise toward the switching point at a rate proportional to the voltage on C1. When the switching point is reached, pin 7 of the IC switches high and initially pulls pin 6 high via capacitor C3. This causes the op-amp to temporarily turn on hard. But capacitor C1 quickly recharges through diode D2, causing the voltage on pin 5 to fall below the switching point and causing the op-amp to switch off once again.

The positive pulse output from the op-amp puts a fixed amount of charge into C2, slightly raising the potential of pin 6. This causes the potential on pin 6 to rise and assist the sharp switch-off of the op-amp. Also

R4 and C2 delay the rise on pin 6 just long enough to get a good output pulse.

The cycle then repeats itself once again. However, during the C1 discharge cycle the rate of charge of C3 is lower with each repetition of the oscillator; this is because the control voltage becomes lower and lower after each cycle, and the output frequency is correspondingly lower. During the C1 charge cycle the reverse applies. The output pulses are buffered by a second op-amp at U1:b, then the current is applied to a driver transistor at Q1. The output waveform has a low duty-cycle but gives a surprisingly loud sound from the speaker.

The siren circuit can be built on either a perf-board with point-to-point wiring or on a printed circuit board. The siren circuit was built on a small printed circuit board measuring 2½ inches × 3½ inches. You're probably anxious to begin constructing the Siren, but before we begin, we are going to take a few moments to prepare. First, you will need to locate a clean, well-lit and well-ventilated work area to build the project. Next, you will want to provide a large table or workbench to have enough room to spread out all the components, diagrams, charts and tools for the project. Locate a 27 to 33-watt pencil tip soldering along with a spool of 22 ga 60/40 rosin core solder. You will also want to secure some small hand tools, such as a pair of needle-nose pliers and a small pair of end-cutters or diagonals. A small flat-blade and Phillips screwdrivers would add nicely to your tool list, along with a magnifying glass and a pair of tweezers. Go to your local Radio Shack store and obtain an anti-static wrist strap and a small jar of "Tip Tinner."

The anti-static wrist strap is used to protect your integrated circuits from high voltage static buildup caused by moving around your work area or getting up and down from your chair. The anti-static strap is ground to a nearby outlet and is good insurance against damage. The "Tip Tinner" is a soldering iron tip cleaner/dresser. After soldering a few solder joints, you place the iron in the "Tip Tinner" and rotate the tip in the jar to clean and dress the tip in order to keep the soldering iron tip in good shape. We're almost ready now, so place the project components in front of you along with the schematics, pictorial diagram and the charts needed for the project. Heat up your soldering iron and we will begin.

Refer to the chart in Table 12-1, which illustrates the resistor color chart. Each resistor will have three to four color bands on the body of the resistor which begin at one end of the resistor body. The first column on the left represents the first digit of the resistor's color code. The digits run from 0 to 9. The second digit of the resistor color code corresponds to the second color band on the resistor and it runs from 0 to 9 as well. The third color

band on the resistor corresponds to the third column on the color chart, and it represents the resistor's multiplier value. If the resistor has a fourth band which is colored silver then the resistor will have a 10% tolerance value, while a gold color band would represent a 5% tolerance value. If the resistor has no color band then the resistor will have a 20% tolerance value. Look through the parts pile and try to locate a resistor which has a (yellow) band, followed by a (violet) band and then an (orange) band. This will be resistor R1 which has a 47 k or 47000 ohm value. Install this resistor onto the PC board, then identify the remaining resistors using the chart and then install them on the printed circuit board. Solder all the resistors to the circuit board, and then follow-up by using your end-cutters to trim the excess resistor leads from the PC board. Cut the excess leads flush to the edge of the circuit board.

Next, we are going to install the capacitors to the circuit board. The Siren Project has three capacitors, two ceramic disk types and an electrolytic capacitor. Many capacitors will have their value printed on them; however, small capacitors often do not have room to print the value on them, so a three-digit code is used to represent their value. Look at the chart shown in Table 12-2 which depicts the three-digit capacitor code.

Table 12-1
Resistor Color Code Chart

Color Band	1st Digit	2nd Digit	Multiplier	Tolerance
Black	0	0	1	
Brown	1	1	10	1%
Red	2	2	100	2%
Orange	3	3	1000 (K)	3%
Yellow	4	4	10000	4%
Green	5	5	100000	
Blue	6	6	1000000 (M)	
Violet	7	7	10000000	
Gray	8	8	100000000	
White	9	9	1000000000	
Gold			0.1	5%
Silver			0.01	10%
No color				20%

Table 12-2

Capacitance Codebreaker Information

This table is designed to provide the value of alphanumeric coded ceramic, mylar and mica capacitors in general. They come in many sizes, shapes, values and ratings; many different manufacturers worldwide produce them and not all play by the same rules. Most capacitors actually have the numeric values stamped on them; however, some are color coded and some have alphanumeric codes. The capacitor's first and second significant number IDs are the first and second values, followed by the multiplier number code, followed by the percentage tolerance letter code. Usually the first two digits of the code represent the significant part of the value, while the third digit, called the multiplier, corresponds to the number of zeros to be added to the first two digits.

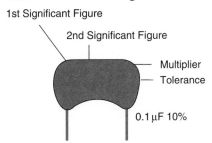

1st Significant Figure

2nd Significant Figure

Multiplier

Tolerance

0.1 µF 10%

Value	Type	Code	Value	Type	Code
1.5 pF	Ceramic		1000 pF/0.001 µF	Ceramic/Mylar	102
3.3 pF	Ceramic		1500 pF/0.0015 µF	Ceramic/Mylar	152
10 pF	Ceramic		2000 pF/0.002 µF	Ceramic/Mylar	202
15 pF	Ceramic		2200 pF/0.0022 µF	Ceramic/Mylar	222
20 pF	Ceramic		4700 pF/0.0047 µF	Ceramic/Mylar	472
30 pF	Ceramic		5000 pF/0.005 µF	Ceramic/Mylar	502
33 pF	Ceramic		5600 pF/0.0056 µF	Ceramic/Mylar	562
47 pF	Ceramic		6800 pF/0.0068 µF	Ceramic/Mylar	682
56 pF	Ceramic		0.01	Ceramic/Mylar	103
68 pF	Ceramic		0.015	Mylar	
75 pF	Ceramic		0.02	Mylar	203
82 pF	Ceramic		0.022	Mylar	223
91 pF	Ceramic		0.033	Mylar	333
100 pF	Ceramic	101	0.047	Mylar	473
120 pF	Ceramic	121	0.05	Mylar	503
130 pF	Ceramic	131	0.056	Mylar	563
150 pF	Ceramic	151	0.068	Mylar	683
180 pF	Ceramic	181	0.1	Mylar	104
220 pF	Ceramic	221	0.2	Mylar	204
330 pF	Ceramic	331	0.22	Mylar	224
470 pF	Ceramic	471	0.33	Mylar	334
560 pF	Ceramic	561	0.47	Mylar	474
680 pF	Ceramic	681	0.56	Mylar	564
750 pF	Ceramic	751	1	Mylar	105
820 pF	Ceramic	821	2	Mylar	205

(Continued)

Table 12-2 (*Continued*)

PicoFarad (pF)	NanoFarad (nF)	MicroFarad (mF, µF or mfd)	Capacitance Code
1000	1 or 1n	0.001	102
1500	1.5 or 1n5	0.0015	152
2200	2.2 or 2n2	0.0022	222
3300	3.3 or 3n3	0.0033	332
4700	4.7 or 4n7	0.0047	472
6800	6.8 or 6n8	0.0068	682
10000	10 or 10n	0.01	103
15000	15 or 15n	0.015	153
22000	22 or 22n	0.022	223
33000	33 or 33n	0.033	333
47000	47 or 47n	0.047	473
68000	68 or 68n	0.068	683
100000	100 or 100n	0.1	104
150000	150 or 150n	0.15	154
220000	220 or 220n	0.22	224
330000	330 or 330n	0.33	334
470000	470 or 470n	0.47	474

Check through the parts and look for a capacitor which has a (102) marked on it. This (102) code represents a value of 1 nF or 0.001 µF; this capacitor will be C2. Now look for a capacitor with a code marking of (103); this represents a value of 10 nF or 0.01 µF and this will be capacitor C3. This project has one electrolytic capacitor at C1. Electrolytic capacitors, as you remember, have polarity, and this must be observed when installing the capacitor, otherwise the circuit is not likely to work properly. These types of capacitors will have a white or black band on the body of the capacitor along one edge. This band will have either a plus (+) or minus (–) marking next to or inside the color band which indicates the polarity. Refer to the schematic and pictorial diagram in Figure 12-3 when installing the electrolytic capacitor and make sure that when you

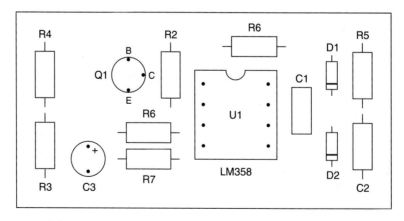

Figure 12-3 *Siren pictorial diagram*

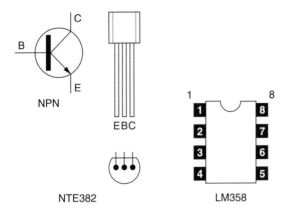

Figure 12-4 *Electronic Siren semiconductors*

install the capacitor the plus (+) side of the electrolytic connects to the junction of R1, R2 and R3. Install all the capacitors, then go ahead and solder the capacitors to the circuit board. Remember to trim the excess capacitor leads from the PC board.

The Siren Project incorporates two silicon diodes, which are shown at D1 and D2. Silicon diodes are another component which has polarity. The symbol for a diode is an arrow or triangle pointing to a line. The arrow will be the anode, while the straight line is the cathode end of the diode. The diode will either have a black or white band which denotes the diode's polarity. The colored band represents the diode's cathode lead. Note that diode D1 has its cathode connected to pin 2 of the op-amp, while the cathode of D2 is connected to the junction of R5 and C2. Install the diodes and then solder them to the PC board. Follow-up by cutting the excess diode leads from the board.

Next we will install the NPN transistor at Q1. Remember that transistors have three leads, a Base lead, a Collector lead and an Emitter lead. The Base lead is at the center of the symbol while the Collector and Emitter join the Base lead. The Emitter lead will always have the arrow on it. An arrow pointing toward the Base lead will represent a PNP type transistor, while an Emitter pointing away from the Base will represent a NPN type transistor. The semiconductor pin-out diagram is illustrated in Figure 12-4 showing the transistor pin-outs, and note that the transistor is shown from the bottom view. TO-92 transistor packages will have the Emitter at one end, with the Base lead in the center and the Collector lead at the opposite end of the package.

This helps orient the transistor when mounting it on the circuit board. You will notice that the Base lead is connected to output of the op-amp at U1:b at pin 7, while the Collector lead is connected to the speaker. Finally, the Emitter lead is connected to the ground or minus (–) bus on the circuit. Go ahead and install the transistor, then solder it to the PC board. Remember to trim the excess component leads from the board.

When constructing the Siren Project circuit, it is recommended to use an integrated circuit socket for the op-amp, in the event of a circuit failure at a later date. It is then a simple matter to replace the op-amp if needed without too much difficulty. Most mortals cannot unsolder an integrated circuit from a circuit board without damaging the PC board, unless you have lots of practice. Most IC sockets will have some form of identification on them, either a cutout or a notch at one end of the socket. Pin 1 will be just to the left of the notch or cutout. When installing the socket on the PC board, make sure that you observe that pin 1 of the IC goes to the junction of C3 and R4 and pin 5 of U1:b, while pin 8 of the IC connects to the plus (+) bus of the circuit. Go ahead and solder the IC socket to the PC board. Locate your anti-static wrist strap and place it on your wrist, while grounding the other end at a nearby power outlet. Now you are ready to insert the integrated circuit into its socket.

When installing the op-amp at U1, be sure to correctly install the device by observing the marking on the op-amp. Most op-amps either have a small round indented circle or a small rectangular cutout or notch at the top of the IC package. Pin 1 of the op-amp is always to the left of the cutout, circle or notch. Be certain when installing the IC to mate pin 1 of the IC to pin of the IC socket, taking an extra moment to re-check the orientation. If you are having difficulties installing the transistor or op-amp, ask a knowledgeable electronics enthusiast for help in determining the pin-outs.

Take a short, well-deserved rest and when we return, we will inspect the circuit board for any possible "cold" solder joints or "short" circuits. Turn the PC board over so that the foil side is facing upwards toward you. Look the circuit board over carefully for possible "cold" solder joints. A good solder joint will look clean, bright, smooth and shiny. If you find a solder joint that looks dull or "blobby," then unsolder the joint, remove the solder from the joint, and then re-solder the joint. Also be alert for "stray" component leads which have stuck to the board after cutting them off the components. The "stray" wire leads, if left in place, could "short" out the circuit once power is applied, and this could damage the Siren circuit.

Once the Siren circuit board has been completed and inspected, you can move on to testing the circuit. Place the Siren circuit board on a clean surface and connect a 2 or 3-inch speaker between the collector lead of Q1, using two lengths of 22 ga stranded wire. Prepare two 6 inch lengths of 22 stranded wire, connect up the pushbutton switch S2, and refer to the schematic. Next, connect the battery clips leads between R2 and the power switch at S1. Be sure to connect the black battery clip lead to the C1/R2 junction and the red battery clip lead to switch S1. Switch will be connected in series with the plus (+) battery clip wire.

Install the 9-volt battery, and turn the power switch S1 to the "on" position. Next, press the momentary switch at S2 for a few seconds and then release the pushbutton. When you first press the pushbutton the Siren will begin to wail upwards in frequency and as you release S2 the Siren will begin to wail downwards. Your portable Siren is now ready for use!

In the event that the Siren circuit does not work correctly, carefully disconnect the battery from the circuit and begin to fully inspect the circuit for any possible mistakes that you may have made during construction. The most likely cause for an inoperative circuit, is the incorrect orientation of the electrolytic capacitors, diodes or transistors. After looking carefully at the circuit you may discover that you made a simple mistake. If so, simply unsolder the component and reinstall it correctly. Another common error is placing the wrong resistor into the circuit board at the wrong location. Carefully check the resistor value and that they are placed in the right location on the PC board. Usually capacitors and diode will not be damaged by reversing them; however, sometimes transistors and integrated circuits can be damaged by incorrect installation. Once you have found your error, you can re-attach the battery and test the circuit once again.

Once your Siren has been tested, you can move on to installing the circuit in chassis box of some type. The Siren circuit can be installed in either a metal or plastic enclosure. The circuit is not critical or prone to outside interference, so a plastic box can be used. Plastic enclosures are often easier to mark-off and drill. A 4 inch × 5 inch × 2 inch plastic box was used to house the Siren circuit. Determine where you wish to mount the switches and the speaker, and drill the proper holes to accommodate these components. The speaker, on/off switch and the pushbutton switch were all mounted on the top front side of the plastic box. A 9-volt battery holder was installed on the bottom side of the chassis box.

Your Siren circuit is now ready! How will you use your new Siren?

Chapter 13

Rolling Dice Project

Parts List

Parts Bin

R1, R3, R4, R5   430 ohm, 5%, ¼-watt resistor

R6, R7, R8   430 ohm, 5%, ¼-watt resistor

R2   470 k ohm, 5%, ¼-watt resistor

R9, R10   150 ohm, 5%, ¼-watt resistor

C1   0.1 μF, 35-volt disk capacitor (104)

Q1, Q2, Q3   2N3904 NPN transistor

Q4, Q5, Q6   2N3904 NPN transistor

U1   LM555 timer IC

U2   CD4017 CMOS IC

LA1–LA6   red LEDs

S1   momentary pushbutton switch (normally open)

B1   9-volt transistor radio battery

Misc   IC socket, PC board, wire, battery clip, enclosure, label, etc.

Push the switch on the Rolling Dice project and bright red dice will begin to "roll" rapidly. Release the button and one of six standard dice patterns stays lit. That's your dice "roll"! If you are playing a game that requires two dice, simply remember the first number and "roll" one more time for the second number. The Rolling Dice utilizes 21 bright LEDs and two integrated circuits, designed to replace the standard pair of dice. The

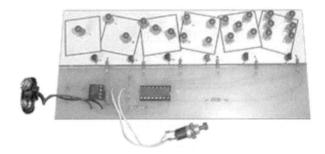

Figure 13-1   *Rolling Dice Project*

Rolling Dice project, shown in Figure 13-1, is easy to build and is powered from a standard 9-volt battery.

The Rolling Dice project depicted in Figure 13-2 produces all six die patterns using two integrated circuits and 21 LEDs. The purpose of the LM555 timer at U1 is to produce "clock" pulses. It is configured as an astable multivibrator or oscillator and provides output pulses on pin 3. With power "on," C1, R1 and R2 set the frequency of the "clock" pulses. Resistor R9 provides a load for the output and Switch S1 couples this output to the CD4017 counter at U2, when the button is pushed. The function of U2, a Johnson Decade counter, is to shift its input to its output pins. Each output pin goes "high" in sequence. As the output pin goes "high," a (+) level is applied to the corresponding Base lead of the transistor connected to that output pin through a limiting resistor. Each of these transistors is an NPN type that have their Collector leads tied to a group of LEDs. The LEDs are grouped into all six of the die patterns. As pushbutton switch S1 is depressed, the "clock" signals cause the output pins of U2 to rapidly light the LED groups. As soon as the button on S1 is released, the output pin on U2 that was "high" remains "high" keeping the corresponding transistor turned "on" and showing the LED die pattern that has been selected. The circuit continues to cycle when S1 is depressed again, and a new die pattern is shown when S1 is released. The battery should be removed when you are finished "rolling" the dice or install an optional on/off switch in series with the battery clip.

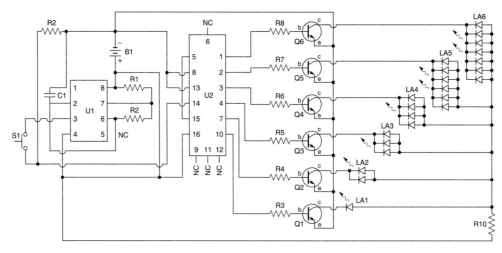

Figure 13-2 *Rolling Dice schematic. Courtesy Chaney (CH)*

The Rolling Dice project was constructed on a 3 inch × 6½ inch circuit board. The circuit is pretty straightforward and can be built in about an hour and half without too much difficulty. The dice are placed along the top of the circuit board, as you notice. A single LED is used to indicate a 1 at LA1. The number two is represented by the two LEDs at LA2, while the number three is shown by LA3. The number four is shown by four LEDs at LA4, five LEDs are illustrated at LA5 and finally the number six is shown by the six diodes which comprise LA6.

Let's begin building the Rolling Dice project. First you will want to locate a suitable table or workbench in a clean and well-lit area where you can spread out your diagrams, components and circuit board. Next, locate a 27–33 watt pencil tip soldering iron with a pointed end or small-flat-edge tip. Find a small roll of #22 ga 60/40

rosin core solder, and a small can of "Tip Tinner," from your local Radio Shack store. You also want to obtain a few small tools for circuit assembly: some small screwdrivers, a wire-cutter, small pair of needle-nose pliers, a magnifying glass and a pair of tweezers. Also try to obtain an anti-static wrist band. Anti-static wrist bands are used to prevent static from damaging integrated circuits when handling them during assembly. You should be able to locate one through you local electronic supply house. Heat up and prepare your soldering iron, cleaning and tinning the solder iron tip.

Place the schematic, Figure 13-2, and layout pictorial diagram, see Figure 13-3, in front of you. Place all ten of the project resistors in front of you and locate Table 13-1, which shows the resistor color code chart. Notice that each resistor will have either three or four color bands beginning at one edge of the resistor. The

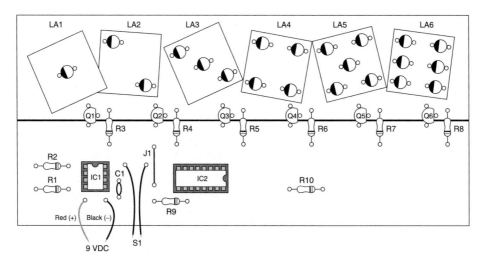

Figure 13-3 *Rolling Dice Project pictorial diagram*

Table 13-1

Resistor Color Code Chart

Color Band	1st Digit	2nd Digit	Multiplier	Tolerance
Black	0	0	1	
Brown	1	1	10	1%
Red	2	2	100	2%
Orange	3	3	1000 (K)	3%
Yellow	4	4	10000	4%
Green	5	5	100000	
Blue	6	6	1000000 (M)	
Violet	7	7	10000000	
Gray	8	8	100000000	
White	9	9	1000000000	
Gold			0.1	5%
Silver			0.01	10%
No color				20%

first color band is the first digit, the second color band is the second digit, and the third color band is the multiplier value for the resistor. The fourth band, if there is one, is the resistor tolerance value. If a resistor does not have a color band then the resistor has a 20% tolerance value, while a silver color band represents a 10% tolerance value. If the resistor has a gold band then the resistor has a 5% tolerance value. You will find that the Rolling Dice project has quite a few resistors that will have a yellow colored first band, followed by a violet color band, and a black third multiplier band. This resistor with a yellow, violet and black color band has a value of 470 ohms. Once you have identified all the resistors, you can install them on the circuit board. When this has been done, you can go ahead and solder them in place. After the resistors have been soldered in, you can cut the excess resistor leads from the board.

Next, take a look at Table 13-2 which illustrates the capacitor values commonly used. The Rolling Dice project has only one capacitor, which has a 0.1 µF value. Note that the capacitor will either have its actual value printed on one side of the capacitor or a (code) value. The 0.1 µF capacitor might have a code marking of (104), this corresponds to a 0.1 uF value. Go ahead and install C1, then solder it in place. Next remove the extra lead length from the capacitor.

The Rolling Dice project employs two integrated circuits and it is advisable to install IC sockets for these two integrated circuits. In the event of a circuit failure, IC sockets will greatly ease the repair of the circuit at a later date. Note that integrated circuits are sensitive to static electricity, so you will need to observe correct IC handling techniques, discussed earlier. Install the IC sockets and solder them in place. Locate the two integrated circuits and place them in front of you. Integrated circuits have locating marks which are used to guide the installation. Most integrated circuits will have either a small indented circle at one end of the IC package or a small cutout or notch. The small circle, notch or cutout will be located just to the right of pin 1 on the IC. Take a look at the IC socket and the underside of the PC board and make sure you can identify which pin is pin 1. Note that IC, U1 has its pin 1 connected to C1 while pin 1 of U2 is connected to resistor R8. If you have difficulties identifying or installing the integrated circuits, seek help from a knowledgeable electronics enthusiast. Once you have identified both pin 1 on the IC and pin 1 on the IC

Table 13-2

Capacitance Codebreaker Information

This table is designed to provide the value of alphanumeric coded ceramic, mylar and mica capacitors in general. They come in many sizes, shapes, values and ratings; many different manufacturers worldwide produce them and not all play by the same rules. Most capacitors actually have the numeric values stamped on them; however, some are color coded and some have alphanumeric codes. The capacitor's first and second significant number IDs are the first and second values, followed by the multiplier number code, followed by the percentage tolerance letter code. Usually the first two digits of the code represent the significant part of the value, while the third digit, called the multiplier, corresponds to the number of zeros to be added to the first two digits.

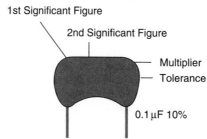

Value	Type	Code	Value	Type	Code
1.5 pF	Ceramic		1000 pF/0.001 µF	Ceramic/Mylar	102
3.3 pF	Ceramic		1500 pF/0.0015 µF	Ceramic/Mylar	152
10 pF	Ceramic		2000 pF/0.002 µF	Ceramic/Mylar	202
15 pF	Ceramic		2200 pF/0.0022 µF	Ceramic/Mylar	222
20 pF	Ceramic		4700 pF/0.0047 µF	Ceramic/Mylar	472
30 pF	Ceramic		5000 pF/0.005 µF	Ceramic/Mylar	502
33 pF	Ceramic		5600 pF/0.0056 µF	Ceramic/Mylar	562
47 pF	Ceramic		6800 pF/0.0068 µF	Ceramic/Mylar	682
56 pF	Ceramic		0.01	Ceramic/Mylar	103
68 pF	Ceramic		0.015	Mylar	
75 pF	Ceramic		0.02	Mylar	203
82 pF	Ceramic		0.022	Mylar	223
91 pF	Ceramic		0.033	Mylar	333
100 pF	Ceramic	101	0.047	Mylar	473
120 pF	Ceramic	121	0.05	Mylar	503
130 pF	Ceramic	131	0.056	Mylar	563
150 pF	Ceramic	151	0.068	Mylar	683
180 pF	Ceramic	181	0.1	Mylar	104
220 pF	Ceramic	221	0.2	Mylar	204
330 pF	Ceramic	331	0.22	Mylar	224
470 pF	Ceramic	471	0.33	Mylar	334
560 pF	Ceramic	561	0.47	Mylar	474
680 pF	Ceramic	681	0.56	Mylar	564
750 pF	Ceramic	751	1	Mylar	105
820 pF	Ceramic	821	2	Mylar	205

(Continued)

Table 13-2 (*Continued*)

PicoFarad (pF)	NanoFarad (nF)	MicroFarad (mF, μF or mfd)	Capacitance Code
1000	1 or 1n	0.001	102
1500	1.5 or 1n5	0.0015	152
2200	2.2 or 2n2	0.0022	222
3300	3.3 or 3n3	0.0033	332
4700	4.7 or 4n7	0.0047	472
6800	6.8 or 6n8	0.0068	682
10000	10 or 10n	0.01	103
15000	15 or 15n	0.015	153
22000	22 or 22n	0.022	223
33000	33 or 33n	0.033	333
47000	47 or 47n	0.047	473
68000	68 or 68n	0.068	683
100000	100 or 100n	0.1	104
150000	150 or 150n	0.15	154
220000	220 or 220n	0.22	224
330000	330 or 330n	0.33	334
470000	470 or 470n	0.47	474

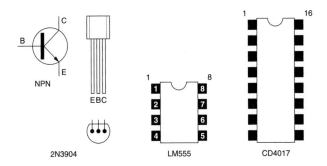

Figure 13-4   *Electronic dice semiconductors*

socket then go ahead and install the integrated circuits into their respective sockets.

Next refer to the transistor and IC mounting diagram shown in Figure 13-4. The Rolling Dice project has six transistors, all of which are 2N3904 NPN types. Most 2N3904 transistors are housed in a TO92 type plastic case as shown. Each transistor will have three leads: a Base lead, a Collector lead, and an Emitter lead. Note that the Emitter lead has the arrow in the schematic diagram and is located at one end of the transistor as shown in the diagram. Once you have identified the transistors and their pin-outs, you can solder them in place. Remember to cut the excess leads from the transistors.

Now locate the template or label and place it over the circuit board on the components side where the LEDs will be mounted. Take a "cut" component lead or slender sharp object and punch through the holes on the label, so that the LED leads can pass through to be soldered. The LEDs are installed in LED arrays from LA1 to LA6 as shown. Refer to the LED mounting diagram depicted in Figure 13-5 and the pictorial or layout diagram. LEDs will generally have a flat edge on one side of the LED package. This flat side of the LED also corresponds to the minus (–) or ground lead of the LED. The dark half moon shown in the diagram also denotes the ground lead of the LED. Install the LEDs one array after the other. Begin with the single LED at

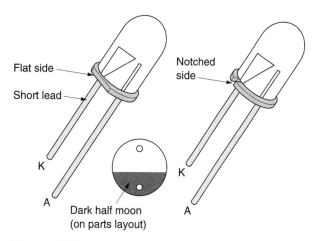

Flat side

Short lead

Notched side

K

A

K

A

Dark half moon
(on parts layout)

Figure 13-5 *LED identification*

LA1, then move on to the two LEDs at LA2, and so forth until you have placed the six LEDs into the LA6 array. Now solder the LEDs to the circuit board, and then cut the excess component leads.

Locate the pushbutton switch S1, and solder two short bare wire leads to the solder tabs on the switch. The free end of the two wires is next soldered to the circuit board across pin 3 of U1 and junction of R9 and pin 14 of U2.

Finally locate and install the 9-volt battery clip. The black or minus (–) wire is soldered to the circuit board at the junction of pin 1 at U1 and pins 8, 13 of U2. You could elect to place an optional on/off SPST toggle or slide switch in series with the red battery clip if desired. The red or plus (+) battery clip lead is soldered to the circuit at pin 8 as shown in the schematic.

Take a well-deserved break for a few minutes and when we return we will inspect the circuit board for "cold" solder joints and possible "short" circuits. Turn the circuit board over so that the foil side of the circuit board is facing you. Now take a look at the solder joints and make sure that they all look clean, smooth and shiny. If the solder joints look dull or "blobby" then you will need to remove the solder from the joint with a solder "sucker" or wick material and then re-solder the joint so that it looks clean, smooth and shiny. Now once again inspect the foil side of the circuit board and look for "cut" component leads which may have "stuck" to the board. Any "stray" wires or component leads can

cause possible "short" circuits which could damage the circuit when power is first applied to the circuit.

Once your inspection has been completed, you can snap the battery clip on a new 9-volt battery. If you installed an optional on/off switch "turn" the switch to the "on" position so current begins to flow into the circuit. Now press the pushbutton at S1. The dice should begin to "roll." Release S1 and the LED array that remains "on" will be your first number. Repeat operation for another number. Note that S1 should be held down for a few seconds to provide a good random number. Press S1 a number of times to make sure that all of the light arrays light up over time, and that all LED arrays work.

If any of the LEDs do not light up, then it is likely that you have installed one of the LEDs backwards. Remove the LEDs and reverse the connections. If you apply power to the circuit and press S1 and no lights come on then you may have a more serious wiring error. If this occurs, then remove the battery and re-inspect the circuit board. You may have installed the wrong resistor in the wrong place. Carefully inspect all the transistor leads to make sure that you connected each of the transistor leads to the correct PC pad. Refer to the schematic diagram and the parts layout diagrams when identifying the transistor leads. Finally, check the integrated circuits to make sure that they have been installed correctly. Note that the pin of the IC must correspond with pin 1 on the socket. Make sure that you understand that pin 1 of U1 connects to the junction of C1 and R9. Also note that pin 1 of the U2 is connected to resistor R8 which drives transistor Q6. This will help you establish the locations of pin 1 on the IC socket. Have a knowledgeable electronics enthusiast help you if you need more assistance.

Once your inspection has detected your error, you can re-attach the battery and again apply power to the circuit and re-try S1 to "roll" the dice. You can use this project with any type of dice games, board games and with some card games as well. Your Rolling Dice project is ready to go! Have some fun with your next board game!

Chapter 14

Game Show Quiz Buzzer

Parts List

Parts Bin

R1   100 ohm, 5%, ¼-watt resistor

R2, R3   10k ohm, 5%, ¼-watt resistor

R4, R5, R6, R7, R8, R9   1k ohm, 5%, ¼-watt resistor

R17, R18, R19, R20   1k ohm, 5%, ¼-watt resistor

R22, R23, R24, R25   1k ohm, 5%, ¼-watt resistor

R9, R10, R11, R12   10k ohm, 5%, ¼-watt resistor

R13, R14, R15, R16   10k ohm, 5%, ¼-watt resistor

C1   100µF, 35-volt electrolytic capacitor

C2, C3, C4, C5   100 nF, 35-volt mica capacitor

D1, D2, D3, D4   1N4148 silicon diode

D5, D6, D7, D8, D9   1N4148 silicon diode

D10, D11, D12, D13   1N4148 silicon diode

L1, L2, L3   red LED

L4, L5   red LED

Q1, Q6   BC557 PNP transistor—2N3906 or NTE159

Q2, Q3, Q4, Q5   BC547 NPN transistor—2N3904 or NTE123AP

Q7, Q8, Q9, Q10   BC547 NPN transistor—2N3904 or NTE123AP

S1, S2, S3, S4, SW5   momentary pushbutton—N/O

S6   SPDT power on/off switch

B1   9-volt transistor radio battery

Misc   PC board, enclosure, battery clip, battery holder, hardware, wire

Ever wonder how they figure out who pushed the answer button first on those Game Shows? This circuit will allow up to four players to compete, or you can daisy chain two of these circuits together for up to eight players. The Game Show Quiz Buzzer circuit is shown in Figure 14-1. This circuit is great for games like Jeopardy or any quiz type games that are timed by a host or monitor. An LED lights up for whoever pushes the button first. You can easily extend the pushbuttons from the circuit board with wires to your own remote switches. This is a great educational tool, or base circuit to build your own game show booth. The circuit must be reset via the quiz master pushbutton to start over again. You can modify this circuit in many ways from remote switches to larger brighter LEDs to AC power.

The Game Show Quiz circuit is centered around the single trigger circuit shown in Figure 14-2. The individual trigger circuit consists of two transistors, two diodes, one capacitor and five resistors as shown. The network of two resistors, capacitor and diode connected to the base of the transistor Qa are used to hold the base of the transistor to ground and to keep the circuit from

Figure 14-1   *Quiz Game. Courtesy Velman (VM)*

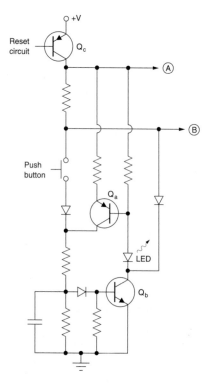

Figure 14-2   *Trigger circuit*

being triggered falsely from noise. When a pushbutton is pressed, voltage is conducted through the switch and diode in series with the momentary switch to the base of transistor Qb through a resistor. Once the pushbutton is pressed, a LED is turned "on" corresponding to the contestant who pressed their pushbutton first.

In the main Game Show circuit illustrated in Figure 14-3, you will notice that there are four identical trigger circuits. Each pushbutton circuit is in effect a trigger circuit which is used to activate the Quiz Show Winner lamp—LD5. The first person to trigger the

circuit by pushing their button first locks out the other contestants. The latch portion of the circuit is centered around the transistor Q1. When the first contestant pushes their button Q1 latches and lights lamp LD5 through transistor Q3. After the question is answered the game host resets the circuit by pressing the reset pushbutton which extinguishes the lamp LD5, and the process is begun again. The circuit is powered from a readily found 9-volt transistor radio battery.

The Game Quiz circuit was built on a 3½ inch × 4½ inch circuit board. Before we begin constructing the Game Show Quiz circuit, you will want to prepare a clean, well-lit worktable or workbench where you can spread out the diagrams, tools and the project parts. We will also need to gather some additional items for building the circuit. Locate a 27 to 33-watt pencil tip soldering iron, either a sharp pointed tip or a small flat-blade type tip. You will also want to locate some #22 ga 60/40 rosin solder, some "Tip Tinner" available at your local Radio Shack, as well as a small wet sponge. The "Tip Tinner" helps clean and dress the soldering tip. After using the wet sponge to clean the tip, you place the soldering iron into the "Tip Tinner" to dress the soldering tip. You will also want to locate a few small tools, such as a small wire-cutter, a pair of small needle-nose pliers, a few small screwdrivers, a magnifying glass and a pair of tweezers. Try to locate an anti-static wrist band to prevent static buildup when handling the semiconductors in the project. Place the schematic layout diagram, shown in Figure 14-4, and Tables 14-1 and 14-2 in front of you on the table.

We are now ready to begin building the Game Show Quiz Buzzer circuit. Heat up your soldering iron and tin the solder iron tip. Refer to Table 14-1, which illustrates the resistor color code chart. Each resistor in the project will have a color code printed on the body of the resistor. The color codes begin at one end of the resistor body. The first digit of the code will represent the first digit of the resistor value, and the second color band will represent the second digit of the resistor value. The third color band on the resistor denotes the multiplier value for the resistor while the fourth band, if any, will represent the tolerance value of the resistor. Arrange the project resistors in front of you and look at the color codes. Look for a resistor with (brown) (black) (brown) on the body of the resistor. Referring to the color code chart you will see that this first resistor has a value of

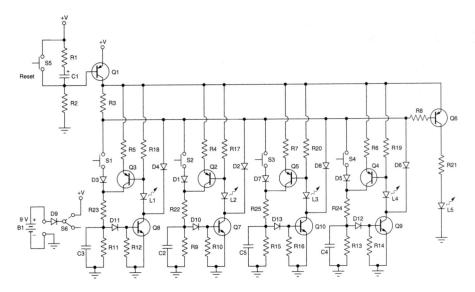

Figure 14-3 *Game Show Quiz circuit*

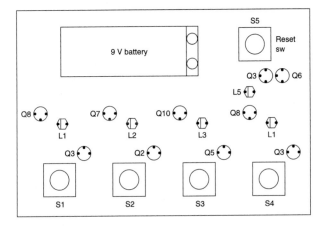

Figure 14-4 *Quiz Game pictorial diagram*

100 ohms; this is resistor R1. Identify each of the resistors using the color code chart, then mount the resistors on the circuit board. Next you can solder in each of the resistors and then use your wire-cutter to trim off the excess component leads, cutting the leads flush with the circuit board.

Next we will look at the capacitors and try to find the values on each of the resistors. Basically there are two types of capacitors in this project. The electrolytic type and the ceramic disk type capacitors. Electrolytic capacitors will have a polarity marking on them. You will find either a black or white stripe on the capacitor body with a plus (+) or minus (–) marking on the body of the component. As mentioned, you will also find ceramic disk capacitors in this project. Sometimes the values of capacitors are not printed on them but a

three-digit code is often used. Refer to the chart in Table 14-2 which shows the capacitor codes and their respective value. Make sure that you can identify each of the capacitors before installing them on the circuit board. Once you have identified all of the capacitors, you can go ahead and install them on the board. Finally solder the capacitors on the circuit board. Then cut the excess component leads flush to the circuit board.

Now we are going to install the silicon diodes on the circuit board. Diodes all have polarity. They will have a black or white colored band on the body of the diode which indicates the diode's cathode lead. Schematically a diode looks like a triangle pointing to a line. The triangle is the anode, while the line is the cathode lead. Make sure that you can identify the diode's leads properly. For example, look at the schematic diagram. Diode D3 has its cathode connected to the Collector of transistor Q3 and resistor R3. The anode of the diode is connected to switch S1. Mount all of the diodes on the circuit board and solder them in place. Finally, cut the excess component leads, flush to the edge of the board.

Locate all of the transistors for the project from the parts pile. There are ten PNP type transistors in this project. Transistors all have three leads, one called the Base, which schematically is the straight line in the middle of the device. Then there is the Collector lead at one end of the transistor, and the Emitter lead which has the arrow on the schematic symbol. Both the Emitter and the Collector originate from the Base lead, as shown in the schematic. Refer to the transistor layout

Table 14-1

Resistor Color Code Chart

Color Band	1st Digit	2nd Digit	Multiplier	Tolerance
Black	0	0	1	
Brown	1	1	10	1%
Red	2	2	100	2%
Orange	3	3	1000 (K)	3%
Yellow	4	4	10000	4%
Green	5	5	100000	
Blue	6	6	1000000 (M)	
Violet	7	7	10000000	
Gray	8	8	100000000	
White	9	9	1000000000	
Gold			0.1	5%
Silver			0.01	10%
No color				20%

diagram shown in Figure 14-5, it depicts the TO-92 plastic case with the Emitter at one end, the Base lead at the center and a Collector lead opposite the Emitter lead. Note, for example, that the Base lead of Q1 connects to capacitor C1 and resistor R2. The Emitter is tied to R3 and Q6, while the Collector of Q1 is connected to the plus (+) 9-volt source. Make sure that you can identify each of the transistor leads before mounting the transistor to the circuit board. Have an electronics enthusiast help you if you are confused by the diagrams. Once the transistors have been mounted, you can go ahead and solder them in place. Remember to cut the excess leads from the transistors.

The Game Show Quiz circuit contains five LEDs. Remember that all LEDs have polarity, which must be observed. Each of the LEDs is in the individual switch or contestant circuit. Refer to the diagram shown in Figure 14-6, which depicts the LED mounting configurations. LEDs, like diodes, have an arrow pointing to the line or cathode on the diode's body. Note that the cathode is the shorter of two leads and is designated as a dark half moon lead on the diagram. For example, LD1 has its anode connected to R18, while its cathode is connected to transistor T8. Go ahead and install the LEDs at their respective locations, then solder

them in place. Don't forget to trim the excess component leads before moving on.

Next we are going to install the pushbutton switches. Switches S1 through S4 are "contestant" pushbuttons, i.e. normally open pushbuttons. RESET switch S5 is at the top left of the schematic diagram. You will want to prepare ten 3-foot lengths of #22 ga pieces of hookup wire, to connect between the "contestant" switches and the circuit board. Solder the free ends of each pair of wires to each of the switches. The remaining switch lead pairs will return to the circuit and connect to the input of each switch position on the circuit board. Switch S6 is the power "on/off" switch which connects in series with the plus (+) battery terminal.

Finally, you can connect a 9-volt battery clip to the circuit board. The black or minus (–) lead will connect to common bus of ground of the circuit board, while the plus (+) or red battery clip lead connects to the anode of diode D9.

Before we apply power to the circuit board we will take a short break, and when we return we will take a look at all the solder joints to make sure that there are no "cold" solder joints which might shorten the life of the circuit. Turn the circuit board over, so that the foil side of the board is upwards and facing you. Look to

Table 14-2

Capacitance Codebreaker Information

This table is designed to provide the value of alphanumeric coded ceramic, mylar and mica capacitors in general. They come in many sizes, shapes, values and ratings; many different manufacturers worldwide produce them and not all play by the same rules. Most capacitors actually have the numeric values stamped on them; however, some are color coded and some have alphanumeric codes. The capacitor's first and second significant number IDs are the first and second values, followed by the multiplier number code, followed by the percentage tolerance letter code. Usually the first two digits of the code represent the significant part of the value, while the third digit, called the multiplier, corresponds to the number of zeros to be added to the first two digits.

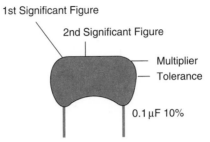

Value	Type	Code	Value	Type	Code
1.5 pF	Ceramic		1000 pF/0.001 μF	Ceramic/Mylar	102
3.3 pF	Ceramic		1500 pF/0.0015 μF	Ceramic/Mylar	152
10 pF	Ceramic		2000 pF/0.002 μF	Ceramic/Mylar	202
15 pF	Ceramic		2200 pF/0.0022 μF	Ceramic/Mylar	222
20 pF	Ceramic		4700 pF/0.0047 μF	Ceramic/Mylar	472
30 pF	Ceramic		5000 pF/0.005 μF	Ceramic/Mylar	502
33 pF	Ceramic		5600 pF/0.0056 μF	Ceramic/Mylar	562
47 pF	Ceramic		6800 pF/0.0068 μF	Ceramic/Mylar	682
56 pF	Ceramic		0.01	Ceramic/Mylar	103
68 pF	Ceramic		0.015	Mylar	
75 pF	Ceramic		0.02	Mylar	203
82 pF	Ceramic		0.022	Mylar	223
91 pF	Ceramic		0.033	Mylar	333
100 pF	Ceramic	101	0.047	Mylar	473
120 pF	Ceramic	121	0.05	Mylar	503
130 pF	Ceramic	131	0.056	Mylar	563
150 pF	Ceramic	151	0.068	Mylar	683
180 pF	Ceramic	181	0.1	Mylar	104
220 pF	Ceramic	221	0.2	Mylar	204
330 pF	Ceramic	331	0.22	Mylar	224
470 pF	Ceramic	471	0.33	Mylar	334
560 pF	Ceramic	561	0.47	Mylar	474
680 pF	Ceramic	681	0.56	Mylar	564
750 pF	Ceramic	751	1	Mylar	105
820 pF	Ceramic	821	2	Mylar	205

(Continued)

Table 14-2 (*Continued*)

PicoFarad (pF)	NanoFarad (nF)	MicroFarad (mF, μF or mfd)	Capacitance Code
1000	1 or 1n	0.001	102
1500	1.5 or 1n5	0.0015	152
2200	2.2 or 2n2	0.0022	222
3300	3.3 or 3n3	0.0033	332
4700	4.7 or 4n7	0.0047	472
6800	6.8 or 6n8	0.0068	682
10000	10 or 10n	0.01	103
15000	15 or 15n	0.015	153
22000	22 or 22n	0.022	223
33000	33 or 33n	0.033	333
47000	47 or 47n	0.047	473
68000	68 or 68n	0.068	683
100000	100 or 100n	0.1	104
150000	150 or 150n	0.15	154
220000	220 or 220n	0.22	224
330000	330 or 330n	0.33	334
470000	470 or 470n	0.47	474

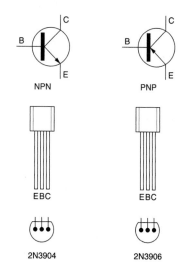

Figure 14-5 *Quiz Game transistors*

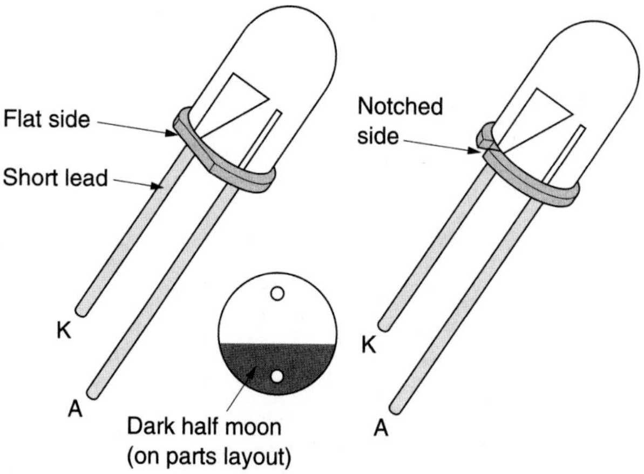

Figure 14-6 *LED identification*

make sure that all the solder joints are clean, smooth and shiny. If you find any of the solder joints look dull or "blobby," you should remove solder from the joint and then re-solder the connection, making sure that the

solder joints all look clean, smooth and shiny, Next we will examine the circuit board for any "short" circuits which may be caused from "stray" component leads which should have been removed when you cut the excess component leads. The component leads often stick to the circuit board from the sticky rosin from the solder. Make sure that you do not see any "bridges" which may connect circuit pads together forming a "short" circuit. Short circuits can damage the circuit's components and the circuit as well.

Now that you have inspected the circuit board, you can connect a 9-volt battery to the terminals and you will be able to test the Game Show Quiz circuit. With power applied to the circuit, press any of the "contestant" pushbutton switches, and you should see one of the LEDs turn on, and it should correspond to the switch that was pressed. If everything is working, the "contestant" lamp will come "on." You should then press the RESET button to reset the circuit. At this point all the LEDs should be "off." Test each of the pushbuttons to make sure that each of the LEDs lights up. Hopefully all the LEDs will work when the switches are pressed, and that the RESET switch should extinguish all the lamps. If everything works, you are all set to start your Game Show career.

In the event that none of the LEDs light up, or if only some of the LEDs light up, then you will have to troubleshoot the circuit. If a particular LED does not light up, then check the circuit around the switch and respective LED. You may have just installed the LED backwards, if you did, simply unsolder the LED. Next, remove the solder from the joint with a solder "wick" or solder "sucker," then rotate the LED and re-solder back into the circuit. Re-attach the battery and you may be all set for your Game Show. However, if the circuit still does not appear to work properly, then you will have to remove the battery and put your troubleshooting skills to work again.

Look for incorrect resistors placed in the wrong location, check the circuit layout diagram against the schematic for each resistor. If you are not sure of a resistor value, use an ohmmeter to verify the part. Refer to the diagrams once again to make sure that you haven't inserted a diode backwards. Also make sure that the electrolytic capacitors have been installed correctly. Now, look at each of the transistors and be sure that each one has been installed correctly in the circuit. Verify the pin-out of the transistors using the transistor pin-out diagram. Note the view of the transistor in the diagram. It is a very common mistake to install a transistor backwards. Look carefully for a problem, and perhaps have a friend look again after you. Sometimes a second pair of eyes can easily find a mistake.

Once you have inspected the circuit board carefully, you can re-install the 9-volt battery and re-test the circuit. Hopefully this time around the circuit will work perfectly. Also note that you can "stack" two of these circuits together for up to eight contestants. Have fun with your new Game Show Quiz circuit.

Chapter 15

Wheel-of-Fortune Game

Parts List

Parts Bin

R1, R5 33 k ohm, 5%,
¼-watt resistor

R2 2.2 megohm, 5%,
¼-watt resistor

R3 82 k ohm, 5%, ¼-watt
resistor

R4 100 k ohm, 5%,
¼-watt resistor

R6 200 ohm, 5%, ¼-watt
resistor

R7 68 ohm, 5%, ¼-watt
resistor

R8 120 ohm, 5%, ¼-watt
resistor

R9 680 ohm, 5%, ¼-watt
resistor

C1, C3 10 µF, 35-volt
electrolytic capacitor

C2 0.47 µF, 35-volt mylar
capacitor

C4 47 µF, 35-volt
electrolytic capacitor

Q1, Q3, Q4 2N3904
transistor

Q2 MU10 UJT transistor
or NTE6410 replacement

U1 CD4017 CMOS Decade
Counter

L1, L2, L3, L4, L5 red
LEDs

L6, L7, L8, L9, L10
red LEDs

S1 momentary pushbutton
(normally open)

B1 9-volt battery

Misc printed circuit
board, IC socket,
battery clip, wire

The Wheel-of-Fortune Game shown in Figure 15-1 is a great game project which can be used for your own version of Wheel of Fortune. Push the "START" button and a bright red "ball" (LEDs) appears to spin around ten numbers, gaining speed as you hold the pushbutton down. When you release the button, the electronic "ball" appears to slow down and finally comes to a stop on a number. As the "ball" spins, a small speaker remits a ticking sound in synchronization with the spinning "ball."

The Wheel-of-Fortune Game shown in schematic Figure 15-2 uses the electronics to give the illusion of a red "ball" spinning around a red and black wheel. To create this effect it is necessary to use three NPN transistors, a single UJT transistor and a CMOS counter chip. Switch S1 is momentarily depressed which turns "on" transistor Q1 which charges up capacitor C1. This charge maintains Q1 in the "on" state for several seconds. When Q1 is "on" transistor Q4 is biased "on" and current flows to the Emitter of the unijunction transistor at Q2. With Q2 having a positive level on its Emitter Q2 begins to oscillate. As Q2 oscillates, it feeds pulses to the base of transistor Q3, which amplifies the pulses. It then feeds them to the speaker and to the input of U1 on pin 14. U1 is known as a CMOS Johnson

Figure 15-1 *Wheel-of-Fortune Game*

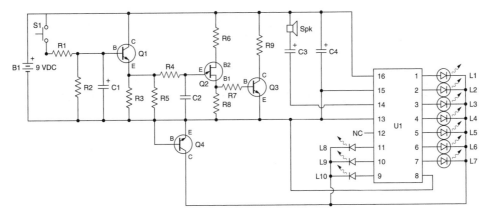

Figure 15-2   *Wheel-of-Forture Game (Courtesy of Cheney Electronics)*

Counter, and as it receives pulses on its input pin, it shifts the output pulse to each of its outputs. The output pins have LEDs connected to them and the LEDs light up in a circular pattern at the same time that the speaker clicks. Transistor Q4 has the function of turning off the LEDs, when transistor Q1 turns "off." If it did not, the LED that is "on" when the circuit stops oscillating would stay "on" until the battery was removed or exhausted. When capacitor C1 discharges, transistor Q1 and Q2 turn "off" and no more pulses reach pin 14 of U1. The integrated circuit now keeps "high" whichever output was originally " high" when the pulses stopped. The LED will "fade out" after several seconds by the action of Q4, as mentioned.

The Wheel-of-Fortune Game was fabricated on a small circuit board as shown in the pictorial diagram in Figure 15-3. Locate a large table or workbench area which you can use to assemble the project. Try to locate a 27 to 33-watt pencil tip soldering iron, along with a roll of #22 ga 60/40 rosin core solder for the project. Go to your local Radio Shack store and buy a small jar of "Tip Tinner," a soldering iron tip cleaner/dresser, along with a pair of tweezers, a magnifying glass and an anti-static strap, which can be used to avoid damaging your integrated circuits when handling them during construction. Getting up and down from your chair or moving around your workbench can generate large static voltages which can easily damage integrated circuits. Procure a pair of small needle-nose pliers, a small pair of end-cutters, a small flat-blade and a small Phillips screwdriver. Prepare your soldering iron by heating it up and cleaning the tip on a wet sponge. If you have "Tip Tinner," place the hot solder iron tip into "Tip Tinner" and rotate the soldering iron tip. Once again clean the tip off with a wet sponge. You

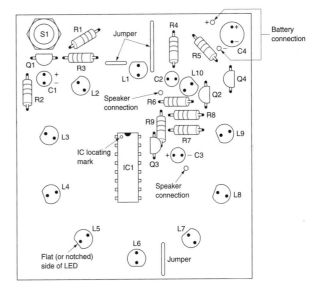

Figure 15-3   *Wheel-of-Fortune Game pictorial diagram*

are now ready to begin placing components on the circuit board.

First begin by inserting an integrated circuit socket for U1 in the center of the circuit board. It is a good idea to use IC sockets, since its cheap insurance against a circuit failure at some later date. You can simply replace the IC chip in just a few minutes. Most people cannot de-solder an IC from a circuit board without damaging the PC board. Pin 1 of the IC socket will connect to LED L1, while pin 16 connects to the plus side of capacitor C4 and the Collector of Q1.

Next, install all the resistors and be sure to observe the color code on the resistor and compare the colors to the resistor color chart in Table 15-1. The first resistor color band closest to the edge is the first digit. The

Table 15-1

Resistor Color Code Chart

Color Band	1st Digit	2nd Digit	Multiplier	Tolerance
Black	0	0	1	
Brown	1	1	10	1%
Red	2	2	100	2%
Orange	3	3	1000 (K)	3%
Yellow	4	4	10000	4%
Green	5	5	100000	
Blue	6	6	1000000 (M)	
Violet	7	7	10000000	
Gray	8	8	100000000	
White	9	9	1000000000	
Gold			0.1	5%
Silver			0.01	10%
No color				20%

second band is the second digit and the third color band is the multiplier value. Often there is a fourth color band which represents the tolerance value. If there is no band then the resistor has a 20% value. If the resistor has a silver band then the resistor has a 10% tolerance value, while a gold band represents a 5% tolerance value. Look through the parts pile for a resistor with its first band colored (orange), its second color band (orange) and its third color band (orange), this resistor will have a value of 33 k or 33000 ohms at R1 and R5. Install these resistors and then identify the remaining resistors using the resistor color chart. Install all the remaining resistors and solder them in place. After soldering the resistors in place, take your end-cutters and snip the excess resistor leads from the circuit board by cutting them flush to the edge of the board.

Now refer to Table 15-2, which illustrates the three-digit capacitor codes used to identify the small capacitors. Look through your capacitors for one marked with a code (474); this capacitor will have a 0.47 µF value at C2. Next insert C2, a mylar capacitor, on the circuit board and solder it in place. Then locate the electrolytic capacitors C1, C2 and C4. These three capacitors have polarity, which must be observed. Electrolytic capacitors will have either a plus or minus marking on the body of the component. Make sure that you pay attention to the markings and insert the capacitors correctly by referring to the schematic and pictorial diagrams as needed. Solder the capacitors onto the PC board, then cut the excess leads from the board.

Next locate transistors Q1, Q3, and Q4. All of these transistors are 2N3904 types. Note that they have three leads: a Base, a Collector and an Emitter. Refer to the pin location diagram in Figure 15-4 to ensure that you insert the transistor pins correctly. The reference diagram shows the transistor view from the bottom of the device. The UJT transistor is different from the other transistors. Observe that Q2 has three leads but they are marked Base 1, Base 2, and the Emitter. Insert the transistors on the circuit board and solder them in place. Remember to remove the excess component leads.

Finally, install the ten red LEDs. Note that the LEDs all have polarity considerations. The cathode band of the LED on the schematic corresponds to the flat side of the LED housing of the shorted lead (see Figure 15-5). Note that the LEDs are all arranged in a circle with L1 at the top of the circuit board. Going counterclockwise from L1, LED L2 is to the left of L1 and so on. Each LED is mounted equidistant from each

Table 15-2

Capacitance Codebreaker Information

This table is designed to provide the value of alphanumeric coded ceramic, mylar and mica capacitors in general. They come in many sizes, shapes, values and ratings; many different manufacturers worldwide produce them and not all play by the same rules. Most capacitors actually have the numeric values stamped on them; however, some are color coded and some have alphanumeric codes. The capacitor's first and second significant number IDs are the first and second values, followed by the multiplier number code, followed by the percentage tolerance letter code. Usually the first two digits of the code represent the significant part of the value, while the third digit, called the multiplier, corresponds to the number of zeros to be added to the first two digits.

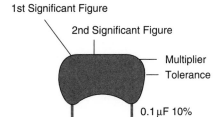

Value	Type	Code	Value	Type	Code
1.5 pF	Ceramic		1000 pF/0.001 µF	Ceramic/Mylar	102
3.3 pF	Ceramic		1500 pF/0.0015 µF	Ceramic/Mylar	152
10 pF	Ceramic		2000 pF/0.002 µF	Ceramic/Mylar	202
15 pF	Ceramic		2200 pF/0.0022 µF	Ceramic/Mylar	222
20 pF	Ceramic		4700 pF/0.0047 µF	Ceramic/Mylar	472
30 pF	Ceramic		5000 pF/0.005 µF	Ceramic/Mylar	502
33 pF	Ceramic		5600 pF/0.0056 µF	Ceramic/Mylar	562
47 pF	Ceramic		6800 pF/0.0068 µF	Ceramic/Mylar	682
56 pF	Ceramic		0.01	Ceramic/Mylar	103
68 pF	Ceramic		0.015	Mylar	
75 pF	Ceramic		0.02	Mylar	203
82 pF	Ceramic		0.022	Mylar	223
91 pF	Ceramic		0.033	Mylar	333
100 pF	Ceramic	101	0.047	Mylar	473
120 pF	Ceramic	121	0.05	Mylar	503
130 pF	Ceramic	131	0.056	Mylar	563
150 pF	Ceramic	151	0.068	Mylar	683
180 pF	Ceramic	181	0.1	Mylar	104
220 pF	Ceramic	221	0.2	Mylar	204
330 pF	Ceramic	331	0.22	Mylar	224
470 pF	Ceramic	471	0.33	Mylar	334
560 pF	Ceramic	561	0.47	Mylar	474
680 pF	Ceramic	681	0.56	Mylar	564
750 pF	Ceramic	751	1	Mylar	105
820 pF	Ceramic	821	2	Mylar	205

(Continued)

Table 15-2 (*Continued*)

PicoFarad (pF)	NanoFarad (nF)	MicroFarad (mF, μF or mfd)	Capacitance Code
1000	1 or 1n	0.001	102
1500	1.5 or 1n5	0.0015	152
2200	2.2 or 2n2	0.0022	222
3300	3.3 or 3n3	0.0033	332
4700	4.7 or 4n7	0.0047	472
6800	6.8 or 6n8	0.0068	682
10000	10 or 10n	0.01	103
15000	15 or 15n	0.015	153
22000	22 or 22n	0.022	223
33000	33 or 33n	0.033	333
47000	47 or 47n	0.047	473
68000	68 or 68n	0.068	683
100000	100 or 100n	0.1	104
150000	150 or 150n	0.15	154
220000	220 or 220n	0.22	224
330000	330 or 330n	0.33	334
470000	470 or 470n	0.47	474

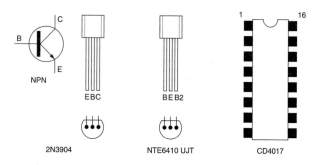

Figure 15-4 *Wheel-of-Fortune semiconductors*

other in a circle pattern with L6 at the bottom opposite L1. Note that the cathodes of the LEDs are all connected to the Collector of Q4. Install and solder the LEDs to the circuit board using the round template shown in Figure 15-6.

Next you can connect the pushbutton switch to the circuit board. Locate two 6 inch pieces of #22 ga stranded-insulated wire. Solder the two wires to the pushbutton terminals. The free ends of the two wires will connect to the circuit board as shown in the pictorial diagram. The switch leads at S1 are connected between the plus (+) end of the battery and resistor R1. Now prepare two 6 inch #22 ga stranded-insulated wires and solder them to the speaker (SPKR) terminals. The free ends of the speaker leads are connected between the minus (−) end of the capacitor C3 and the plus (+) power supply.

At this point in time, you should locate the integrated circuit and prepare to insert it into the circuit board. Integrated circuits all have some sort of locating marks which help to orient the IC in its socket. Take a look at the U1; you will notice either a small indented circle or a small cutout or notch at one end of the IC package. If you see a small indented circle, then pin 1 of the IC will be to the left of it. If you see a cutout or notch on the top of the IC, then pin 1 will again be to the left of the cutout. Once you have observed the correct orientation, you can insert the IC into the socket, using your

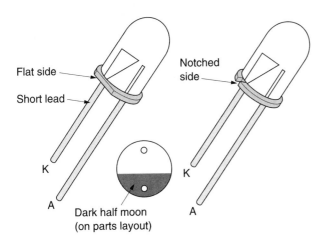

Figure 15-5   *LED identification*

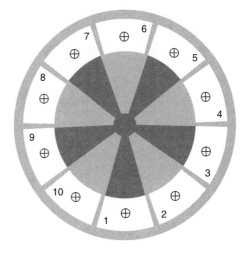

Figure 15-6   *Wheel-of-Fortune template*

anti-static wrist strap. Be sure to line-up pin 1 of the IC socket with pin 1 on the IC chip itself. If you have any difficulties identifying the pin-outs or orientation of the transistors or integrated circuits, you should ask a knowledgeable electronics enthusiast to assist you.

Finally, connect up the 9-volt battery clip leads. The plus (+) or red lead connects to the positive supply bus on the circuit board, i.e. the Collector of Q1, and junction of R6, R9 and C4. The black or minus (–) battery clip lead connects to the ground side of the circuit, i.e. the Emitters of Q1, Q3 and Q4.

This is a good time to take a short break, and when we return you will check over your circuit board. When inspecting your completed circuit board you will be looking first for "cold" solder joints, and then you will be looking for "stray" component leads. First look at your solder joints and see if all the joints, look clean, smooth and shiny. If they do not look shiny and smooth, you will need to clean the joint and re-solder the connection to ensure a good solder joint for reliable operation. Next you will want to inspect the circuit board for "stray" component leads that may have "stuck" to the circuit board when you cut off the component leads. These "stray" leads could "short circuit" the board and cause a failure when power is first applied.

Connect up a 9-volt transistor radio battery and press the pushbutton. If all goes well, then you should see the LEDs begin to light up. You are now ready to play the Wheel-of-Fortune.

If one of the LEDs does not light up, then remove the battery. You may have installed one of the LEDs

backwards. It will then be a simple matter to unsolder the particular LED, rotate it 180 degrees and re-insert it into the circuit board.

If the circuit seems "dead" and the LEDs do not light up, you will need to inspect the entire circuit board for any errors due to mistakes in components orientation. First be sure that each resistor has the correct color code for that location. Next, be sure that all the capacitors have been installed correctly, check the orientation of the electrolytic capacitors. Then inspect the installation of transistors Q1, Q3 and Q4 followed by the UJT at Q2. Refer to the schematic, the pin-out diagrams and the pictorial diagrams. Now re-check the orientation of the integrated circuits and make sure everything was inserted into the sockets properly and that pin 1 of the IC goes to L1. Finally re-check the installation of all ten of the LEDs to make sure they were installed with the flat side of the LED package or the cathode pointing to the half moon or ground bus connection on the circuit board.

Once you have re-checked the circuit board, you can re-apply power and try the circuit again. When you have located your error and everything appears to be working correctly, you can locate the decal template and place it over the LEDs. Punch out the holes for the LEDs on the template and orient the decal over the LEDs so that LED 1 corresponds to the number one on the decal. Your Wheel-of-Fortune is now ready to play. Have fun!

Chapter 16

Skeet-Shooting Game

Parts List

Parts Bin

R1, R5, R6, R7, R8, R9
22 k ohm, 5%, ¼-watt
resistor

R10, R11, R12, R13, R14
22 k ohm, 5%, ¼-watt
resistor

R2   220 k ohm, 5%,
¼-watt resistor

R3   2.2 megohm, 5%,
¼-watt resistor

R4   2.2 k ohm, 5%,
¼-watt resistor

R15   680 ohm, 5%, ¼-watt
resistor

C1, C4   100 nF, 35-volt
green cap

C2   470 nF, 35-volt
green cap

C3   47 nF, 35-volt green
cap

C5   100 µF, 35-volt
electrolytic capacitor

D1   1N4004 silicon diode

Q1, Q2, Q3, Q4, Q5
PN100 NPN transistor—
2N3904 or NTE123AP

Q6, Q7, Q8, Q9, Q10
PN100 NPN transistor—
2N3904 or NTE123AP

U1   CD4093B QUAD
Schmitt NAND IC

U2   CD4017B Decade
Counter IC

L1, L2, L3, L4 green
LEDs

L5, L6, L7, L8, L9
green LEDs

L10   red LED

S1   RESET Pushbutton
(normally open) SPST

S2   FIRE pushbutton
(normally open) SPST

Sx   optional on/off
switch SPST

B1   9-volt transistor
radio battery

Misc   PC board, IC
sockets, battery clip,
hardware, enclosure,
wire, etc.

How good is your hand–eye coordination? Rifle enthusiasts test theirs by trying to shoot down "skeets" or clay targets, hurled up into the sky by a catapult or trap. This electronic game project is a less dangerous way to test your shooting skills indoors at any time (see Figure 16-1).

Trying to shoot down a fast-moving clay target or skeet is a real test of hand–eye coordination, because you have to move your rifle to anticipate the skeet's movement, and then press the trigger at exactly the right instant to make sure the bullet and skeet will arrive at the same place at the same time. This may sound easy but in reality it isn't.

The skeet-shooting game sends a burst-of-light 'skeet' up along a string of LEDs, and you have to try to "hit" it by pressing a FIRE button at exactly the right time as it reaches the red LED at the end. Most of the time you will miss, but the game is a lot of fun. When you do "hit" the skeet though, the game stops with the end LED glowing. After the "joy" of hitting your target,

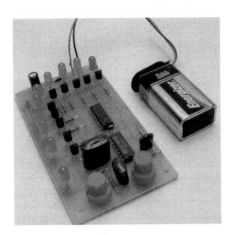

Figure 16-1   *Skeet-Shooting Game*

resistor and a 470 nF capacitor for an astable multivibrator or oscillator around U1:d, running at about 10 Hz. This clocking pulse every 1/10 of a second is fed to pin 14 of U2. The other clock pulse input at pon3 of U2 is not used, but is tied to ground. When the battery is first connected to the circuit, U1:d begins sending pulses into U2 which begins the counting, and the outputs begin to go "high," first Q1, then Q2 and so on.

Because each of the counter's outputs are connected to the base of switching transistors via a 22 k series resistor, this means when each output goes "high" it turns on that particular transistor. As each transistor is turned "on," it begins to draw current and lights up the corresponding LED. It is connected between its Collector and the plus (+) line through a 680 ohm resistor. So as U2 counts the pulses from U1:d, the LEDs flash on in sequence—from LED 1 up to LED 10. This is how the "skeet" flies up.

After the counter's Q10 output has gone "high" briefly, turning on LED 10, the next pulse to arrive from U1:d causes the counter to start again from scratch. So

you can write down your score and start the game again by pressing the RESET button. The Skeet Game is a great way to improve your hand–eye coordination.

The Skeet-Shooting Game schematic, shown in Figure 16-2, is built around two integrated circuits, a CD4093B CMOS QUAD Schmitt NAND Gate and a CD4017B Decade Counter IC. There is a 220 k ohm

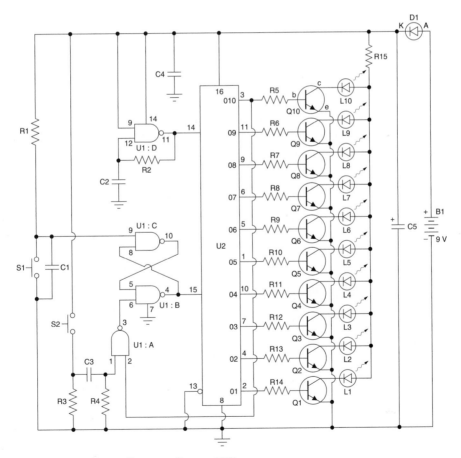

Figure 16-2   *Skeet-Shooting Game. Courtesy Jaycar (JC)*

the counting sequence and the LEDs flashing in turn repeats itself, with new "skeets" moving up the LEDs at a rate of about one every second. So far we have been assuming that this counting can continue because you haven't tried pressing the "FIRE" button at S2. So let's see how this counting sequence is controlled.

Notice that gates (B) and (C) of U1 are connected as a simple flip-flop, with each gate's output connected to one input of the other gate. As a result of this "cross coupling," the flip-flop is able to remain in either of two stable states. One with the output of U1:c high and that of U1:b low. The flip-flop is known as bistable.

Notice that U1:c's input pin 9 is connected up to the plus (+) 9-volts via a 22 k ohm resistor, and also connected down to ground via a 100 nF capacitor. So when we first apply power, pin 9 of U1:c starts off by being pulled down to the negative line by the capacitor, before it charges up via the 22k ohm resistor. Also notice that the output of U1:b at pin 4 is connected to pin 15 of counter U2, as well as to pin 8 of U1:c. This input of U2 is the counter's master RESET input, which makes the counter stop counting when it's taken to the logic "high" level, i.e. 5 volts. When the flip-flop turns "on" and switches the output of U!:b to "high," this pulls the RESET "low," and as a result the counter can start counting immediately sending "skeets" up the string of LEDs.

But how does the circuit tell when you have pressed S2 at exactly the right time? As you can see, the output of U1:a is connected to the pin 6 input of the second flip-flop gate U1:b. One input pin 2 of U1:a is also connected to pin 3 of U2, which is the O10 output. So each time the counter reaches "10" and sends this output pin "high" to turn "on" Q10 and LED 10, it also pulls pin 2 of U1:a "high."

Notice that the FIRE button at S2 is connected between the plus (+) 9-volt line and the 47 nF capacitor which connects to the second input pin 1 of U!:a. Both ends of the capacitor are also connected to ground, so that until S2 is pressed, both sides of the capacitor are pulled down. This means that pin 1 of U1:a is pulled "low," and its output will remain "high" even when pin 2 is "high" each time the counter reaches "10."

When you press S2, this connects the junction of the 4 nF capacitor and the 2.2 M resistor to plus (+) 9 volts. As a result the 47 nF capacitor is able to charge up quickly, via the 2.2 k ohm resistor; this produces a short

pulse of positive voltage across the resistor, pulling up pin 1 of U1:a. So the effect of pressing S2 is to produce a very short pulse on pin 1 of U1:a. This pulse is about 100 µs even if the button is pressed longer. This brief pulse on pin 1 of U1A won't have any effect unless pin 2 happens to be pulled "high" at exactly the same time. This will only happen if you've pressed S2 at exactly the same time U2's O10 output has gone "high," to make LED 10 light up. So this is how the circuit can tell if you've pressed S2 at the "right" time.

When you do press S2 at the same instant as U2's O10 output has gone "high," this makes both inputs of the NAND gate U1: a high" at the same time, so the output pin 3 will go low. This, in turn, will pull down input pin 6 of U1:b, which makes the flip-flop switch states. Pin 4 of U1:b will flip "high" pulling pin 15 of U2 "high" as well, thus immediately stopping the counter with LED 10 glowing to register a "hit."

The circuit will stay in this state until you either disconnect the battery or press the "RESET" button at S1. As you can see, pressing S1 connects pin 9 of U1:c to ground, making the flip-flop switch back into its original state with pin 10 "high" and pin 4 "low," this removes the "high" from U2's RESET pin, and allows it to start counting again.

Well, let's go ahead and build the Skeet-Shooting Game. Prepare a large working table where you can spread out all of your diagrams and parts and tools for the project preparation. First you will need a 27 to 33-watt soldering iron or pencil iron, some 60/40 rosin core solder, "Tip Tinner" and a wet sponge to clean your soldering iron tip. You will also want to locate an anti-static wrist band to safely handle the integrated circuits. We chose to build the Skeet-Shooting Game on a 2¼ inch × 4¼ inch printed circuit board for most reliable results. We also chose to use integrated circuit sockets to ensure easy replacement of an IC in the event of a circuit failure at some possible later date. The Skeet-Shooting Game circuit shown in the pictorial diagram in Figure 16-3 uses two integrated circuits, a 14 pin CD4093B and a 16 pin CD4017B IC.

Let's take a few moments to secure a well-lit worktable or workbench. Locate a 27 to 33-watt pencil tip soldering iron as well as a roll of 22 ga 60/40 rosin core solder. Go to your local Radio Shack store and pick up a small jar of "Tip Tinner," a soldering iron tip cleaner/dresser, and an anti-static wrist strap which will

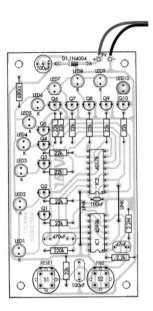

Figure 16-3 *Skeet-Shooting Game pictorial*

help to prevent damage to the integrated circuits from high voltage static charges. Static charges can build up from just getting in and out of your chair or moving around your worktable. Plug in the wrist strap to wall outlet which ground the strap. A few tools, such as a pair of needle-nose pliers, a pair of end-cutters or diagonals, a magnifying glass and a pair of tweezers can be very helpful during construction.

Finally, locate a small flat-blade and Phillips screwdriver.

With all the diagrams and parts in front of you let's begin. Locate the resistors and refer to Table 16-1 which lists the resistor color codes. Most resistors will have at least three or four colored bands on the body of the resistor. The first band, at one end of the resistor body, will be the first digit, the second color band corresponds to the second digit of the code, and the third color band represents the resistor multiplier value. If there is not a fourth band the resistor will have a 20% tolerance value, but if the resistor has a silver band, then the resistor has a 10% tolerance value. If the fourth color band is gold then the resistor has a 5% tolerance value. Look through the resistors and see if you can find a resistor with a red first band, red in the second band, with a third color band of an orange color. This would represent a 22k ohm resistor at R1. Repeat this process to identify each of the resistors. When you have identified all of the resistors, you can insert them into the circuit board and solder them in place. Take your end-cutters and cut the excess resistor leads. Cut the leads flush to the edge of the circuit board.

Next refer to Table 16-2, which lists the capacitor codes, and which can be used to identify the capacitors in your project. This project has five capacitors; some

Table 16-1
Resistor Color Code Chart

Color Band	1st Digit	2nd Digit	Multiplier	Tolerance
Black	0	0	1	
Brown	1	1	10	1%
Red	2	2	100	2%
Orange	3	3	1000 (K)	3%
Yellow	4	4	10000	4%
Green	5	5	100000	
Blue	6	6	1000000 (M)	
Violet	7	7	10000000	
Gray	8	8	100000000	
White	9	9	1000000000	
Gold			0.1	5%
Silver			0.01	10%
No color				20%

Table 16-2

Capacitance Codebreaker Information

This table is designed to provide the value of alphanumeric coded ceramic, mylar and mica capacitors in general. They come in many sizes, shapes, values and ratings; many different manufacturers worldwide produce them and not all play by the same rules. Most capacitors actually have the numeric values stamped on them; however, some are color coded and some have alphanumeric codes. The capacitor's first and second significant number IDs are the first and second values, followed by the multiplier number code, followed by the percentage tolerance letter code. Usually the first two digits of the code represent the significant part of the value, while the third digit, called the multiplier, corresponds to the number of zeros to be added to the first two digits.

1st Significant Figure

2nd Significant Figure

Multiplier

Tolerance

0.1 µF 10%

Value	Type	Code	Value	Type	Code
1.5 pF	Ceramic		1000 pF/0.001 µF	Ceramic/Mylar	102
3.3 pF	Ceramic		1500 pF/0.0015 µF	Ceramic/Mylar	152
10 pF	Ceramic		2000 pF/0.002 µF	Ceramic/Mylar	202
15 pF	Ceramic		2200 pF/0.0022 µF	Ceramic/Mylar	222
20 pF	Ceramic		4700 pF/0.0047 µF	Ceramic/Mylar	472
30 pF	Ceramic		5000 pF/0.005 µF	Ceramic/Mylar	502
33 pF	Ceramic		5600 pF/0.0056 µF	Ceramic/Mylar	562
47 pF	Ceramic		6800 pF/0.0068 µF	Ceramic/Mylar	682
56 pF	Ceramic		0.01	Ceramic/Mylar	103
68 pF	Ceramic		0.015	Mylar	
75 pF	Ceramic		0.02	Mylar	203
82 pF	Ceramic		0.022	Mylar	223
91 pF	Ceramic		0.033	Mylar	333
100 pF	Ceramic	101	0.047	Mylar	473
120 pF	Ceramic	121	0.05	Mylar	503
130 pF	Ceramic	131	0.056	Mylar	563
150 pF	Ceramic	151	0.068	Mylar	683
180 pF	Ceramic	181	0.1	Mylar	104
220 pF	Ceramic	221	0.2	Mylar	204
330 pF	Ceramic	331	0.22	Mylar	224
470 pF	Ceramic	471	0.33	Mylar	334
560 pF	Ceramic	561	0.47	Mylar	474
680 pF	Ceramic	681	0.56	Mylar	564
750 pF	Ceramic	751	1	Mylar	105
820 pF	Ceramic	821	2	Mylar	205

Table 16-2 (*Continued*)

PicoFarad (pF)	NanoFarad (nF)	MicroFarad (mF, μF or mfd)	Capacitance Code
1000	1 or 1n	0.001	102
1500	1.5 or 1n5	0.0015	152
2200	2.2 or 2n2	0.0022	222
3300	3.3 or 3n3	0.0033	332
4700	4.7 or 4n7	0.0047	472
6800	6.8 or 6n8	0.0068	682
10000	10 or 10n	0.01	103
15000	15 or 15n	0.015	153
22000	22 or 22n	0.022	223
33000	33 or 33n	0.033	333
47000	47 or 47n	0.047	473
68000	68 or 68n	0.068	683
100000	100 or 100n	0.1	104
150000	150 or 150n	0.15	154
220000	220 or 220n	0.22	224
330000	330 or 330n	0.33	334
470000	470 or 470n	0.47	474

capacitors may have their actual value printed on the component body, while other capacitors may have a three-digit code marked on them and you will have to use the chart to identify them. For example, a code marked (103) would indicate a 10 nF or 0.01 μF capacitor. Only one of the capacitors in this project will have polarity issues. The capacitor at C5 is an electrolytic type and will have a black or white band at either end of the component body, which denotes the polarity. Make sure that you observe the color band before inserting the electrolytic capacitor into the PC board. Once you have identified the capacitors, you can go ahead and install them on the circuit board and solder them in place. Remember to cut the excess capacitor leads from the PC board.

Next, locate the single silicon diode at D1. Diodes also have polarity marking which will help you in installing the component. Diodes will have either a black or white color band at one end of the diode body. The colored band denotes the diode's cathode. On the schematic the anode is the arrow or triangle which points to the cathode, which is represented by a flat line.

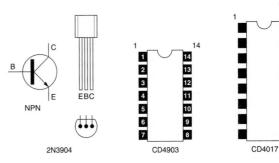

Figure 16-4 *Skeet-Shooting semiconductors*

The Skeet-Shooting Game has a number of transistors, one for each output on the decade counter, from one to ten. The transistors used in this project are all NPN types. Transistors generally have three leads: a Base lead, a Collector lead and an Emitter lead. Refer to the diagram in Figure 16-4, which depicts the pin-outs for the transistor and the integrated circuits used in this project. The transistor symbol has a flat line representing the Base lead with both a Collector and Emitter connected to the Base. The Emitter lead will have an arrow pointing away from the Base lead if the

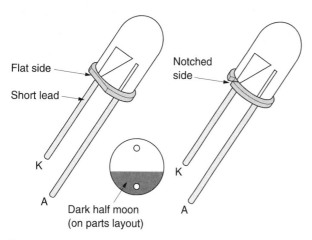

Flat side

Short lead

Notched side

K

A

Dark half moon
(on parts layout)

K

A

Figure 16-5 *LED identification*

transistor is a NPN type. The pictorial diagram for the transistor shows the three leads in a row with the Emitter at one end, the Base lead in the center and the Collector lead at the opposite end of the transistor body. Note that the transistor is referenced from the bottom view. Once you have identified the transistors you can insert them into the circuit board and solder them in place. Remove the excess transistor leads from the PC board.

As mentioned, the Skeet-Shooting Game has two integrated circuits; we advise using IC sockets for easy removal in the event of circuit failure. Most people have a very difficult time removing integrated circuits from a circuit board and often damage the board in the process of removal. So to make life a bit easier, use IC sockets when building the circuit. All integrated circuits will have some sort of identification marking on the IC package to help in installing the device. Look at the IC package and you will see either a small indented circle or a small cutout or notch at one end of the IC's body. The pin of the IC will be just to the left of these markings. Now, you will need to make sure that you can identify which pin on the IC socket is pin 1, before inserting the IC into the socket. Take your time and get it right the first time, to prevent destroying the IC when power is first applied. Have an adult or friend check your work! If you are sure that you have identified pin 1 on the IC and pin 1 on the socket, then go ahead and install the IC into its respective socket.

We are making great progress! Now look for all the LEDs; there should be nine green LEDs and one red LED in the parts pile. Refer to the diagram shown in

Figure 16-5, which shows the pin-out diagram for an LED. Light emitting diodes or LEDs will also have a type of polarity marking on them. LEDs must be installed correctly or they will not light up. Note on the diagram that the LED will have a flat edge; you can confirm this by looking at the actual LED. Also note that the symbol for an LED is nearly the same as that of a diode. The cathode is the flat line, with the anode as a triangle pointing to the cathode. The cathode lead of the LED is usually the shorter lead and is often represented with a dark half moon. Refer to the pictorial diagram and you will see that the first nine LEDs are green, which lead up to the final LED, which is red. Mount the LEDs onto the circuit board and solder them in place. Remember to cut the excess component leads.

At this point you can look for both pushbutton switches, S1 and S2. These PC mounted switches can be installed now. Note that you should also elect to use pushbutton switches with longer leads if you are going to mount the PC board remote from the circuit board. Note that S2, the FIRE switch, is installed between R3/C3 and pin 16 of U2. Switch S1, the RESET pushbutton, is connected between the junction of R3/R4 and resistor R1.

Finally, you can locate the 9-volt battery clip and solder to the circuit board. Note that the minus (–) or black lead is the common point of the circuit or ground which connects to pin 8 of U2 and pin 7 of U1. The red or plus (+) battery clip lead can be wired in series with an optional SPST power switch if desired. The red or plus lead would connect to one end of the power switch and the other end of the switch is connected to pin 16 on U2 and pin 14 of U1.

Why don't you take a short well-deserved break and when we return we will inspect the circuit board for any possible "cold" solder joints or "short" circuits. Turn the circuit board, so that the foil side of the circuit board faces up toward you. Take a look at the solder joints to make sure that they all look clean, smooth and shiny. "Cold" solder joints can cause premature circuit failures and poor operation from your circuit, so it's best to discover them before power is applied. If you find a solder joint that looks dull, "blobby" or lumpy, remove the solder from the joint and re-solder it until it looks good. Next, look the circuit board over again to make sure that there are no "shorts." A "short" or "bridge" can cause your circuit to fail immediately and perhaps damage it. So it is best to find a "short" before power is applied to the circuit for the first time. Look for any "stray" component leads which may have stuck to the underside of the PC board after the excess component leads were cut or removed. Sometimes these excess leads can stick or get stuck from solder flux. Make sure that there are no extra wires or "bridges" between circuit pads.

Once the board looks clean, you can attach a 9-volt battery, to see if the circuit works. Snap on the battery clip and if you installed the optional switch, turn the switch to the "on" position. You will be rewarded by seeing a small flash of light "travel" up the string of LEDs, starting with LED 1 at the lower left and going through each LED in turn until it reaches the red LED L10 at the top right. Then another "skeet" should follow it up the LEDs and so on, at about one per second.

Now it's time to see if you can press the FIRE button S2 at the exact time that one of these light "skeets" reaches LED10. This may take quite a while to achieve, but when you do get the timing for your button press right, all movement will stop and LED L10 will stay lit to confirm you had a "hit."

You can restart the game by pressing the RESET button S1. Let one of your friends try the next round and see if he/she can "hit" the "skeet." If all goes well and the circuit works correctly, you can challenge your sister/brother or you mom/dad or your friends to a game of "Skeet."

In the event that your Skeet-Shooting Game does not work, which sometimes does happen, you will have to remove the battery and re-inspect the circuit for any errors you might have made during construction.

If one or more of the LEDs does not seem to light up, you may have installed them backwards. It is a simple matter to unsolder the LED and remove the solder from the joint. Then rotate the LED 180 degrees, install

it back into the circuit board, then solder it in place. In the event that the circuit is "dead" and does not light up at all, you will have to "test" your troubleshooting skills.

First, re-check the color codes on each of the resistors to make sure that they are in their respective places. Next, look for diodes and capacitors which may have been installed backwards, this can easily happen even to the most sophisticated builders. Next, look at all the transistors to make sure that you have installed them correctly. Carefully compare the drawing with the component to make sure that you can identify which lead is truly the Base. Many times transistors are installed backwards with Emitter reversed from the Collector. Have a friend or adult check your work; often a different pair of eyes will catch an error. Re-check the placement of the integrated circuits in their respective sockets and then make sure that each of the pins of the IC socket is connected to the correct component in the circuit. It's always possible that you may have received a defective IC or that you damaged it while handling it. Replace the IC if necessary.

You are almost finished now. Refer once again to the installation diagram for the LEDs. Many project errors are a result of LEDs installed backwards. If most of the green LEDs light up in sequence but one lamp is out, the chances are that you installed an LED backwards. Re-check your LED polarity and re-install an LED if necessary.

When you have checked all of these "fine" points you are ready to apply power once again to see if the circuit will now work correctly. Hopefully you have corrected your errors and the circuit should now work fine. Happy hunting. Have fun!

Chapter 17

Grab-the-Gold Game

Parts List

Parts Bin

R1, R2 1k ohm
  resistor, 5%, ¼-watt
  resistor

R3, R4 120 ohm
  resistor, 5%, ¼-watt
  resistor

R5 5k ohm
  potentiometer

C1 10 µF, 35-volt
  electrolytic capacitor

C2, C3 0.01 µF, 35-volt
  disc capacitor

D1 1N4001 silicon diode

U1 LM555 timer IC

U2 4017 CMOS IC

L1, L2, L3, L4 red
  LEDs

L5, L6, L7, L8, L9 red
  LEDs

L10 orange LED

S1 momentary pushbutton
  switch (normally open)

B1 9-volt transistor
  radio battery

Misc PC circuit board,
  wire, IC sockets,
  battery clip, etc.

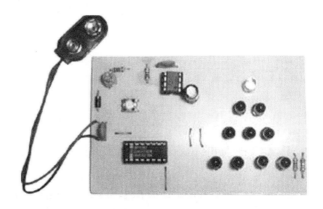

Figure 17-1 *Grab-the-Gold Game*

The Grab-the-Gold Game shown in Figure 17-1 is an easy to play but hard to master game. This game features a bright red LED pyramid with a clear LED at the top of the pyramid. When power is applied, each of the ten LEDs lights up in a sequence pattern. If you press and hold the pushbutton, just before the top clear LED is to light, this LED will stay on illuminating its bright orange glow, indicating that you made it to the top and picked up the gold. If your timing is off, one of the red LEDs will light up, indicating "you lost." To make things more interesting, there is a speed control to move things along.

The Grab-the-Gold Game uses two integrated circuits and a handful of electronic components. The game uses nine red LEDs and one orange LED as the game display which are arranged as a pyramid. The heart of the Grab-the-Gold Game is the LM555 timer and U1 is configured as an astable oscillator or free running oscillator (see the schematic diagram in Figure 17-2). The frequency of the timing pulses created at U1 is determined by the RC time constant set up by the resistors R1 and R2 and capacitor C1 and the potentiometer at P1. The output of U1 is presented at pin 3 and is coupled directly to the clock input of U2 on pin 14. The LM555 timer provides the clock pulses to the integrated circuit at U2. Integrated circuit U2 is a CMOS Johnson Counter. As the clock pulses from U1 are sent one at a time to the Johnson Counter, the counter shifts a (+) output sequentially to each output. Each time an output pin goes high, it lights up the next LED in sequence. When switch S1 is pressed, it shorts out capacitor C1. When the capacitor C1 is shorted it stops the clock pulses from reaching the input of U2, thereby freezing the LED display.

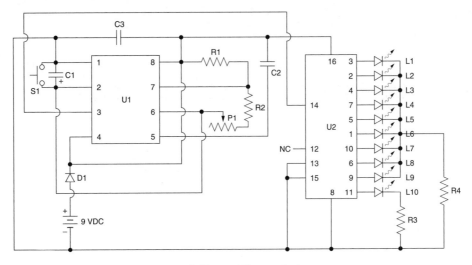

Figure 17-2 *Grab-the-Gold Game. (Courtesy of Cheney Electronics)*

The Grab-the-Gold Game is powered by a commonly found 9-volt transistor radio battery. You can elect to place the game circuit in a plastic enclosure or just cover the bottom of the circuit with tape to protect the circuit from shorting out.

The Grab-the-Gold Game circuit was fabricated on a 4 inch × 2½ inch circuit board (refer to the pictorial diagram shown in Figure 17-3). Circuit board construction ensures a more reliable circuit operation as everything is soldered in place with good solid connections. Construction of the game is very straightforward and can be built in a few hours' time.

Before we actually begin building the project, let's take a few minutes to secure a well-lit worktable or workbench in order to spread out all the diagrams, charts, components and tools required. You will want to obtain a small 27 to 33-watt pencil tipped soldering iron, a wet sponge, a small jar of "Tip Tinner", and a solder iron prep and cleaner from your local Radio Shack store. Try to locate an anti-static wrist band for use when handling the integrated circuits. Powerful static charges can damage sensitive integrated circuits as you move around when handling and building the circuit, from getting up and down from your chair or moving back and forth on a rug, etc. You will also want to secure a pair of diagonal cutters, a pair of needle-nose pliers, a pair of tweezers, a magnifying glass and a few small screwdrivers. Heat up your soldering iron and we will begin!

Next, refer to the chart in Table 17-1, which illustrates the resistor color code values. Locate the resistors from the parts pile and take a look at one of them and notice the color bands, which start at one end of the resistor body. From the resistor color code chart you will notice that there are different columns in the chart. The first color represents the resistor's first digit, a number from 0 to 9. The second color represents the resistor's second number value, and the third color band represents the resistor's multiplier value. A fourth color band depicts the tolerance of the resistor. A silver band indicates a 10% tolerance value, while a gold band indicates a 5% tolerance. If there is no fourth color band then the resistor will have a 20% tolerance value. Try to find a resistor with a (brown), a (black) and a (red) color band; this resistor will have a 1000 ohm value and will represent R1 and R2. Next refer to the pictorial diagram, shown in Figure 17-3. Mount these resistors in their respective locations and solder them in place on the circuit board. Identify the remaining resistors and install them on the circuit board in their correct locations, then solder them in place on the PC board.

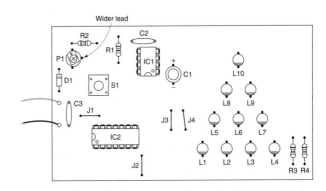

Figure 17-3 *Grab-the-Gold pictorial diagram*

Table 17-1

Resistor Color Code Chart

Color Band	1st Digit	2nd Digit	Multiplier	Tolerance
Black	0	0	1	
Brown	1	1	10	1%
Red	2	2	100	2%
Orange	3	3	1000 (K)	3%
Yellow	4	4	10000	4%
Green	5	5	100000	
Blue	6	6	1000000 (M)	
Violet	7	7	10000000	
Gray	8	8	100000000	
White	9	9	1000000000	
Gold			0.1	5%
Silver			0.01	10%
No color				20%

Take your end-cutters or diagonal cutters and cut the excess resistor leads from the circuit board. Cut the excess leads flush to the edge of the circuit board.

Next, you can locate and install the potentiometer P1. Note that trimmer potentiometers will have either three leads in-a-row or two leads with an offset center lead which represents the wiper or adjustable arm. Install the potentiometer on the circuit board, then solder it in place. Remember to cut the excess component leads.

Now, we can go ahead and install the capacitors. Refer now to the chart in Table 17-2, which illustrates the capacitor codes found on smaller disks and mylar capacitors. Often capacitors will have their values printed directly on the component, but sometimes, on small-sized capacitors, there is just no room for printing of the full value, so a code is used to represent the actual value. Capacitors C2 and C3 will both have a code marked (103), and from the chart you will see that the value should be 0.01 μF. Once you have identified these capacitors, you can go ahead and install them on the circuit board. These capacitors have no polarity markings, so they can be installed in either direction. Capacitor C1 is an electrolytic capacitor, so it will have polarity markings on it. Be sure to observe the polarity

markings on capacitor C1; to ensure that you install it properly, refer to the schematic and pictorial diagram as needed. Note, the negative side of the capacitor is often marked with a color stripe along the edge or side. You will see either a black or white color band and near it you will find either a minus (–) or plus (+) marking; this will indicate the polarity. Install C1 on the circuit board at its proper location, then remove the excess leads from the board.

Identify the only silicon diode for the project at D1. Remember that diodes have polarity markings on them. Look for either a black or white color band on the diode at one end of the diode body. The anode lead of the diode will be the triangle or arrow which points to the cathode lead. The color band will usually indicate the diode's cathode lead. Observe that the anode lead connects to the battery plus (+) terminal, and the cathode of the diode connects to both pin 4 and pin 8 of U1.

Next we need to deal with installing the two integrated circuits, so refer now to the semiconductor pin-out diagram in Figure 17-4. Integrated circuit sockets are highly recommended in the event of a circuit failure at a later date, since most people cannot unsolder an IC from

Table 17-2

Capacitance Codebreaker Information

This table is designed to provide the value of alphanumeric coded ceramic, mylar and mica capacitors in general. They come in many sizes, shapes, values and ratings; many different manufacturers worldwide produce them and not all play by the same rules. Most capacitors actually have the numeric values stamped on them; however, some are color coded and some have alphanumeric codes. The capacitor's first and second significant number IDs are the first and second values, followed by the multiplier number code, followed by the percentage tolerance letter code. Usually the first two digits of the code represent the significant part of the value, while the third digit, called the multiplier, corresponds to the number of zeros to be added to the first two digits.

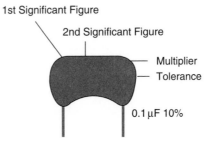

Value	Type	Code	Value	Type	Code
1.5 pF	Ceramic		1000 pF/0.001 μF	Ceramic/Mylar	102
3.3 pF	Ceramic		1500 pF/0.0015 μF	Ceramic/Mylar	152
10 pF	Ceramic		2000 pF/0.002 μF	Ceramic/Mylar	202
15 pF	Ceramic		2200 pF/0.0022 μF	Ceramic/Mylar	222
20 pF	Ceramic		4700 pF/0.0047 μF	Ceramic/Mylar	472
30 pF	Ceramic		5000 pF/0.005 μF	Ceramic/Mylar	502
33 pF	Ceramic		5600 pF/0.0056 μF	Ceramic/Mylar	562
47 pF	Ceramic		6800 pF/0.0068 μF	Ceramic/Mylar	682
56 pF	Ceramic		0.01	Ceramic/Mylar	103
68 pF	Ceramic		0.015	Mylar	
75 pF	Ceramic		0.02	Mylar	203
82 pF	Ceramic		0.022	Mylar	223
91 pF	Ceramic		0.033	Mylar	333
100 pF	Ceramic	101	0.047	Mylar	473
120 pF	Ceramic	121	0.05	Mylar	503
130 pF	Ceramic	131	0.056	Mylar	563
150 pF	Ceramic	151	0.068	Mylar	683
180 pF	Ceramic	181	0.1	Mylar	104
220 pF	Ceramic	221	0.2	Mylar	204
330 pF	Ceramic	331	0.22	Mylar	224
470 pF	Ceramic	471	0.33	Mylar	334
560 pF	Ceramic	561	0.47	Mylar	474
680 pF	Ceramic	681	0.56	Mylar	564
750 pF	Ceramic	751	1	Mylar	105
820 pF	Ceramic	821	2	Mylar	205

(Continued)

Table 17-2 (*Continued*)

PicoFarad (pF)	NanoFarad (nF)	MicroFarad (mF, µF or mfd)	Capacitance Code
1000	1 or 1n	0.001	102
1500	1.5 or 1n5	0.0015	152
2200	2.2 or 2n2	0.0022	222
3300	3.3 or 3n3	0.0033	332
4700	4.7 or 4n7	0.0047	472
6800	6.8 or 6n8	0.0068	682
10000	10 or 10n	0.01	103
15000	15 or 15n	0.015	153
22000	22 or 22n	0.022	223
33000	33 or 33n	0.033	333
47000	47 or 47n	0.047	473
68000	68 or 68n	0.068	683
100000	100 or 100n	0.1	104
150000	150 or 150n	0.15	154
220000	220 or 220n	0.22	224
330000	330 or 330n	0.33	334
470000	470 or 470n	0.47	474

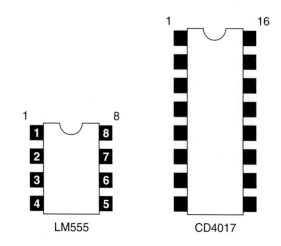

Figure 17-4 *Grab-the-Gold semiconductors*

connects to LED L6. Pin 16 of U2 is connected to the (+) plus supply voltage or the battery. It's generally a good idea to install the integrated circuit sockets before installing all the other components.

Using your grounded anti-static wrist band, pick up the integrated circuits carefully and insert them into their respective sockets. Bend all the pins along one edge at the same time, by pressing the IC to a table edge slowly and gently. Generally integrated circuits will have either a small indented circle, a notch or cutout at one end of the IC package. Just to the left of the notch or cutout, you will find pin 1 on the IC. Make sure to line-up pin 1 on the IC with pin 1 on the socket before installing them.

Next, install the orange LED, L10, at the top of the circuit board. Make sure you observe the correct polarity, so that the cathode or the flat side of the LED lens corresponds to the half moon on the pictorial diagram. Basically there are three ways to identify the cathode of an LED (see the LED pin-out diagram in Figure 17-5). Once the L10 is installed, you can move

the circuit board without damaging it. Usually an IC socket will have a notch or cutout at one end of the package. Just to the left of the notch or cutout will be pin 1 of the socket. Pin 1 of U1 will connect to the junction of capacitor C1, C3 and switch S1, while pin 8 of U1 will connect to the plus (+) 9-volt supply. Pin 1 of U2

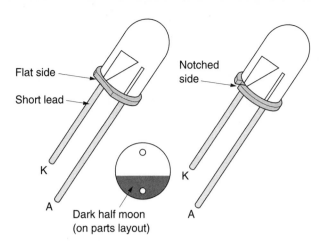

Flat side

Short lead

Notched side

K

A

K

A

Dark half moon
(on parts layout)

Figure 17-5 *LED identification*

on to installing the remaining red LEDs, L1 through R9. Refer to the circuit layout diagram for the LED placement. Solder all of the LEDs to the circuit board, then cut the excess leads by flush cutting the leads to the edge of the board.

At this time you can solder the 9-volt battery clip to the circuit board. The black or minus (−) battery clip lead will connect to the ground bus of the circuit, while the red or plus (+) battery clip lead wire will connect to the anode of diode D1. You can elect to place an SPST slide or toggle switch in series with the plus (+) or red battery clip lead if you wish to have an "on/off" switch.

Before applying power to the Grab-the-Gold Game, you should inspect the circuit board for any "cold" solder joints. The solder joints should all look smooth and clean. If any of the solder joints look dull or lumpy, re-solder them to ensure good connections. Now is a good time for a second visual inspection: this time you will be looking for any "stray" component leads which may have stuck to the circuit board during construction. These "stray" leads could "short" out the circuit when power is applied, thus destroying the circuit. These quick inspections will save you much time and grief, when first applying power to the circuit.

You are now ready to test and play the Grab-the-Gold Game! First, turn the potentiometer to the fully clockwise position, then install a 9-volt battery.

At this time, the LEDs should light up in a sequential pattern. Study the time it takes to light the orange LED

at the top of the pyramid. Your goal is to press and hold the pushbutton at the precise moment that L10 is lit up. If your timing is "great," you will have "Grabbed the Gold." When you think you have succeeded in the correct timing, try and challenge yourself by rotating the potentiometer counterclockwise to speed up the game. Things should get more challenging now!!

In the event that your Grab-the-Gold Game does not work or one of the LEDs does not light-up, you will have to remove the battery and inspect the circuit board for any errors. If one of the LEDs does not light, you may have installed an LED backwards; this is a common mistake. Unsolder the connection and reverse the leads to the LED and then re-solder the connection. If the circuit does not light-up at all, then things might be a bit more complicated and you may have a more serious problem. First, you will want to make sure that each of the resistors has the correct value at a particular location; this is also a common mistake. Now make sure that the diode has been installed correctly; check the polarity marking once more. Next, you will want to make sure that the capacitor at C1 was installed correctly. Observe the polarity to make sure that the plus (+) lead of C1 is connected to pin 2 of U1. Finally, you will want to make sure that the integrated circuits have been installed correctly on the circuit board. Remember that each IC will have a notch or cutout and to the left of the cutout you will find pin 1. Remember to line-up pin 1 of the IC with pin of the socket at U1, and that pin 1 of both IC and socket should connect to the junction of C1, C3 and S1. Remember that pin 1 of U1 connects to LED L6. If you are still not sure of the IC placements, check with a knowledgeable electronics friend. When you are sure that all is well and that you have potentially discovered your problem, re-connect the battery and turn "on" the circuit.

You will have lots of fun with the Grab-the-Gold Game. Challenge your sisters or brothers to the "next" level by changing the position of the potentiometer control. You can pass the game around like a hot potato from one person to the next. When you have finished playing the game be sure to remove the battery, or turn the power switch to "off". Have fun with your new Grab-the-Gold Game.

Chapter 18

Stairway-to-Heaven Game

Parts List

Parts Bin

R1, R2, R5, R11 470 ohm, 5%, ¼-watt resistor

R3 33 k ohm, 5%, ¼-watt resistor

R4 22 k ohm, 5%, ¼-watt resistor

R6 10 k ohm, 5%, ¼-watt resistor

R7 150 ohm, 5%, ¼-watt resistor

R8 270 ohm, 5%, ¼-watt resistor

R9 330 ohm, 5%, ¼-watt resistor

R10 390 ohm, 5%, ¼-watt resistor

R12 560 ohm, 5%, ¼-watt resistor

R13, R14 1 k ohm, 5%, ¼-watt resistor

R15, R16 2.2 k ohm, 5%, ¼-watt resistor

R17 3.3 k ohm, 5%, ¼-watt resistor

R18 4.7 k ohm, 5%, ¼-watt resistor

C1 10 µF, 35-volt electrolytic capacitor

C2 22 µF, 35-volt electrolytic capacitor

C3 470 µF, 35-volt electrolytic capacitor

D1, D2 1N4148 silicon diode

D3, D4, D5 Red LEDs

D6, D7, D8 Red LEDs

D9 Bi-color LED

Q1, Q2, Q3, Q4 BC547 NPN transistor—2N3904 or NTE123AP replacement

Q5, Q6, Q7 BC547 NPN transistor—2N3904 or NTE123AP replacement

Q8 BC557 NPN transistor—2N3906 or NTE159 replacement

U1 74C14 HEX Schmidt inverter

S1 SPST on-off power switch

S2 momentary pushbutton switch

B1 9-volt transistor radio battery

Misc PC circuit board, enclosure, battery clip, battery holder, wire, etc.

The Stairway-to-Heaven game, shown in Figure 18-1, is lots of fun for kids of all ages. This game will test your skill and your hand–eye coordination, and it takes quite a bit of practice to master. The game is compact, so you can take it anywhere. Practice at home, then thrash the pants off all your friends! Press the switch when the bi-color LED is green and the chain of LEDs will gradually light up. But press it when the bi-color LED is red and all your hard work is undone, and you start

Figure 18-1 *Stairway-to-Heaven Game*

When pin 6 of the 74C14 is high, the bi-colored LED is green. Pressing the pushbutton switch at S2 will allow capacitor C3 to charge up via resistor R4. Transistor Q8 is now turned "off." The charging of C3 is set at an exponential rate. So that many more pushes of the switch at S2 at the correct time are needed to get the first 10%.

When pin 6 of U1:C is low, then the following occur: if switch S2 is pressed the charge in C3 will rapidly discharge via the resistor at R5; then, if switch S2 is not pressed any charge in C3 is placed across the seven transistors Q through Q7, and a diode. At each there is an approximate .6 volt drop. So the number of LEDs in the chain D3 through D8, which can be turned "on" is determined by the charge in C3; the greater the charge the more LEDs will be able to turn "on." If the charge in C2 is 1.5 volts then Q1 will be partially "on"; there is a .6 volt drop across the diode and Q7 leaving 3 volt to partially turn "on" transistor Q1; also, if transistor Q8 is turned "on." This allows some of the LEDS to turn "on," depending on the charge in C3 as discussed previously. The bi-colored LED will light-up red.

The Stairway-to-Heaven game was constructed on a $2\frac{1}{2} \times 4\frac{1}{4}$ inch circuit board, which can be built in about two hours. Before we get into building the project, let's take a few minutes to get organized. First, you will want to locate a suitable table or workbench to build the project. Obtain a table large enough to be able to spread

all over again! The chain of illuminated LEDs then goes out. This project teaches a number of electronics concepts and is a lot of fun to play.

The heart of the Stairway-to-Heaven game is the three HEX Schmidt trigger inverter IC, as seen in the schematic diagram shown in Figure 18-2. Sections U1:a through U1:c form an oscillator which cycles at a rate by the 22 µF capacitor at C2 and the value of the R3. The Schmidt trigger provides noise free inputs which give sharp switching transitions in the circuits operation from one state to another.

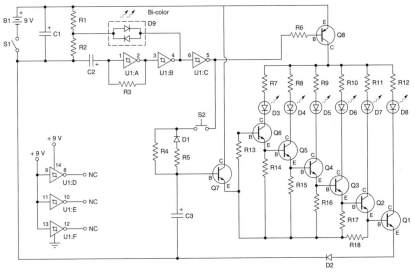

Figure 18-2 *Stairway-to-Heaven Game*

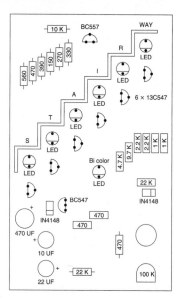

Figure 18-3 *Stairway-to-Heaven pictorial*

out all your project parts, schematics, layout diagrams and charts. Make sure your work area is well lit and ventilated in order to remove solder fumes. Next, you will want to procure a low wattage soldering. A 27 to 33-watt pencil tip soldering iron will work nicely for this project. A pointed or small flat wedge tip is the most preferable tip to use when soldering the components to the circuit board. Locate a spool of 60/40 tin/lead rosin core solder, along with a small tin of "Tip Tinner" from your local Radio Shack store. "Tip Tinner" works great to clean and prepare the soldering tip for soldering. Every few solder joints, you should clean the tip with "Tip Tinner"; you should also have a small wet sponge to help clean the tip as well. You will also want to have some small hand tools for the project. Locate a small pair of needle-nose pliers, a small pair of wire-cutters or end-cutters, a few small screwdrivers, tweezers and a magnifying glass.

Obtain an anti-static wrist band for handling integrated circuits. It's very easy to damage integrated circuits when building electronic circuits, so it's advisable to use an anti-static wrist band to avoid static buildup when moving around your workbench or building table. Getting up and sitting down or moving across the carpet when building a circuit can easily build up static charges. When building electronic circuits with integrated circuit projects, try to discharge yourself especially during the brief time that you are handling the integrated circuit packages. The anti-static

wrist strap is wrapped around your wrist and a long grounding wire is plugged into a wall power outlet for grounding.

Place the schematic Figure 18-2 and pictorial diagram in Figure 18-3 in front of you and plug in your soldering iron. Locate and refer to Table 18-1, which depicts a chart showing the resistor color code chart. Most resistors will have three or four color bands on the body of the resistor. The first 0 to 9 number on the color code chart illustrates the resistor's first digit. The second color band illustrates the second digit of the resistor code, and the third color band denotes the resistor's multiplier value. The color bands start at one end of the resistor body, and go towards the center of the resistor body. If there is no fourth color band then the resistor has a 20% tolerance value, while a fourth silver color band denotes a 10% tolerance value. A fourth band which is gold in color notes a 5% tolerance value. So look in the resistor parts pile and try to find a resistor that has a (yellow) (violet) and (brown) color bands. The first digit is yellow or the digit (4), the second color is violet or the digit (7), and the third multiplier band is brown with a value of 100, so the first resistor at R1 is a 470 ohm resistor. Take this first resistor and place it on the circuit board at its proper location. Note that 470 ohm resistors are also used at R2, R5, R11. Install these resistors on the circuit board and solder them at their respective locations. Identify the remaining resistors and place them in their proper locations and solder them to the circuit board. Use your end-cutters to remove the excess resistors from the circuit board. Flush cut the resistor leads close to the circuit board, so that there are no long leads left on the board.

Next, let's move on to installing the capacitors for the project. The Stairway-to-Heaven game utilizes a number of electrolytic capacitors. Remember that these types of capacitors have polarity considerations when installing them. Electrolytic capacitors all have either a black or white color band along the side of the capacitor body. Next to the color band or inside the color band, you will find a plus (+) or minus (–) marking which indicates the polarity of the nearest lead. When installing these capacitors, refer to both the schematic and the pictorial diagram for the positioning of the capacitor within the circuit and how it is placed on the circuit board. Note that capacitor C1 is a 10 µF electrolytic capacitor; in the circuit diagram the plus (+) lead of this capacitor

Table 18-1

Resistor Color Code Chart

Color Band	1st Digit	2nd Digit	Multiplier	Tolerance
Black	0	0	1	
Brown	1	1	10	1%
Red	2	2	100	2%
Orange	3	3	1000 (K)	3%
Yellow	4	4	10000	4%
Green	5	5	100000	
Blue	6	6	1000000 (M)	
Violet	7	7	10000000	
Gray	8	8	100000000	
White	9	9	1000000000	
Gold			0.1	5%
Silver			0.01	10%
No color				20%

connects to the both R1 and the plus (+) side of the battery at B1. When installing C1 on the circuit board, make sure that the plus (+) lead of the capacitor is nearest the connection to R1. Solder capacitor C1 to the circuit board and then remove the excess component leads, cutting them flush to the circuit board. Identify the remaining capacitors, install them on the circuit board and solder them in place. Don't forget to cut the excess component leads after soldering the capacitors on the circuit board.

The Stairway-to-Heaven game project also incorporates a few silicon diodes and, as you know, these components have polarity issues. Silicon diodes have polarity concerns and they must be installed correctly if the circuit is going to work properly. On the diode body, you will find either a black or white color band. Referring to the schematic diagram, you will notice that the symbol for a diode is an arrow pointing to a line. The arrow part of the diode is the anode, while the line that it points to is the cathode. When installing diode D1, make sure that the cathode lead connects to the switch at S2, and that the anode connects to resistor R5; note also that the anode of diode D2 connects to transistor Q1, while the cathode of D2 is connected to capacitor C3.

Next, we are going to install the eight NPN transistors in this project. Transistors, as you will remember, have three leads. There will be a Base lead, a Collector lead and an Emitter lead. Refer to the diagram shown in Figure 18-4, which illustrates the pin-out for the transistors and the integrated circuits. The small TO-92 plastic transistor package has its three leads beginning at one end of the transistor body. Seen from the bottom of the transistor package the leads start from one end. First you will see the Emitter lead, then the Base lead in the center with the Collector lead opposite the Emitter lead. Referring to the schematic, you will see that the Base lead is the straight line to which both the Emitter and Collector connect to. The Emitter lead has the arrow pointing away from the transistor body, which indicates the transistor is a NPN type. The Collector of each of the transistors is connected directly to the LEDs of the display.

Handle the transistors carefully when installing them on the circuit board. Solder the transistors in place on the circuit board and then remove the excess component leads.

When installing integrated circuits in electronics circuits, it is a good idea to first install an integrated

Table 18-2

Capacitance Codebreaker Information

This table is designed to provide the value of alphanumeric coded ceramic, mylar and mica capacitors in general. They come in many sizes, shapes, values and ratings; many different manufacturers worldwide produce them and not all play by the same rules. Most capacitors actually have the numeric values stamped on them; however, some are color coded and some have alphanumeric codes. The capacitor's first and second significant number IDs are the first and second values, followed by the multiplier number code, followed by the percentage tolerance letter code. Usually the first two digits of the code represent the significant part of the value, while the third digit, called the multiplier, corresponds to the number of zeros to be added to the first two digits.

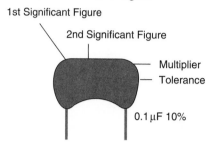

1st Significant Figure
2nd Significant Figure
Multiplier
Tolerance
0.1 µF 10%

Value	Type	Code	Value	Type	Code
1.5 pF	Ceramic		1000 pF/0.001 µF	Ceramic/Mylar	102
3.3 pF	Ceramic		1500 pF/0.0015 µF	Ceramic/Mylar	152
10 pF	Ceramic		2000 pF/0.002 µF	Ceramic/Mylar	202
15 pF	Ceramic		2200 pF/0.0022 µF	Ceramic/Mylar	222
20 pF	Ceramic		4700 pF/0.0047 µF	Ceramic/Mylar	472
30 pF	Ceramic		5000 pF/0.005 µF	Ceramic/Mylar	502
33 pF	Ceramic		5600 pF/0.0056 µF	Ceramic/Mylar	562
47 pF	Ceramic		6800 pF/0.0068 µF	Ceramic/Mylar	682
56 pF	Ceramic		0.01	Ceramic/Mylar	103
68 pF	Ceramic		0.015	Mylar	
75 pF	Ceramic		0.02	Mylar	203
82 pF	Ceramic		0.022	Mylar	223
91 pF	Ceramic		0.033	Mylar	333
100 pF	Ceramic	101	0.047	Mylar	473
120 pF	Ceramic	121	0.05	Mylar	503
130 pF	Ceramic	131	0.056	Mylar	563
150 pF	Ceramic	151	0.068	Mylar	683
180 pF	Ceramic	181	0.1	Mylar	104
220 pF	Ceramic	221	0.2	Mylar	204
330 pF	Ceramic	331	0.22	Mylar	224
470 pF	Ceramic	471	0.33	Mylar	334
560 pF	Ceramic	561	0.47	Mylar	474
680 pF	Ceramic	681	0.56	Mylar	564
750 pF	Ceramic	751	1	Mylar	105
820 pF	Ceramic	821	2	Mylar	205

Table 18-2 (*Continued*)

PicoFarad (pF)	NanoFarad (nF)	MicroFarad (mF, µF or mfd)	Capacitance Code
1000	1 or 1n	0.001	102
1500	1.5 or 1n5	0.0015	152
2200	2.2 or 2n2	0.0022	222
3300	3.3 or 3n3	0.0033	332
4700	4.7 or 4n7	0.0047	472
6800	6.8 or 6n8	0.0068	682
10000	10 or 10n	0.01	103
15000	15 or 15n	0.015	153
22000	22 or 22n	0.022	223
33000	33 or 33n	0.033	333
47000	47 or 47n	0.047	473
68000	68 or 68n	0.068	683
100000	100 or 100n	0.1	104
150000	150 or 150n	0.15	154
220000	220 or 220n	0.22	224
330000	330 or 330n	0.33	334
470000	470 or 470n	0.47	474

circuit socket prior to installing the actual IC. Integrated circuit sockets are a form of low cost insurance. In the event of a circuit failure at some possible later date, the IC socket will make life a lot easier. Most people cannot unsolder an integrated circuit from a circuit board without damaging the circuit board. Integrated circuit are low cost and readily obtained. Install the IC socket onto the circuit board, noting that there is a small notch or cutout at one end of the socket. This is an orientation device which helps install the socket. The notch will be in the center at one end of the plastic package. Pin 1 of the socket will be to the left of the notch; this will help you install the IC socket, since pin 1 of the IC socket should be next to capacitor C2.

When installing the actual integrated circuit, you will note that it too has a notch, cutout or small indented circle at one end of the IC case. Pin 1 of the IC will be just to the left of the notch or cutout. When installing the actual IC in its socket, make sure that you are wearing your anti-static wrist band to avoid damage to the IC during installation. Note that there are three unused sections of the 74C14 IC; the input pins of these

sections should be connected to plus (+) 9-volts. The outputs of sections U1:D, U1:E and U1:F can be left unconnected. Also note that U1 has its power connections on pins 7 and 14. The ground connection is on pin 7, while the plus (+) voltage connection is made at pin 14 of the chip. Make sure that you align pin 1 of the IC with pin 1 of the IC socket when placing the IC into its socket. Observe that pin 1 of the IC should be connected to the plus (+) side of capacitor C2, and that pin 7 of the IC is connected to the ground connection. If you are having any difficulty installing the transistors or integrated circuit, then contact a knowledgeable electronics enthusiast for help.

Finally, we are going to install the LEDs for the project onto the circuit board. LEDs, as you remember, will have two leads. Referring to the schematic, you will find that a LED looks much like a silicon diode on the schematic diagram, with one difference being that the LED also has two arrows next to the package indicating light is given off from the LED. Refer to the diagram shown in Figure 18-5, which depicts the LED package and pin-out diagram. The LED package will

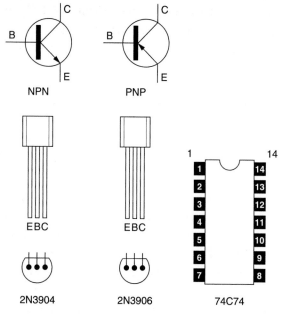

Figure 18-4 *Stairway-to-Heaven semiconductor diagram*

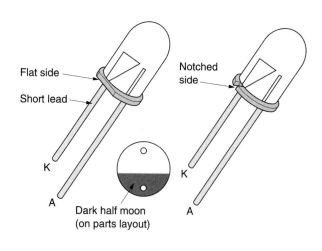

Figure 18-5 *LED identification*

generally have a flat edge at one side of the diode package, this will indicate the cathode lead of the LED. You can also identify the cathode lead, since it is usually the shorter lead of the two. Sometimes you can also identify the cathode lead by a notch on the lens of the plastic LED package. Remember that the anode lead of the LED is an arrow pointing to a straight line, and that the line is the cathode lead. In this Stairway-to-Heaven game the cathode leads of the LEDs are connected to the Collector leads of the transistors, and the anode of the LEDs are connected together and to the Collector of transistor Q8. Once you have identified the LEDs and when they should be placed on the circuit board, then you can go ahead and install them on the circuit board. Solder the LEDs to the circuit board and then cut the excess LED leads flush to the circuit board.

The Stairway-to-Heaven game is powered from a single 9-volt transistor radio battery. We chose to incorporate an "on/off" switch in series with the battery connected to the circuit. Take the plus (+) red battery clip wire and wire it in series with one of the switch leads. The remaining switch lead is then connected to the circuit board at the plus bus or where resistor R1 and C1 and Q7 meet. The black or minus (–) battery clip lead is connected directly to the minus (–) or ground bus of the circuit.

Now, before we power up the circuit to test it out, take a short well-deserved break, then when we return

we will inspect the circuit board for any possible "short" circuits or "cold" solder joints. Pick up the circuit board and place the foil side up toward you. First, we will inspect the circuit board for any possible "cold" solder joints. Cold solder joints could lead to premature circuit failure, so it's best to try to avoid them or locate them quickly. Look over the solder joints carefully, they should all look clean, smooth and bright and shiny. If any of the solder joints look dull or "blobby" then remove the solder from that particular joint and re-apply solder to the joint, this time making sure the joint looks clean, smooth and shiny.

Once again we will inspect the circuit board, but this time we are going to look for possible "short" circuits which could be caused by "stray" component leads that were trimmed. Because rosin core can often leave a sticky residue after many components have been soldered to a circuit board, it is necessary to inspect the PC board for component leads that get "stuck" to the board from the rosin core residue. Look the board over carefully for any wires or component leads that may be lying across PC copper lands or pads that could "short" out a line of PC traces on the circuit board. Circuit board "shorts" can cause circuit damage when power is first applied, so it is best to eliminate the problem before the circuit is turned "on."

Now that you have inspected the circuit board for "shorts" or "cold" solder joints, we can move on to connecting the 9-volt battery and testing the circuit. Once the battery has been connected, you can turn on the switch at S1. Next, you can press the momentary pushbutton switch at S2. If you press the switch while the bi-color LED is green, then the array of red LEDs will begin lighting up. If you press S2 when the bi-color LED is red, the array of red LEDs will jump back down to the bottom of the red LED array.

If your Stairway-to-Heaven game appears to work and the bi-color LED lights up in both red and green color and the red LEDs count up to the top of the LED array, then your circuit appears to be working correctly. If your Stairway-to-Heaven game does not light up at all, or if only some of the LEDs light up, then you will have to disconnect the battery from the circuit and troubleshoot the circuit to try to find an error or defective component.

First, you will want to try and determine what is wrong with the circuit based on what it is or is not doing. If some of the LEDs are not working, then you may have just installed an LED backwards. In that case, you can simply reverse the leads of the LED. One step further, if the LED does not light up after being reversed, then you may have a defective LED or you may have installed one of the transistors incorrectly. Check your installation of the transistor associated with that particular LED. You may have installed a transistor incorrectly! Mistakes can occur and do occur! If the circuit does not appear to work correctly or erratically, then you may want to check further into the circuit.

Check the installation of the resistors first; it is a common mistake to get the wrong value resistor in the wrong place on the circuit board. Make sure all the resistors are in the right place on the circuit board. If you are in doubt, you can check each resistor with an ohmmeter; however, you may have to disconnect one end of the resistor to do so. Next you will want to make sure the capacitors have been installed correctly by observing the color polarity band on the capacitor body. So check the two silicon diodes in the circuit to make sure that they have been installed correctly; you may want to compare the schematic with the pictorial or layout diagram. Finally, you will want to make sure that the integrated circuit has been installed correctly in the IC socket. Remember, that pin 1 of the IC socket has to align with pin 1 of the actual integrated circuit chip, and that pin 1 of the socket must be connected to the plus (+) side of capacitor C2 and that pin 7 is connected to ground, while pin 14 is connected to the battery supply voltage as shown in the schematic. If you are confused by either the integrated circuit installation or the installation of the transistors, check with a friend who is an electronics enthusiast for help. Once you have checked over the circuit carefully for any errors or have replaced any parts, then you can go ahead and connect up the 9-volt battery and re-check the operation of the circuit to see if it will work as it should. With the battery reinstalled and the power switch turned to the "on" position, you can examine the circuit again to see if it works as it should.

Press the momentary switch S2 during the time that the bi-color LED is lighted green, then the chain of LEDs in the array should begin to light up in sequence. But if you pressed the pushbutton when the bi-color LED was red, then the lamps will fall back down to the bottom and you will have to try again. The game is easy to play but hard to master. Challenge your friends to play and have fun!

Chapter 19

Electronic Cricket

Parts List

Parts Bin

R1 4.7k ohm, 5%,
¼-watt resistor

R2, R9 2.2k ohm, 5%,
¼-watt resistor

R3, R5, R7 1 megohm,
5%, ¼-watt resistor

R4 10 megohm, 5%,
¼-watt resistor

R6 100k ohm, 5%,
¼-watt resistor

R8, R11 10k ohm, 5%,
¼-watt resistor

R10 22k ohm, 5%, ¼-watt
resistor

R12 100 ohm, 5%,
¼-watt resistor

C1, C2, C3 100 nF,
35-volt green-cap

C4 47nF, 35-volt
greencap

C5 220 µF, 35-volt
electrolytic capacitor

D1 1N4148 silicon diode

D2 1N4004 silicon diode

L1, L2 red LEDs

Q1, Q2 PN100
NPN transistor—
2N3904 or NTE123AP
replacement

U1 TL082 dual op-amp IC

U2 CD4093 quad Schmitt
NAND IC

SPK 8-ohm speaker

B1 9-volt transistor
radio battery

Misc PC board, IC
sockets, battery
clips, wire, etc.

The Electronic Cricket project, illustrated in Figure 19-1, is an intriguing novelty project, that just sits there in the dark, until it hears a sound! When the Electronic Cricket hears a noise it suddenly starts making a chirping sound like a cricket, and at the same time its LED eyes start blinking. Your friends will wonder how its works, your parents will go insane trying to figure out where the noise is coming from!

The Electronic Cricket circuit depicted in Figure 19-2 appears to be a simple circuit, but it's actually somewhat complicated. The first part of the dual op-amp at U1 amplifies the weak signals picked up by the electret microphone at M1. The signals are fed into U1:a through a 100 nF capacitor at C1. The op-amp amplifies the incoming signal about 500 times based on the feedback path and the ratio of input resistor (R2) and the gain resistor at R3.

Figure 19-1 *Electronic Cricket project*

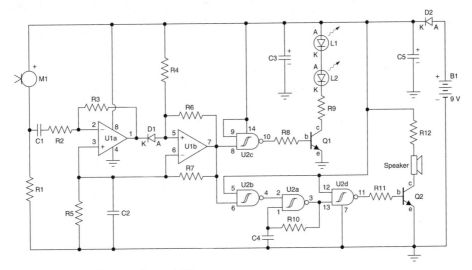

Figure 19-2   *Electronic Cricket. Courtesy Jaycar (JC)*

The pre-amplifier gets its forward bias from the output of U1:b, the second half of U1. When the "Cricket" is waiting to hear a sound, the output of U1:b is switched up very close to + 9 volts. This provides a bias voltage of about +4.5 volts at the positive input pin 2 of U1:a, based on the voltage divider formed by the two 1 megohm resistors between pin 7 of U1:b and ground. The output at pin 1 of U1:a rests at very close to the same voltage, because of the way the amplifier works.

At the same time the negative input of U1:b at pin 6 is also held at 4.5 volts, since it is connected to the same divider. But the positive input of U1:b is held slightly higher in voltage at about +5 volts, because of the voltage drop in diode D1. That's why the output of U1:b rests at very close to +9 volts, because this amplifier acts as a comparator and in this situation its positive input is more positive than it is negative.

When the microphone picks up some sound, U1:a amplifies the AC signals from the microphone and its output voltage swings up and down with amplified AC. When the voltage swings negative, this pulls down the voltage at pin 5 of U1:b via diode D1. As a result, the output of comparator U1:b suddenly switches low, which removes the +9 volts from the top of the voltage divider feeding pins 3 and 6. So the bias voltage for both U1:a and U1:b starts dropping, as the 100 nF capacitor across the lower 1 megohm resistor starts discharging, and the pre-amplifier stops working.

But this situation doesn't remain for long. Essentially U1:a and U1:b together form a low frequency oscillator and a series of short pulses appear at the output of U1:b before its flies back up to +9 volts again. Everything returns to the original "waiting" state, if the sounds that were picked up by the microphone have stopped. However, if the mic picks up more sounds when U1:a begins working again as a pre-amplifier, the whole process starts again.

The rest of the "cricket" senses these changes in the voltage at U1:b at pin 7, and uses them to generate the cricket chirps and makes the LED "eyes" blink. As you can see, gate U2:b is connected as an inverter, with input pin 6 connected to the output of U1:b. So when the output of U1:b is "waiting" at +9 volt, the output of U1:b switches low, the output of U2:b switches up to +9 volts instead.

Now gate U2:a has a 22 k ohm resistor connected from its output back to its input at pin 1, and a 47 nF capacitor connected from the same input to ground. This provides positive feedback around the gate, which therefore tends to oscillate at about 2 kHz. But it can only oscillate when its second input at pin 2 is pulled up to + 9 volts. When the second input is held low the gate's output is switched high and the gate cannot oscillate.

When the output of U1:b pulses low in response to a sound being picked up by the mic, this causes the output of U1:b to pulse high. Each time it pulses high

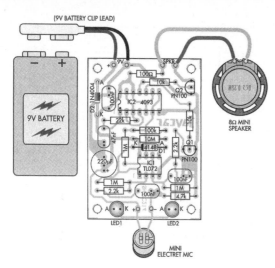

Figure 19-3 *Cricket pictorial diagram*

this allows U2:a to oscillate, producing bursts of 2 kHz square waves signals at its output at pin 3. The bursts of square wave signals are fed through U2:d, used here as an inverting buffer, and then fed to the base of transistor Q2, via a 10 k ohm resistor.

As a result Q2 is turned "on" and "off" in bursts, at a rate of 2 kHz. And since the mini speaker is connected from the Collector of Q2 to the + 9 volt line, via the 100 ohm resistor, this causes bursts of current pulsing at 2 kHz to pass through the speaker.

The output of U1:b is also connected to the pin 8 input of gate U2:c, which is also used as an inverter, just like U2:b. So when the output of U1:b pulses low in response to sounds being picked up by the mic, the output of U2:c pulses high. Since this output is connected to the base of transistor Q1, via another 10 k ohm resistor, this means that Q1 is turned "on" during each pulse. And the LEDs are connected in series between the Collector of Q1 and the +9 volt line, via a 2.2 k current limiting resistor, so as Q1 conducts it allows pulses of current to flow through the LEDs. That is how the "cricket's" LED flash at the same time as the chirping.

By now you are probably eager to start building the Electronic Cricket, so let's begin.

The "cricket" was constructed on a 2¼ inch × 4¼ inch circuit board. Before we begin building the circuit, take a few minutes to prepare the work area. First locate a large work surface area which you can use to spread out all the project parts. Locate the schematic and pictorial diagram, shown in Figure 19-3, and all the

necessary tools for the project. Locate a 27 to 33-watt pencil type soldering iron with a sharp point or flat edge. Locate some small diameter 60/40 rosin core solder. Try to locate an anti-static wrist band for handling the integrated circuits. Anti-static wrist bands prevent static shocks from damaging your delicate IC, so it is a good idea to use them. Next, locate all the project parts and place them in front of you. Locate the schematic, the layout diagram or pictorial and the resistor and capacitor code charts.

So let's begin: locate the resistor color code chart in Table 19-1, this will help you determine the correct resistor values from the colors marked on the body of the resistors. Place the project resistors in front of you now. Resistors will have three or four color bands on the body of the resistor. The first color corresponds to the first digit of the resistor code, while the second color code depicts the second digit of the resistor's code. The third color band on the resistor is the multiplier value. Resistors often also have a fourth color band which denotes the resistor's tolerance value. No fourth band indicates a 20% tolerance value, while a silver band indicates a 10% tolerance and a gold band denotes a 5% tolerance value. The first color band of the code should be close to one edge of the resistor body. So, look for a resistor with a yellow band, followed by a violet band and an orange band; this will be the 47 k ohm resistor at R1. Place R1 on the circuit board now. Identify the remaining resistors and mount them in their respective locations using both the schematic and the pictorial diagram shown in Figure 19-3. Once the resistors have been placed on the circuit board, you can go ahead and solder them in place, then cut the excess component leads, flush to the edge of the circuit board.

Next look through the components and try to locate the capacitors for the project. There are five capacitors in this project with one of them an electrolytic. Look at the chart in Table 19-2, which illustrates the capacitor codes which are often marked on capacitors instead of their actual value. A three-digit code marked (104), for example, denotes a capacitor value of 0.1 µF or 100 nF. Identify the capacitors; you should find three 100 nF capacitors and one 47 nF capacitor. The project also contains a single electrolytic capacitor at C5, a 220 µF value. Note that electrolytic capacitors have polarity and this must be observed when installing them to avoid

Table 19-1

Resistor Color Code Chart

Color Band	1st Digit	2nd Digit	Multiplier	Tolerance
Black	0	0	1	
Brown	1	1	10	1%
Red	2	2	100	2%
Orange	3	3	1000 (K)	3%
Yellow	4	4	10000	4%
Green	5	5	100000	
Blue	6	6	1000000 (M)	
Violet	7	7	10000000	
Gray	8	8	100000000	
White	9	9	1000000000	
Gold			0.1	5%
Silver			0.01	10%
No color				20%

damaging the component or circuit. Electrolytic capacitors usually have a white or black band on the body of the component. Near the colored band, you will see a plus (+) or a minus marking (–). Refer to the schematic and layout diagram when installing the electrolytic to make sure you will place or mount the capacitor correctly. Once you have identified all of the capacitors and their respective locations, you can install them on the circuit board and solder them in, and then cut the excess lead lengths.

This project also utilizes silicon diodes, which are also polarity sensitive and must be installed with respect to their markings. Diodes will have either a black or white band at one end of the diode body. The colored band denotes the diode's cathode lead, or most minus lead. A diode's anode is shown schematically as an arrow or triangle and the cathode of the diode is shown as the flat line to which the arrow is pointing. Refer to the layout diagram and schematic when installing the diodes to make sure you are installing them correctly. Insert the diodes carefully and solder them in place. Cut the excess component lead after soldering the diodes in place.

Next we will go ahead and install the transistors; this project has two NPN type transistors. Transistors generally have three leads coming from the bottom of the device, see Figure 19-4 for transistor IC pin-outs. You will find a Collector lead at one end of the transistor, a Base lead is usually the center lead, and an Emitter lead opposite the Collector lead. Schematically, the Base lead is the flat line in the diagram. The Collector and Emitter leads go into the Base lead, with the Emitter lead having a small arrow on it. If the arrow points away from the transistor, then the transistor is a NPN type device, but if the arrow points inwards toward the transistor then it is a PNP type transistor. Refer to the schematic and pictorial or layout diagram and make sure you can identify each of the transistor leads before attempting to install them. If you are in doubt or are having difficulties, then ask a parent or electronically knowledgeable friend for help. Once you have identified the transistors and their pin-outs and where they will be mounted on the circuit board, then you can go ahead and install them on the circuit board. Next, remove the excess component leads from the circuit board.

When building electronic circuits, it is often a good idea to use integrated circuit sockets. Integrated circuit sockets are very helpful in the event of a circuit failure. It is much simpler to remove an IC from a socket than trying to unsolder an IC from the circuit. Most people

Table 19-2

Capacitance Codebreaker Information

This table is designed to provide the value of alphanumeric coded ceramic, mylar and mica capacitors in general. They come in many sizes, shapes, values and ratings; many different manufacturers worldwide produce them and not all play by the same rules. Most capacitors actually have the numeric values stamped on them; however, some are color coded and some have alphanumeric codes. The capacitor's first and second significant number IDs are the first and second values, followed by the multiplier number code, followed by the percentage tolerance letter code. Usually the first two digits of the code represent the significant part of the value, while the third digit, called the multiplier, corresponds to the number of zeros to be added to the first two digits.

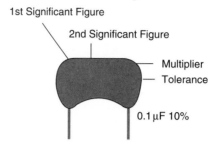

1st Significant Figure
2nd Significant Figure
Multiplier
Tolerance
0.1 µF 10%

Value	Type	Code	Value	Type	Code
1.5 pF	Ceramic		1000 pF/0.001 µF	Ceramic/Mylar	102
3.3 pF	Ceramic		1500 pF/0.0015 µF	Ceramic/Mylar	152
10 pF	Ceramic		2000 pF/0.002 µF	Ceramic/Mylar	202
15 pF	Ceramic		2200 pF/0.0022 µF	Ceramic/Mylar	222
20 pF	Ceramic		4700 pF/0.0047 µF	Ceramic/Mylar	472
30 pF	Ceramic		5000 pF/0.005 µF	Ceramic/Mylar	502
33 pF	Ceramic		5600 pF/0.0056 µF	Ceramic/Mylar	562
47 pF	Ceramic		6800 pF/0.0068 µF	Ceramic/Mylar	682
56 pF	Ceramic		0.01	Ceramic/Mylar	103
68 pF	Ceramic		0.015	Mylar	
75 pF	Ceramic		0.02	Mylar	203
82 pF	Ceramic		0.022	Mylar	223
91 pF	Ceramic		0.033	Mylar	333
100 pF	Ceramic	101	0.047	Mylar	473
120 pF	Ceramic	121	0.05	Mylar	503
130 pF	Ceramic	131	0.056	Mylar	563
150 pF	Ceramic	151	0.068	Mylar	683
180 pF	Ceramic	181	0.1	Mylar	104
220 pF	Ceramic	221	0.2	Mylar	204
330 pF	Ceramic	331	0.22	Mylar	224
470 pF	Ceramic	471	0.33	Mylar	334
560 pF	Ceramic	561	0.47	Mylar	474
680 pF	Ceramic	681	0.56	Mylar	564
750 pF	Ceramic	751	1	Mylar	105
820 pF	Ceramic	821	2	Mylar	205

(Continued)

Table 19-2 (Continued)

PicoFarad (pF)	NanoFarad (nF)	MicroFarad (mF, μF or mfd)	Capacitance Code
1000	1 or 1n	0.001	102
1500	1.5 or 1n5	0.0015	152
2200	2.2 or 2n2	0.0022	222
3300	3.3 or 3n3	0.0033	332
4700	4.7 or 4n7	0.0047	472
6800	6.8 or 6n8	0.0068	682
10000	10 or 10n	0.01	103
15000	15 or 15n	0.015	153
22000	22 or 22n	0.022	223
33000	33 or 33n	0.033	333
47000	47 or 47n	0.047	473
68000	68 or 68n	0.068	683
100000	100 or 100n	0.1	104
150000	150 or 150n	0.15	154
220000	220 or 220n	0.22	224
330000	330 or 330n	0.33	334
470000	470 or 470n	0.47	474

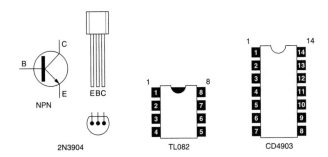

Figure 19-4   *Cricket semiconductor diagram*

are not skillful at removing integrated circuits and will often damage the circuit board as a result.

Now, locate the two integrated circuits from the parts. Integrated circuits will generally have some form of markings on them to help orient them on the circuit board. Look for a small indented circle at one end of the IC package, or you might find a notch or a cutout at one end of the IC package. If you look just to the left of the notch, cutout or indented circle you will find pin 1 of the IC. Once you have identified pin 1 of the IC package, you will have to identify which pin on the socket is pin 1, so

you can align the IC with the socket. Refer to the schematic and parts layout diagram to find pin 1's location. Once you are sure of the orientation of the socket and you know which is pin 1, you can insert the IC into the socket. Once again if you have difficulty locating pin 1 on the schematic, ask a knowledgeable electronics friend. Solder the IC socket to the circuit board.

Find the two red LEDs from the remaining parts. LEDs have polarity so you must determine which lead is the anode and which lead is the cathode, since the LED's schematic symbol looks much the same as a diode. Refer to Figure 19-5, which depicts the pin-outs for an LED. Notice that one edge of the LED will have a flat edge, the lead nearest to the flat edge is the cathode lead. Refer to the diagram and you will also notice that there is a half moon on the LED symbol; the half moon also denotes the cathode lead of the LED. Once you have identified the LED leads, then you will have to refer to the schematic and parts layout diagrams to see where the cathode pins of the LEDs are placed. From the schematic, you will note that the anode of LED 1 is connected to the plus (+) or the junction of D2 and C5. The cathode of L1 is connected to the anode of L2 and the cathode of L2 is

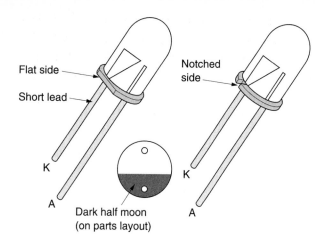

Flat side

Short side

Short lead

Notched side

K

A

K

A

Dark half moon
(on parts layout)

Figure 19-5 *LED identification*

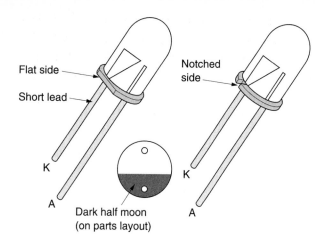

connected to resistor R9. Solder the LEDs to the circuit board, then cut the extra lead lengths.

Locate two 6-inch pieces #22 ga stranded-insulated hookup wire. Solder the two free ends of the wire to the 8-ohm speaker. The remaining free end of the two wires will now be soldered to the circuit board. One wire will be soldered to the Collector of transistor Q2 and the other speaker wire will be connected to resistor R12. Once the speaker has been connected, you can next solder the 9-volt battery clip to the circuit board. The red or plus (+) battery clip lead will connect to the cathode of diode D2. The black or minus (–) battery clip lead will be connected to the ground bus of the circuit board.

Finally locate the electret microphone and solder it to the circuit board. An electret mic does have polarity, so it must be observed when installing it into the circuit board. The electret microphone or mic will have one terminal marked with a plus (+) marking. The plus (+) on the mic will be connected to the plus (+) bus of the circuit or the anode end of diode D2. The remaining mic lead will be connected to capacitor C1.

Well, your Electronic Cricket is now just about ready for testing, but before we do that, take a short well-deserved break. After the short break, we will inspect the circuit board before we apply power to the circuit. First, we are going to inspect the circuit board for "cold" solder joints, which could lead to premature circuit failure. Turn the circuit board over, so the foil side of the circuit board is upwards and facing you. Make sure that all of your solder joints look clean,

smooth and shiny. If you find any of the solder joints look dull or "blobby" then you should remove the solder from the solder joint and solder the joint over again.

Next, we are going to inspect the circuit board for "short" circuits, which could cause the circuit to "short-out" and possibly destroy the circuit when power is first applied. Often when cutting excess component leads, sometimes the "stray" component leads will "stick" to the circuit board possibly from excess rosin from the solder. Inspect the circuit board for any wire "bridges" which short between the circuit pads. Remove any loose wires or component leads from the circuit board. Once the board is clean, we can move on to testing the "cricket" circuit.

Locate a 9-volt transistor radio battery and attach the battery clip to the battery. The "cricket's" eye LEDs should briefly flash and the speaker may give a quick chirp. After the blink and chirp, the circuit should go quiet. Now try speaking a few words near the microphone, the "cricket" should "come to life" and begin chirping and flashing. When the noise goes away, the "cricket" should go back to sleep.

If the circuit does not respond to noise, or if the LEDs do not flash or no chirping sounds come from the speaker, then you most likely made a mistake when building the circuit. Do not fear yet! Remove the battery from the battery clip and you will have to re-inspect the circuit board for any possible errors.

Place the circuit board in front of you with the component side upwards facing you. First, you will want to make sure that you placed the correct resistor, with the correct value into the correct hole on the board. Mistakes are made, often the wrong resistor is placed in the wrong place on the circuit board. Check resistor color codes for each resistor and make sure you have the right resistor in the right place. Then, move on to inspect the capacitors are in their respective locations. Make sure that the electrolytic capacitors have been inserted correctly on the circuit board. They are polarity sensitive! Now locate the two diodes and make sure that they have been inserted correctly and that their cathode bands are facing the right direction on the circuit.

Now look at the transistors, to make sure that they have been inserted correctly. Commonly builders often mix up the Collector lead with the Emitter lead and the transistors are connected backwards. So carefully check each transistor for correct placement. Next, look to make sure that the integrated circuits have been placed in the socket in the right direction. Pin of the IC must match pin on the IC socket. Finally check to make sure that the two LEDs have been installed correctly. Once you have inspected all of the component placement, you can re-attach the 9-volt battery and re-test the circuit once again.

Your insidious little "cricket" can now be hid in your living room or bedroom and see if it will drive someone crazy. You can replace the 22 k resistor for a different cricket chirp; changing the resistor will change the pitch of the chirping sounds. The circuit is best left "naked" with no enclosure around the circuit. You could mount the circuit board on standoffs to elevate the circuit, you could also cover the bottom of the circuit board with "duct" tape to protect the circuit from "shorting out" if it touches a metal table or surface.

Have fun with your new Electronic Cricket. Place it in your sister's room, or your living room, or at a friend's house for some real fun!

Chapter 20

Filtered Blacklight Project

Parts List

Parts Bin

R1, R3 10 ohm, 5%, ¼-watt resistor

R2 680 ohm, 5%, ¼-watt resistor

P1 5k trimmer potentiometer

C1 0.0047µF, 35 volt mylar disk capacitor (472)

C2 0.1µF, 35-volt disk capacitor (104)

D1 1N4007 silicon diode

FTB filtered blacklight tube 5¼ inch

U1 LM555 timer IC

Q1 RCA 309 PNP power transistor

T1 Inverter transformer

S1 power switch SPST slide or toggle switch

Misc IC socket, PC board, wire, hardware, case, etc.

The portable Filtered Blacklight Project depicted in Figure 20-1 features a 5¼ inch long filtered ultraviolet long-wave blacklight tube, which is ideal for "lighting up" posters, invisible inks, mineral samples, insects, glow-in-the-dark stars and other fun blacklight sensitive objects. The portable blacklight project operates on a 6-volt lantern battery or four "D" cells for a completely portable device.

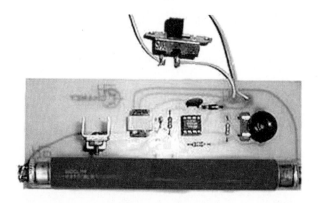

Figure 20-1 *Blacklight Project*

The portable blacklight project depicted in Figure 20-2 uses one integrated circuit and one transistor and operates from a 6-volt power source. The project converts 6 volts to over 150 volts AC to power the fluorescent blacklight tube. The integrated circuit is an LM555 timer IC and is configured as an astable multivibrator or oscillator. This AC signal has a frequency controlled by capacitor C2 and resistor R1. The pulse width of the IC timer is controlled by potentiometer P1, and as it is adjusted the brightness of the lamp will go from dim to bright. The output of U1 appears on pin 3 of the IC and is coupled by resistor R3 and capacitor C1 to the Base of transistor Q1. Transistor Q1 is a power PNP type which functions to amplify the output power from the IC. Also Q1 directly drives the

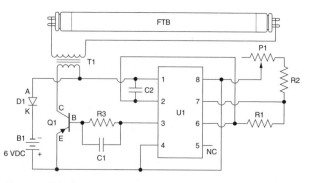

Figure 20-2 *Blacklight*

inverter transformer T1. The transformer, through its turns ratio, converts the low voltage AC to over 150 volts AC. This voltage is applied to a special ultraviolet blacklight tube that causes UV sensitive inks and objects to glow brightly. Power for the project is supplied from a 6-volt lantern battery or from four "D" cells. Diode D1 prevents damage to the circuit from reversed polarity connections to the battery. The special blacklight tube that is used has a black filter coating on the inside of the tube that prevents most of the visible light spectrum from being emitted, therefore this allows the brilliance of the filtered fluorescent tube to be more effective than a normal blacklight tube.

Are you ready to begin building the Portable Blacklight Project? First prepare a clean work area, where you can spread out you parts and diagrams for the project. Locate a 27 to 30-watt solder iron and some solder, a clean wet sponge and some "Tinner." Locate an anti-static wrist band which will be used to handle the integrated circuit. Warm up the soldering iron and we will begin. The Filtered Blacklight Project was constructed on a printed circuit board. Printed circuit boards will guarantee the most reliable operation. You could also elect to build the circuit using point-to-point wiring or "quick board," which are available at your local Radio Shack store.

Place all the project parts in front of you as well as the schematic and pictorial diagram, shown in Figure 20-3, for the blacklight project. Now refer to the resistor color code chart listed in Table 20-1. Take note that each resistor will have at least three or four color coded bands on the body of the resistor. The first color band, which will be at one edge of the resistor, is the first band or will represent the first digit in the color code chart. The second color band is the second digit of the resistor code. The third color band on the resistor corresponds to the code multiplier. If there is no fourth color band then the resistor has a 20% tolerance value, but if the fourth band is silver then the resistor has a 10% tolerance value. If the fourth color coded band is gold then the resistor will have a 5% tolerance. Look through the project resistors and note that R1 and R3 will have a first color band which is brown, representing a digit number one. The second color band will be black, this represents a zero, and the third color band will be black and therefore the resistor value is 10 ohm. Now look for the resistor R2; it will have a blue first

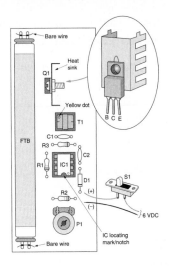

Figure 20-3 *Blacklight pictorial diagram*

band, a gray second band and a brown third band; this resistor will have a 680 ohm value.

Now go ahead and install these three resistors and solder them in place on the printed circuit board. After soldering the resistors to the circuit board, cut the excess component leads flush to the edge of the circuit board. Next locate the trimmer potentiometer, it will have three leads, either all in a row or two outside leads with an offset middle lead. The center lead of the potentiometer is the adjustable or wiper lead. You will want to make sure that you install this component correctly. Take a look at the schematic and layout diagram and see where the wiper lead goes in the circuit before installing the trimmer. Once installed, you can solder the trimmer in place. Cut the excess component leads from the potentiometer.

Now let's move on to locating and installing the capacitors. The Filtered Blacklight Project has three capacitors; note that capacitors in this project have no polarity, so orientation is not critical. Refer to the capacitor code chart in Table 20-2. You will have to identify the marking on the capacitors before installing them. Capacitors generally will have their value printed on them or some form of three-digit code marked on them. Note that the capacitor at C1 will be marked with a (472) code, you can look up this code and you will find that the value for the capacitor is 0.0047 μF. Capacitor C2 will have a (104) code marked on it, which corresponds to a value of 0.1 μF. Place the capacitors into the correct holes on the printed circuit

Table 20-1

Resistor Color Code Chart

Color Band	1st Digit	2nd Digit	Multiplier	Tolerance
Black	0	0	1	
Brown	1	1	10	1%
Red	2	2	100	2%
Orange	3	3	1000 (K)	3%
Yellow	4	4	10000	4%
Green	5	5	100000	
Blue	6	6	1000000 (M)	
Violet	7	7	10000000	
Gray	8	8	100000000	
White	9	9	1000000000	
Gold			0.1	5%
Silver			0.01	10%
No color				20%

board, and solder them in place, follow up by removing the excess capacitor leads.

The blacklight project has a single silicon diode which is connected between the minus (–) side of the battery and pin 1 of the integrated circuit at U1. Diodes are polarity sensitive devices and will have to be mounted correctly for the circuit to function. Diodes generally will have a white or black band at one side of the device. The colored band corresponds to the cathode of the diode which is shown as the flat line on the schematic. The anode of the diode is the triangle in the diagram. Install the diode and solder it in place; remember to cut the excess diode leads flush to the edge of the circuit board.

Now look for the transistor, which will be installed at Q1. Transistors are generally three-lead devices and will be either a PNP device or a NPN device. Refer to the transistor identification diagram shown in Figure 20-4. Transistors will have three leads marked as follows: the flat side on the transistor symbol in the mounting diagram is the Base lead marked (B). The Collector lead comes off the Base lead at an angle. The third transistor lead is the Emitter which is opposite the Collector lead. The Emitter lead will have a small arrow which will point toward the interior of the transistor if the device is

a PNP transistor as is Q1 in this project. If the transistor is a NPN device then the arrow on the Emitter will point away from the transistor. The Base lead of the transistor is connected to the junction of R3 and C1. The Emitter lead of Q1 is connected to the plus (+) power supply bus at the battery, while the Collector is connected to one end of the primary of transformer T1. Place a heatsink over the transistor, using some heat-sink compound between the transistor and the heatsink. Fasten the heatsink to the transistor, solder the transistor to the printed circuit board and then cut the excess component leads flush to the edge of the PC board.

The Filtered Blacklight Project utilizes a single integrated circuit, an LM555 timer IC. An integrated circuit socket is highly recommended, in the event of a circuit failure at some later date. If the circuit fails, then it is a simple matter to replace the integrated circuit. Most people cannot unsolder an integrated circuit without damaging the printed circuit board. The IC socket will have a notch at one end of the socket. The pin just to the left of the notch will be pin 1. Solder the IC socket in place. Using the anti-static wrist band to avoid damage while handling the integrated circuit, take the IC and insert it into the IC socket. At one end of the IC, you will notice that there will be a small indented

Table 20-2

Capacitance Codebreaker Information

This table is designed to provide the value of alphanumeric coded ceramic, mylar and mica capacitors in general. They come in many sizes, shapes, values and ratings; many different manufacturers worldwide produce them and not all play by the same rules. Most capacitors actually have the numeric values stamped on them; however, some are color coded and some have alphanumeric codes. The capacitor's first and second significant number IDs are the first and second values, followed by the multiplier number code, followed by the percentage tolerance letter code. Usually the first two digits of the code represent the significant part of the value, while the third digit, called the multiplier, corresponds to the number of zeros to be added to the first two digits.

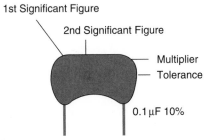

Value	Type	Code	Value	Type	Code
1.5 pF	Ceramic		1000 pF/0.001 μF	Ceramic/Mylar	102
3.3 pF	Ceramic		1500 pF/0.0015 μF	Ceramic/Mylar	152
10 pF	Ceramic		2000 pF/0.002 μF	Ceramic/Mylar	202
15 pF	Ceramic		2200 pF/0.0022 μF	Ceramic/Mylar	222
20 pF	Ceramic		4700 pF/0.0047 μF	Ceramic/Mylar	472
30 pF	Ceramic		5000 pF/0.005 μF	Ceramic/Mylar	502
33 pF	Ceramic		5600 pF/0.0056 μF	Ceramic/Mylar	562
47 pF	Ceramic		6800 pF/0.0068 μF	Ceramic/Mylar	682
56 pF	Ceramic		0.01	Ceramic/Mylar	103
68 pF	Ceramic		0.015	Mylar	
75 pF	Ceramic		0.02	Mylar	203
82 pF	Ceramic		0.022	Mylar	223
91 pF	Ceramic		0.033	Mylar	333
100 pF	Ceramic	101	0.047	Mylar	473
120 pF	Ceramic	121	0.05	Mylar	503
130 pF	Ceramic	131	0.056	Mylar	563
150 pF	Ceramic	151	0.068	Mylar	683
180 pF	Ceramic	181	0.1	Mylar	104
220 pF	Ceramic	221	0.2	Mylar	204
330 pF	Ceramic	331	0.22	Mylar	224
470 pF	Ceramic	471	0.33	Mylar	334
560 pF	Ceramic	561	0.47	Mylar	474
680 pF	Ceramic	681	0.56	Mylar	564
750 pF	Ceramic	751	1	Mylar	105
820 pF	Ceramic	821	2	Mylar	205

(Continued)

Table 20-2 (*Continued*)

PicoFarad (pF)	NanoFarad (nF)	MicroFarad (mF, μF or mfd)	Capacitance Code
1000	1 or 1n	0.001	102
1500	1.5 or 1n5	0.0015	152
2200	2.2 or 2n2	0.0022	222
3300	3.3 or 3n3	0.0033	332
4700	4.7 or 4n7	0.0047	472
6800	6.8 or 6n8	0.0068	682
10000	10 or 10n	0.01	103
15000	15 or 15n	0.015	153
22000	22 or 22n	0.022	223
33000	33 or 33n	0.033	333
47000	47 or 47n	0.047	473
68000	68 or 68n	0.068	683
100000	100 or 100n	0.1	104
150000	150 or 150n	0.15	154
220000	220 or 220n	0.22	224
330000	330 or 330n	0.33	334
470000	470 or 470n	0.47	474

circle, a notch or a small cutout in the plastic package. Just to the left of the cutout or notch will be pin 1 of the actual IC. You will have to make sure that pin 1 of the actual integrated circuit is plugged into pin 1 of the socket. Note that pin 1 of the IC should be connected to the anode of D1 and C2, while pin 8 is opposite pin 1, and is connected to the plus supply voltage and P1. If you are having difficulty identifying and/or installing the transistor of integrated circuit then ask a knowledgeable electronics enthusiast to help you.

Next you will want to mount the filtered blacklight to the circuit board. The easiest way to mount the lamp to the circuit board is to take a 2 to 3 inch length of bare solid core #20 ga wire and wrap each of the lamp-posts/pins with about four to five turns of the bare wire. Note that there are two lamp-posts at either end of the lamp. The two posts on each end of the lamp are wired together. Once all the posts have been wrapped with bare wire, you can line-up the lamp-posts with the hole on the PC board and insert the bare wires into the circuit board at their respective locations. Solder the wire to the circuit board and then remove the excess component leads.

Once all the components have been placed on the circuit board, you can attach a 6-volt battery clip to the circuit board. You can power the filtered blacklight from any 6-volt source; a "wall wart" power supply could be used, but for portable operation, you will want to power the circuit from batteries. You can select a two "C" or two "D" cell battery holder to power the circuit. Wire the two battery holders in series so that the black or minus (−) lead from the battery holder is connected to the cathode end of diode D1. The plus (+) lead from the battery holder is then connected to the junction of Emitter of the transistor Q1 and U1 pin 4 and pin 8. You can connect a power switch in series with the plus (+) or red power battery clip lead.

Now that your blacklight circuit board is completed, you can take a short, well-deserved rest and when we return we will inspect the circuit board for any possible "short" circuits or "cold" solder joints. First we will inspect the circuit board for any possible "cold" solder joints. Pick up the circuit board and turn the PC board so that the foil side of the board is facing upwards toward you. Carefully inspect the solder joints, to make

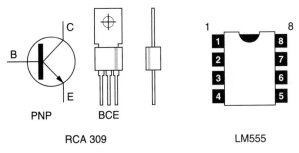

Figure 20-4 *Blacklight semiconductors*

sure that all the solder joints look clean, bright, smooth and shiny. If you find any of the solder joints look dull or "blobby" then you will want to unsolder that particular joint, remove the solder and then re-solder the joint, so that the joint looks clean, smooth and shiny. Next we will inspect the circuit for any "short" circuits, which may be caused from any "stray" wires or component leads which may have attached themselves to the circuit board. Rosin core solder often leaves a sticky residue on the circuit board and this can sometime cause a "cut" component to stick to the circuit board after it was cut from the board. "Stray" leads and wires can "short" between the copper circuit pads or traces and cause the circuit to "short" out and thus damage the circuit when the circuit is first powered-up. Look the circuit board over carefully and remove any wires or flecks of solder. Once your PC board has been inspected and you are satisfied with the results, you can then attach the battery clip and then turn the power switch to the "on" position to test the filtered blacklight.

At this point you should see the lamp light up; you may have to darken the room a little to see the lamp if your room lighting is too bright. You may have to adjust the potentiometer control at P1 for optimum results if needed.

In the event that your blacklight does not immediately work, you will have to remove the battery clip and re-inspect the circuit board for any possible errors or faulty components. First look over the PC board to make sure that each resistor at each location has the correct value. It is a common mistake to insert the wrong resistor into the wrong location. Look over

the color codes carefully and use an ohmmeter if you are unsure of yourself. Next, you will want to make sure that the silicon diode has been installed correctly. Remember the anode is the triangle or arrow which points to the cathode. The cathode of the diode is connected to the minus (−) side of the battery. Now, make sure that you have installed the transistor correctly. Re-check the schematic and compare it with the transistor pin-out diagram. The pin-outs are usually referenced from the bottom view of the transistor. The plastic TO-92 transistor package will usually have all the leads in a row with the Emitter at one end, with the Base lead in the center and the Collector lead opposite the Emitter lead. In TO-5 transistor packages, the Base lead will be offset from the other two leads. The case may have a tab on it and it will generally correspond to the Emitter lead.

Finally you will need to carefully inspect the installation of the integrated circuit. Pin 1 of the IC socket, if you remember, is connected to capacitor C2 and the anode of the diode, while pin 8 of the IC is connected to pin 4 and to the wiper of P1. If you are confused with the pin-outs of either the transistor or integrated circuit, then consult with an electronics enthusiast for help. Make sure that you have correctly installed the transformer, note that the primary of the transformer is connected to D1 and transistor Q1. The primary will have the shorter number of turns and lowest resistance. You can use an ohmmeter to test the primary resistance vs. the secondary resistance. The secondary winding will step up the voltage so it will have more winding and a higher resistance.

Now that you have inspected the circuit board carefully and taken whatever steps to correct your problem, you can re-connect the battery lead and re-test the circuit once more. Apply power to the circuit and you should see the blacklight come to life. You can use your new filtered blacklight to light up your party posters. Check for counterfeit money, or to look at beautiful phosphorescent rocks that get stimulated with a blacklight such as Calcite or Willamite. Have fun with your new blacklight.

Chapter 21

Mini-Electronic Organ

Parts List

Parts Bin

Main Board

R1, R4 4.7 k ohm, 1%,
⅛-watt resistor

R2, R3, R5, R7 100 k
ohm, 1%, ⅛-watt
resistor

R6 47 k ohm, 1%,
⅛-watt resistor

R8 51 k ohm, 1%,
⅛-watt resistor

R9 150 ohm, 1%,
⅛-watt resistor

P1 100 k ohm trimmer
potentiometer

C1, C2 2.2 µF,
35-volt electrolytic
capacitor

C3 47 µF, 35-volt
electrolytic capacitor

C4 10 nF, 35-volt green
capacitor

C5 470 µF, 35-volt
electrolytic capacitor

C6 1000 µF, 35-volt
electrolytic capacitor

D1 1N4004 silicon diode

Q1, Q2 PN100 NPN
transistor—2N3904 or
NTE123AP replacement

Q3 PN200 PNP
transistor—2N3906 or
NTE159 replacement

U1 LM555 timer IC

SPK 8 ohm mini speaker

S1 Power on/off switch
(SPST)

S2 Vibrato switch
(SPST)

J1 2 circuit mini jack
(keyboard)

P1 2 circuit mini plus
(keyboard)

B1 9-volt transistor
radio battery

Misc PC board,
IC socket, wire,
stylus, battery clip,
etc.

Keyboard

PC-1 PC keyboard

R1, R2 11 k ohm, 1%, ⅛-
watt resistor

R3, R4 10 k ohm, 1%,
⅛-watt resistor

R5 9.1 k ohm, 1%,
⅛-watt resistor

R6, R7 8.2 k ohm, 1%,
⅛-watt resistor

R8, R9 7.5 k ohm, 1%,
⅛-watt resistor

R10 6.8 k ohm, 1%,
⅛-watt resistor

R11, R12 6.2 k ohm,
1%, ⅛-watt resistor

R13, R14 5.6 k ohm,
1%, ⅛-watt resistor

R15 5.1 k ohm, 1%,
⅛-watt resistor

R16, R17 4.7 k ohm, 1%,
⅛-watt resistor

R18 4.3 k ohm, 1%,
⅛-watt resistor

R19, R20 3.9 k ohms,
1%, ⅛-watt resistor

R21 3.6 k ohm, 1%,
⅛-watt resistor

R22, R23 3.3 k ohm, 1%,
⅛-watt resistor

R24 3.0 k ohm, 1%,
⅛-watt resistor

ST-1 Pen stylus, made
from discarded meter
probe

Misc 2 feet #22 ga
stranded-insulated
wire

Optional Audio Interface

R1 10 k ohm, 5%,
¼-watt resistor

P1 10 k ohm Linear
potentiometer

R3 100 k ohm, 5%,
¼-watt resistor

R3 10 k ohm, 5%,
¼-watt resistor

C1 1 nF, 35-volt mylar
capacitor

C2 47 nF, 35-volt mylar
capacitor

Misc PC board, wire,
shielded cable,
connectors

The Mini Electronic Organ will allow you to play and practice your favorite tunes, musical compositions, etc. This circuit will entertain children and adults as well for many hours. The Mini Electronic Organ features an integrated circuit oscillator, a piano-like keyboard with audio output for external amplification, see Figure 21-1.

The heart of the Mini Electronic Organ, shown in schematic diagram Figure 21-2 and the pictorial/layout diagram Figure 21-3, is the timer IC at U1. The multipurpose LM555 timer IC is used as an oscillator to produce the musical notes. But it only oscillates when the stylus is touching the key pads, because the oscillator's feedback path is from U1's output pin 3

Figure 21-1 *Electronic Organ Project*

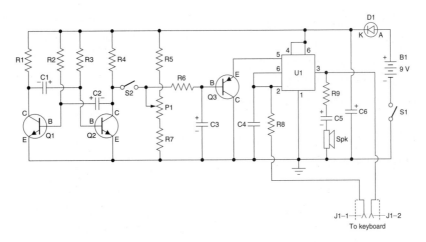

Figure 21-2 *Mini Organ. Courtesy Jaycar (JC)*

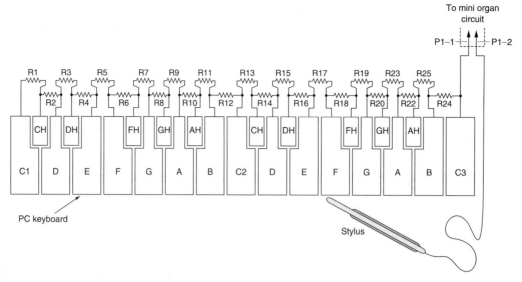

Figure 21-3 *Mini Organ keyboard*

back to the input sensing pins 2 and 6, via the stylus, the 51 k series resistor and whichever other resistors are in series with the keypad you touch.

The oscillator's frequency is set by this feedback resistance and the 10 nF capacitor connected from pins 2 and 6 to the negative line, which is why we can vary the organ's note simply by touching the stylus to each of the various key pads. If you work out the total feedback resistance for each of the keypads, you will find that the resistance decreases from left to right by almost exactly the same factor (1.0595), from one note to the next. This makes the frequency or pitch of the notes increase by the same factor or amount, which happens to be the ratio between notes for modern "equal temperament" musical scales.

Because the speaker is also connected between U1's output pin 3 and the negative line, via the 100 ohm series resistor to limit the current and the 470 µF electrolytic capacitor, i.e. to block DC, some of the IC's output is fed to the speaker. As a result the speaker produces sound whenever the oscillator is working, and at the oscillation frequency.

We produce the vibrato effect, by using transistors Q1 and Q2 in another oscillator circuit, to generate a very low frequency (about 0.5 Hz). This is then used to vary the frequency of note oscillator U1, so that the note "wavers" at this rate.

Note that transistors Q1 and Q2 are connected in a cross-coupled multivibrator circuit, where the Base of

each transistor is connected to the other's Collector. This means that they can't conduct at the same time, but instead flip back and forth—with the first one conducting and then the other, and so on. The frequency they oscillate at is set mainly by the two 2.2 µF capacitors and the 100 k ohm Base resistors. The oscillator produces a square wave signal at the Collector of Q2, which switches back and forth between +9 volts and ground.

If we were to feed this square wave signal directly to U1, it would not produce a musical vibrato effect but simply sound as if each note had been separated into two. So, instead, when S1 is closed, we first feed the signal through a simple "low-pass" filter circuit formed by the 47 k ohm resistor and the 47 µF capacitor. This removes most of the "sudden changes" or harmonics from the vibrato signal, leaving a voltage which still varies up and down at 0.5 Hz rate, but now does so quite smoothly. This smoothed vibrato signal is then fed through transistor Q3 and into the control voltage input pin 5 of U1, so it can vary the note frequency up and down by a small amount, and the result is a nice vibrato effect.

When switch S2 is opened to turn off the vibrato, we instead feed a small DC voltage from trimpot P1 through the filter and Q3 into U1. Potentiometer P1 is used to adjust the substitute DC voltage so that it matches the average DC level of the vibrato signal, to make sure that U1's note frequencies are the same whether the vibrato is turned on or off.

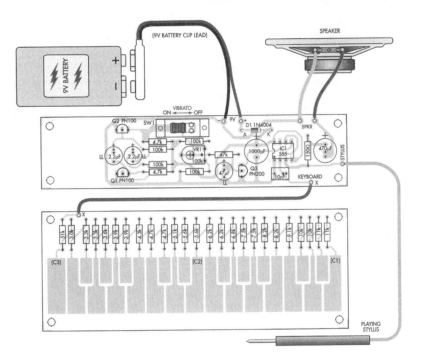

Figure 21-4 *Electronic Organ pictorial diagram*

We chose to build the Mini Electronic Organ on two circuit boards for most reliable operation. The main electronics board measures 5¼ inches × 1½ inches. The musical keyboard was constructed on a second printed circuit board which measures 5¼ inches × 2 inches, but you could elect to build both circuits on a single larger circuit board. The circuit layout diagram is shown in Figure 21-4. Although it is possible to build the circuit on a perf-board, a conventional circuit board is highly recommended for reliable operation.

If you are ready, let's begin constructing the Mini Electronic Organ. First, take a few moments to find a well-lit work area, with large table or workbench, so you can spread out all the diagrams, charts, components and tools. Locate a 27 to 30-watt solder iron and prepare the iron by heating it up and cleaning the tip. Get a can of "Tip Tinner" from you local Radio Shack store. Then insert the heated iron tip into the "Tinner" can. Next clean the tip on a wet sponge and your iron should now be ready to begin assembling your circuit. Locate a spool of 22 ga 60/40 rosin core solder, a pair of needle-nose pliers, a pair of end-cutters or diagonals, a magnifying glass and a pair of tweezers. It is also a good idea to have a small flat-blade and Phillips screwdriver set. You should also locate an anti-static wrist strap to help protect the integrated circuits as they are being handled. High voltage static charges can readily build up when

building the circuit, due to you moving around and getting up and down from your chair, so it's best to use a grounded anti-static wrist band. The wrist band plugs into a regular household outlet for grounding.

Locate an integrated circuit socket for U1. Integrated circuit sockets are a good idea in the event of circuit failure at a later date. You can easily replace the IC without trying to de-solder the IC from the circuit. Because most people cannot unsolder IC from a PC board without damaging the board, it is best to just use sockets from the beginning. Integrated circuit sockets have a small notch or cutout at one end of the package. To the left of the cutout, you will find pin 1. You will want to make sure that pin 1 of the socket is connected to the circuit ground bus, while the opposite pin 8 is connected to the cathode of diode D1. Using your anti-static wrist strap place the IC into its socket, noting that the IC will have a notch, cutout or small indented circle at one end of the IC package. Line-up pin 1 of the IC with pin 1 of the IC socket and insert the IC into the socket.

Next find the resistors for the project. Refer to the resistor color code chart and make sure you understand how the color code works before installing the resistor—see Table 21-1. The first color band on the end of the resistor is the first digit, the next color band is the second digit. The third color band represents the multiplier. Often there is a fourth band which represents the resistor

Table 21-1

Resistor Color Code Chart

Color Band	1st Digit	2nd Digit	Multiplier	Tolerance
Black	0	0	1	
Brown	1	1	10	1%
Red	2	2	100	2%
Orange	3	3	1000 (K)	3%
Yellow	4	4	10000	4%
Green	5	5	100000	
Blue	6	6	1000000 (M)	
Violet	7	7	10000000	
Gray	8	8	100000000	
White	9	9	1000000000	
Gold			0.1	5%
Silver			0.01	10%
No color				20%

tolerance value. If there is no band the resistor is 20% resistor. If the fourth color band is silver then the resistor is 10% tolerance value and if the fourth band is gold then the resistor has a 5% tolerance value. Look for a resistor with a (yellow) band, a (violet) band and a (red) band; this will be resistor R1, a 4.7 k ohm resistor. Now, install the resistor R1 into the circuit board. Identify and install the remaining resistors. Next, you can solder all of the resistors to the circuit board, then remove the excess leads from the board. Use your end-cutters and cut the leads flush to the edge of the PC board.

Potentiometers are often configured with three leads in a row or two outside pins with an offset center pin, which represents the center wiper of the potentiometer. Next install the potentiometer into the circuit board, and solder it into place, then remove the excess leads.

The electronic organ circuit has a single silicon diode, at D1. Note that diodes are polarity sensitive and must be installed correctly in order for the circuit to work correctly. Diodes will have a black or white band at one end of the diode's body which denotes the polarity. Schematically the anode of a diode is the arrow or triangle which points to the cathode which is the symbol of a line. The colored band denotes the diode's cathode side. When installing the diode make sure that

you understand how to mount the diode into the circuit. In the electronic organ circuit the anode of the diode faces the battery's plus (+) terminal and the cathode faces toward pin 4 and pin 8 of integrated circuit U1. Now insert the silicon diode D1 into the circuit board and solder it in place onto the circuit board. Remember to cut the excess component leads.

Next, we will install the capacitors onto the circuit board. Refer now to the chart in Table 21-2, which illustrates the capacitors' codes and how they are used. Often capacitors will have their actual value marking printed on the capacitors, but many times, small disk or mylar capacitors will have a three-digit code printed instead of a value, so you must understand how these codes work. For example, if you find a capacitor with a code marking of (103), you will see from the chart that this capacitor has a value of 0.01 μF or 10 nF and this will be C4. Install C4 on the circuit board and solder it in place. The electronic organ circuit has six capacitors, five of which are electrolytic types. You will remember that electrolytic capacitors will have polarity markings, and therefore they must be installed correctly for the circuit to work properly. Note that installing capacitors backwards could damage the circuit, so you should try to install them correctly the first time. Capacitors will have either a black or white band at one end of the capacitor body, in that

Table 21-2

Capacitance Codebreaker Information

This table is designed to provide the value of alphanumeric coded ceramic, mylar and mica capacitors in general. They come in many sizes, shapes, values and ratings; many different manufacturers worldwide produce them and not all play by the same rules. Most capacitors actually have the numeric values stamped on them; however, some are color coded and some have alphanumeric codes. The capacitor's first and second significant number IDs are the first and second values, followed by the multiplier number code, followed by the percentage tolerance letter code. Usually the first two digits of the code represent the significant part of the value, while the third digit, called the multiplier, corresponds to the number of zeros to be added to the first two digits.

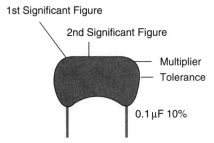

1st Significant Figure
2nd Significant Figure
Multiplier
Tolerance
0.1 µF 10%

Value	Type	Code	Value	Type	Code
1.5 pF	Ceramic		1000 pF/0.001 µF	Ceramic/Mylar	102
3.3 pF	Ceramic		1500 pF/0.0015 µF	Ceramic/Mylar	152
10 pF	Ceramic		2000 pF/0.002 µF	Ceramic/Mylar	202
15 pF	Ceramic		2200 pF/0.0022 µF	Ceramic/Mylar	222
20 pF	Ceramic		4700 pF/0.0047 µF	Ceramic/Mylar	472
30 pF	Ceramic		5000 pF/0.005 µF	Ceramic/Mylar	502
33 pF	Ceramic		5600 pF/0.0056 µF	Ceramic/Mylar	562
47 pF	Ceramic		6800 pF/0.0068 µF	Ceramic/Mylar	682
56 pF	Ceramic		0.01	Ceramic/Mylar	103
68 pF	Ceramic		0.015	Mylar	
75 pF	Ceramic		0.02	Mylar	203
82 pF	Ceramic		0.022	Mylar	223
91 pF	Ceramic		0.033	Mylar	333
100 pF	Ceramic	101	0.047	Mylar	473
120 pF	Ceramic	121	0.05	Mylar	503
130 pF	Ceramic	131	0.056	Mylar	563
150 pF	Ceramic	151	0.068	Mylar	683
180 pF	Ceramic	181	0.1	Mylar	104
220 pF	Ceramic	221	0.2	Mylar	204
330 pF	Ceramic	331	0.22	Mylar	224
470 pF	Ceramic	471	0.33	Mylar	334
560 pF	Ceramic	561	0.47	Mylar	474
680 pF	Ceramic	681	0.56	Mylar	564
750 pF	Ceramic	751	1	Mylar	105
820 pF	Ceramic	821	2	Mylar	205

(Continued)

Table 21-2 (*Continued*)

PicoFarad (pF)	NanoFarad (nF)	MicroFarad (mF, µF or mfd)	Capacitance Code
1000	1 or 1n	0.001	102
1500	1.5 or 1n5	0.0015	152
2200	2.2 or 2n2	0.0022	222
3300	3.3 or 3n3	0.0033	332
4700	4.7 or 4n7	0.0047	472
6800	6.8 or 6n8	0.0068	682
10000	10 or 10n	0.01	103
15000	15 or 15n	0.015	153
22000	22 or 22n	0.022	223
33000	33 or 33n	0.033	333
47000	47 or 47n	0.047	473
68000	68 or 68n	0.068	683
100000	100 or 100n	0.1	104
150000	150 or 150n	0.15	154
220000	220 or 220n	0.22	224
330000	330 or 330n	0.33	334
470000	470 or 470n	0.47	474

band will be a plus (+) or minus (–) marking. When installing the capacitors onto the circuit board, you will have to observe the polarity marking on the capacitor, as well as understand that the circuit polarity must match the component polarity. Look at the circuit schematic diagram, and observe where capacitor C1 is placed. Note that the minus (–) end of the capacitor is connected to the Collector of Q1 and the plus (+) side of C1 is connected to the Base of Q2. Once you have discovered how this works, you can go ahead and install the capacitor at C1 and solder it in place. Repeat the procedure for all of the other electrolytics. Finally solder all of the electrolytic capacitors in place. Remove the excess capacitor leads from the PC board.

Next, let's install the three transistors. Note that transistors generally have three leads, a Base, a Collector, and an Emitter. Refer to the semiconductor mounting diagram shown in Figure 21-5. The TO-92 type transistor housing will have all three leads in a row, with the Emitter at one end. The Base lead will be in the center, with the Collector lead at the opposite end of the transistor. The pin-out diagram shows the transistor from bottom view as a reference, and this is important to note. Observe that the transistors Q1 and Q2 are NPN types, while transistor Q3 is a PNP type. In NPN types the arrow on the Emitter will point away from the Base and with the PNP types the Emitter arrow will point toward the Base lead. Take a look at the schematic and you will see that the Emitter of Q1 is connected to the Emitter of Q2 and to the system ground, this will help you insert the leads into the circuit board for orientation purposes. Note that the Base lead of Q1 is connected to capacitor C2 and resistor R2, while the Collector of Q1 is connected to C1 and R1. When you understand the orientation of all of the transistors, you can go ahead and install them, trimming the leads and soldering them in place.

Next we are going to install the IC at U1. It is recommended that you use an integrated circuit socket. Integrated circuit sockets will greatly aid servicing at a later time if the circuit fails. Before installing the IC, note that it will either have a small indented circle or a notch cutout at the top of the IC case. Pin 1 of the IC and pin of the socket will be just to the left of the indented circle or notch. You will also have to observe

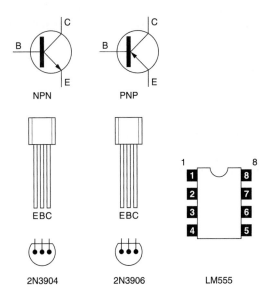

Figure 21-5   *Electronic Organ semiconductor diagram*

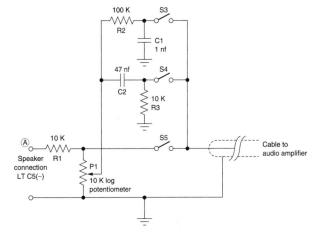

Figure 21-6   *Mini Organ amplifier output connections*

where pin 1 of the IC should be placed with respect to the circuit. Note that pin 1 in the circuit is connected to ground, while pin 8 is connected to the cathode of D1. Be sure to use your anti-static wrist strap when handling the IC. Refer to the schematic and the pictorial diagrams to help you install the IC. When you understand and can visualize how to mount U1, go ahead and install U1 into its socket, if you are still not sure or confused, ask a parent or knowledgeable friend.

Now locate the 8-ohm speaker, locate two 6-inch lengths of #22 ga stranded-insulated hookup wire and solder the two free ends to the speaker tabs. Then take the remaining wire ends and solder the wires to the circuit board. One speaker wire will connect to ground, and the other lead will connect to the minus (–) end of capacitor C5.

Finally connect up the 9-volt battery clip to the circuit. The black or minus (–) battery clip lead is connected to the system ground. The red or plus (+) battery lead is wired in series with the power switch at S1. The other switch lead is connected to the plus (+) 9-volt system bus or to pins 4 and 8 of U1 as shown.

Now let's construct the keyboard, shown in Figure 21-3. The keyboard or key pad is constructed on a piece of printed circuit board; refer to the schematic and pictorial diagrams. There are 25 keys made from copper foil left on the circuit board. You could also take a blank sheet of phenolic and use sticky copper foil tape

to make a keyboard if desired. Notice that the keyboard has many resistors connected in series; all of these resistors are 1%, ⅛-watt types, which begin with the 11 k value at (C1) and end with the 3 k ohm resistor at (C3). Note that the keyboard assembly is connected to the main circuit board via a two-circuit jack and plug arrangement; you can use a ⅛ inch phone jack or RCA type jack for this purpose. The keyboard is connected to the main board by a single wire from key (C3) to the 51 k resistor at pin 2 of U1.

The stylus which is used to activate the keyboard and "play" the organ was made from a discarded multimeter probe, but you can improvise and use any metal probe or wire end.

The Mini Electronic Organ creates sound through its 8-ohm speaker, but you can also elect to send the output from the electronic organ to a mono or stereo amplifier system if desired. The circuit shown in Figure 21-6 illustrates a circuit which can be used for this purpose. This simple interface circuit is used by removing the speaker from the circuit and connecting the resistor at R1 or point (A) to the minus (–) lead of capacitor C5. Potentiometer P1 adjusts the output of the Mini Electronic Organ circuit's audio level. The potentiometer is fed to a two-filter network which can be switched in and out of the circuit for different sound effects as desired. The values given for the C1/R2 network and the C2/R3 network are examples or "starting" points; you can experiment with these values for the desired sound. The output of the interface circuit should be a shielded cable to the amplifier of your choice.

Now is a good time to check over you circuit board before applying power to the circuit. First you will want to look the board over to make sure that all the solder joints look smooth and clean and shiny. If you find a solder joint that does not look smooth, simply clean the solder joint and re-solder the connection. Clean solder joints ensure long reliable operation. Next you will want to inspect the circuit board to make sure that there are no "cut" or "stray" component leads which may have stuck to the board after cutting the component leads. You will want to make sure that there are no extraneous leads left to "short" out the circuit once power is applied.

Once the IC has been installed and the battery clip attached to the battery, turn switch S1 to the "on" position, and your Mini Electronic Organ is now ready to perform its magic. With the main circuit board connected to the keypad via P1/J1, bring the stylus to the keypad and touch the stylus point to one of the keys on the keypad and you should hear a sound note from the speaker. If so, all is well. You are now ready to practice some songs for your mom, dad or friends!

If you do not hear any sounds, you will have to disconnect the battery and do another visual inspection of the circuit board. First, remove the battery and inspect the circuit board over again. This time you will need to verify the color code on each of the resistors to make sure that they are in the right location. If they are in the right location, then you can move on to inspecting to make sure that the diode at D1 has been installed correctly with the black band of cathode pointing toward pins 4 and 8 of U1. Next make sure that all of the transistors are oriented correctly. Finally, make sure the IC has been installed correctly, so that pin 1 of the IC is connected to pin 1 of the IC socket or the system ground. Check your battery connections, making sure the black lead connects to the power ground bus of the circuit, and that the red battery lead is connected to the plus (+) and pins 4 and 8 of U1. If all looks good, you can attach the keyboard to the main circuit board, apply power and try the circuit again. Place the stylus on one of the keys on the keyboard, and you should hear a sound. Next try each of the pads and make sure that you hear sound when each pad is touched. Finally, you can make any fine "tuning" adjustments to the potentiometer P1 on the main board.

Your Mini Electronic Organ is now ready to entertain your family and friends!

Chapter 22

3-Channel Color Organ

Parts List

Parts Bin

R1 3.3 k ohm, 5%, ¼-watt resistor

R2, R5, R12 18k ohm, 5%, ¼-watt resistor

R3, R6, R11 220 k ohm, 5%, ¼-watt resistor

R4, R8, R13 100 ohm, 5%, ¼-watt resistor

R7 1.8 k ohm, 5%, ¼-watt resistor

R9 680 ohm, 5%, ¼-watt resistor

R10 2 k ohm, 5%, ¼-watt resistor

P1, P2, P3, P4 10 k ohm trim potentiometers

C1, C4 10 µF, 35-volt electrolytic capacitor

C3, C6, C8 33 µF, 35-volt electrolytic capacitor

C2, C5, C7 0.047 µF, 35-volt disk capacitor (473)

Q1, Q2, Q3 C458 NPN transistor or 2N3904 or NTE123AP

D1, D2, D3 T106B— Silicon Controlled Rectifier or NTE5456

T1, T2, T3 600-600 mini-transformer— Mouser 553TY304P

L1, L2, L3 110-volt color lamps—or color lamp strings

B1 9-volt battery or 9-volt power supply

S1 SPST power "on/off" switch

F1, F2, F3 110-volt, .5 amp fuse

P1 110-volt wall plug

Misc PC board, wire, enclosure, hardware, etc.

The 3-Channel Color Light Organ highlighted in Figure 22-1 is an entertaining project, which will allow you to pulse three different color flood lamps or three different banks of colored lamps, to the beat of your audio source. An audio source such as a radio, stereo system or karaoke system can be used to drive the color lamps to your music. The 3-Channel Color Organ divides the sound spectrum into three distinct channels: a high frequency channel, a medium and low frequency channel. Each will drive up to 800 watts of lamps

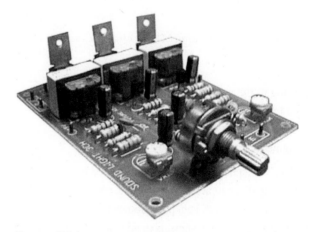

Figure 22-1 *Color Organ Project*

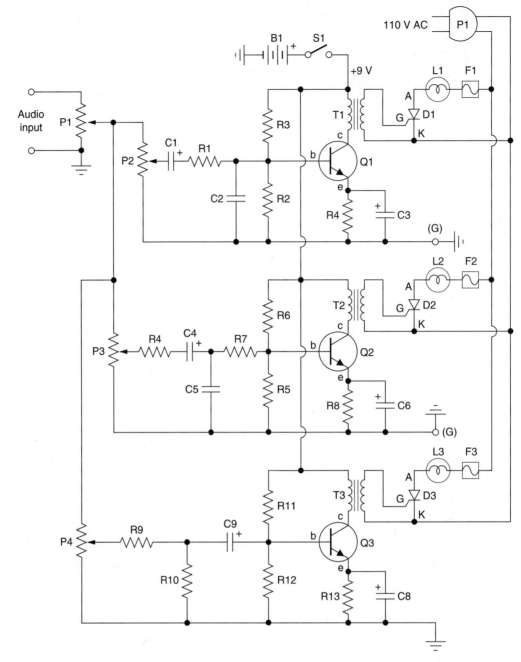

Figure 22-2 *3-Channel Color Organ. Courtesy Future Kit (FK)*

pulsing to the music, the effect can be awesome. You could use three colored lamps or strings of separate color lamps strung around the room: let your imagination run wild.

The 3-Channel Color Organ is illustrated in the schematic diagram at Figure 22-2. The audio signal enters the circuit via overall gain potentiometer at P1. From this main level control the signal is divided into three channels with a gain potentiometer for each channel, i.e. P2, P3 and P4. Resistor capacitor filter networks at the input of each

of the channels divides up the audio spectrum so that a high, medium and low channel are provided. Notice that C1, R2 and C2 for the low frequency or Bass channel, and that R4, C4, C5 and R7 comprise the medium frequency channel filter. The high frequency filter is formed by components R9, R10 and C7.

Each of the frequency filters in channels 1, 2 and 3 are fed into a transistor, which is used to drive the audio transformers at T1, T2 and T3. Each of the transformers in turn is used to trigger the SCR based on the audio

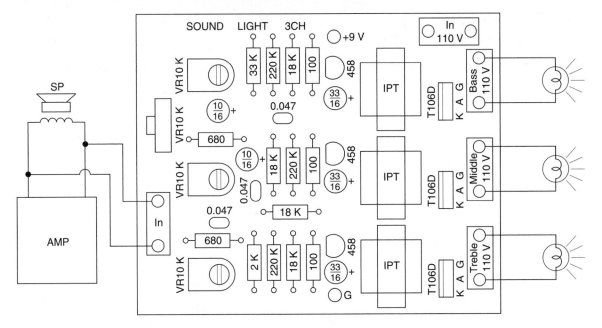

Figure 22-3   *Color Organ pictorial diagram*

input from the transformers. The transformers are used to modulate the input to the SCR gate marked (G) on the schematic. Each lamp is connected to the anode of the SCR which is connected in series with a 110-volt source. The cathodes of the SCRs are connected to the common 110-volt circuit as shown. Note that the input circuit on the primary circuit is isolated from the secondary circuit by the transformer. The primary side of the circuit is powered from a 9-volt battery source while the secondary part of the circuit is kept separate from the primary through the isolation of the transformer. Remember that voltage on the secondary part of the circuit is 110 volts AC, which can be dangerous, so have an adult help you when it is time to connect the circuit to the AC line.

The 3-Channel Color Organ circuit was fabricated on a 3 inch × 3 inch glass epoxy circuit board, and should be housed in a plastic case, so that the secondary part of the circuit is not exposed to human body parts, i.e. hands, head, feet, etc. The circuit is pretty straightforward and can be built in less than two hours.

Before we begin building the color organ, take a few moments to locate a clean, well-lit table or workbench, so you can spread out all the materials you will need to build the project. First you will want to procure a 27 to 33-watt pencil tip soldering iron with a sharp pointed tip or a small flat or wedge tip. You will also want to obtain some 60/40 rosin core solder, a wet sponge and a small

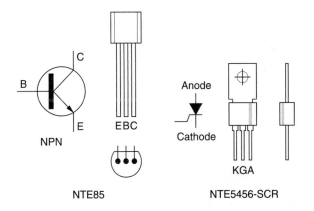

Figure 22-4   *Color Organ semiconductors*

tin of "Tip Tinner" from your local Radio Shack store. You should also locate some small tools such as a small pair of edge-clipping wire-cutters, a small pair of needle-nose pliers, some small screwdrivers, a magnifying glass, and a pair of tweezers. Place all the project parts in front of you along with the schematic and the pictorial diagram, shown in Figure 22-3, as well as the tables and semiconductor pin-out diagram, shown in Figure 22-4. Now you can plug in your soldering iron and we will begin constructing the color organ circuit.

Refer to the resistor color chart shown in Table 22-1, this will aid you in identifying the resistors. On each of the resistors you will see three or four color bands starting at one end of the resistor body. The first color

Table 22-1

Resistor Color Code Chart

Color Band	1st Digit	2nd Digit	Multiplier	Tolerance
Black	0	0	1	
Brown	1	1	10	1%
Red	2	2	100	2%
Orange	3	3	1000 (K)	3%
Yellow	4	4	10000	4%
Green	5	5	100000	
Blue	6	6	1000000 (M)	
Violet	7	7	10000000	
Gray	8	8	100000000	
White	9	9	1000000000	
Gold			0.1	5%
Silver			0.01	10%
No color				20%

band represents the first digit of the resistor value, the second color band represents the second digit of the color code, and the third color band denotes the resistor multiplier value. A fourth silver color band represents a 10% tolerance value for the resistor, while a gold band denotes a 5% tolerance (see chart). So, let's try and locate the first resistor R1, look for a resistor with (orange) (orange) (red) color bands. This resistor will be a 3.3 k ohm resistor which is placed at R1. Identify the remaining resistors, place them on the circuit board in the correct locations. After installing the resistors, solder each of them in place using your solder iron and your wire-cutters to cut the excess component leads flush to the circuit board.

Locate the four trimmer potentiometers in the parts pile. Potentiometers will generally have three leads either all in a row or two outside leads with a center offset lead, which is the wiper or adjustable lead of the potentiometer. After identifying the potentiometers, go ahead and install them on the circuit board. Solder the trim potentiometers to the circuit board, then cut the excess component leads flush to the circuit board. Note that all of the potentiometers have a 10 k value.

Next, we will identify all of the capacitors and install them. There are two types of capacitor in this project,

disk capacitors and electrolytic capacitors. Electrolytic capacitors will have their values printed on the body of the component, while the disk capacitors may not have their value on them, but may have a code printed on them. Let's start with the disk capacitors: refer to the chart in Table 22-2, which lists the common capacitor codes. Look for three capacitors with a code marking of (473); this represents a capacitor value of 0.047 µF. These capacitors will be placed at C2, C5 and C7. Install these disk capacitors in their respective locations. Next, locate the electrolytic capacitors in the parts pile. Electrolytic capacitors are polarity sensitive and must be installed correctly for the circuit to work. Installing electrolytic capacitors backwards can cause damage to the component and possibly to the circuit itself. On the body of the electrolytic capacitor you should see a black or white color band. On or near the color band you should see a plus (+) or minus (–) marking, this demotes the polarity. Look at the schematic and make sure you can identify the plus (+) marking diagram and match it up when placing the electrolytic capacitor on the circuit board. When you have identified all of the electrolytic capacitors, you can install them on the circuit board. After placing the capacitors on the board, you can solder them in place, then remove the excess component leads.

Now we will install the three NPN transistors. Generally transistors have three leads, a Base, a Collector and an Emitter lead. Looking at the schematic you will notice that the Base lead is the flat line with the Emitter and Collector pointing toward the Base lead. The Emitter lead shows an arrow either pointing toward the Base lead, this is a PNP transistor, or an arrow pointing away from the transistor, this is an NPN type transistor. Refer to the diagram depicted in Figure 22-4, which illustrates the transistor package and pin-out information. When installing the transistor, make sure you can identify the leads correctly. A plastic TO-92 transistor package will have the Emitter lead at one end, the Base lead in the center and the Collector lead opposite the Emitter. Refer to the schematic when installing the transistors. The Base lead of Q1 connects to the junction of resistor R2 and R2, while the Base lead of Q2 connects to the junction of R5 and R5. The Base lead of transistor Q3 is connected to the junction of R11 and R12. The Collector of each of the transistors is used to drive the transformers at T1, T2 and T3. The Emitter of Q1 is fed to resistor R4, while the Emitter of Q2 is fed to R8 and the Emitter of Q3 is connected to R13. If you are having difficulties identifying the transistor pin-outs seek a knowledgeable electronics friend to help you. Solder the transistors in their respective locations, then cut the excess leads.

Locate the three transformers; these are isolation transformers with a 1:1 ratio so the primary and secondary are exactly the same. The primary winding is wired to the transistors, while the secondary windings are fed to the colored lamps. Mount the transformers on the circuit board, then cut the component leads flush to the board.

Finally you will need to install the power SCRs. Refer to the diagram shown in Figure 22-4, which illustrates the SCR pin-outs. Note that the SCRs look similar to diode symbols, but with a gate lead connected to the cathode. The transformers are fed to the gate of the SCRs, while the cathodes are connected to the lamp commons. The anode of the SCR is connected directly to the lamps. If you intend to drive large loads with the SCR, then you should plan on using heatsinks on each of the SCRs to prevent overheating. Place some heat transfer compound between the SCR and heatsink when securing the parts together. Solder the SCR to the circuit board, then remove the excess component leads.

Look at the schematic diagram and you will see a (G) symbol on the minus side of C3, C6 and C8, this is common ground point for the battery supply voltage. An optional power switch at S1 is wired in series with the red or plus (+) lead of the 9-volt power source to power the circuit. The minus (–) or black lead of the battery clip is connected to the system ground at (G).

Note that the secondary or display circuits are isolated from the primary driver circuits via the transformer. The 9-volt power source only connects to the primary circuit, which is separate from the secondary circuit for safety. The secondary circuit is "live" or "hot," that is to say it is at a 110-volt AC potential, so be very careful when wiring and testing the secondary circuit. Ask a knowledgeable friend or parent to help you with final wiring!

The 3-Channel Color Organ is now complete and ready to go. Take a short, well-deserved break and when we return we will inspect the circuit board for any possible "shorts" or "cold" solder joints which could cause premature circuit failure. Turn the circuit board over, so that the foil side is facing up toward you. Now take a look at the solder joints, to make sure they all look clean, smooth and shiny. If you find a solder joint that looks dull or "blobby," then you will want to unsolder that particular joint and re-solder it, so that it looks clean, smooth and shiny.

Next you will inspect the circuit board for any possible "short" circuits. Often when removing cut component leads, they will stick to the underside of the circuit board. Sometimes the rosin from the solder is left on the board, it can be sticky and make the cut leads adhere to the circuit board. These "stray" component leads can sometimes stay on the board and form a "short" circuit across the copper lines or circuit lands.

Once you are satisfied with the inspection, you can now connect up the different colored lights that you want to use. Connect up an audio source to the input of the color organ, then connect up a 9-volt power source and then connect the lamps to the line voltage and switch on the color organ with S1. Adjust the volume on the radio or stereo, and adjust the main gain control on the color organ to midway through its range. Make sure that the individual gain controls P2, P3 and P4 are also adjusted to midrange. You should see the lamps pulsate along with the music.

If for some reason the color organ does not work, unplug the circuit from the 110-volt power line and

Table 22-2

Capacitance Codebreaker Information

This table is designed to provide the value of alphanumeric coded ceramic, mylar and mica capacitors in general. They come in many sizes, shapes, values and ratings; many different manufacturers worldwide produce them and not all play by the same rules. Most capacitors actually have the numeric values stamped on them; however, some are color coded and some have alphanumeric codes. The capacitor's first and second significant number IDs are the first and second values, followed by the multiplier number code, followed by the percentage tolerance letter code. Usually the first two digits of the code represent the significant part of the value, while the third digit, called the multiplier, corresponds to the number of zeros to be added to the first two digits.

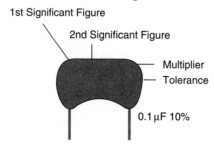

Value	Type	Code	Value	Type	Code
1.5 pF	Ceramic		1000 pF/0.001 µF	Ceramic/Mylar	102
3.3 pF	Ceramic		1500 pF/0.0015 µF	Ceramic/Mylar	152
10 pF	Ceramic		2000 pF/0.002 µF	Ceramic/Mylar	202
15 pF	Ceramic		2200 pF/0.0022 µF	Ceramic/Mylar	222
20 pF	Ceramic		4700 pF/0.0047 µF	Ceramic/Mylar	472
30 pF	Ceramic		5000 pF/0.005 µF	Ceramic/Mylar	502
33 pF	Ceramic		5600 pF/0.0056 µF	Ceramic/Mylar	562
47 pF	Ceramic		6800 pF/0.0068 µF	Ceramic/Mylar	682
56 pF	Ceramic		0.01	Ceramic/Mylar	103
68 pF	Ceramic		0.015	Mylar	
75 pF	Ceramic		0.02	Mylar	203
82 pF	Ceramic		0.022	Mylar	223
91 pF	Ceramic		0.033	Mylar	333
100 pF	Ceramic	101	0.047	Mylar	473
120 pF	Ceramic	121	0.05	Mylar	503
130 pF	Ceramic	131	0.056	Mylar	563
150 pF	Ceramic	151	0.068	Mylar	683
180 pF	Ceramic	181	0.1	Mylar	104
220 pF	Ceramic	221	0.2	Mylar	204
330 pF	Ceramic	331	0.22	Mylar	224
470 pF	Ceramic	471	0.33	Mylar	334
560 pF	Ceramic	561	0.47	Mylar	474
680 pF	Ceramic	681	0.56	Mylar	564
750 pF	Ceramic	751	1	Mylar	105
820 pF	Ceramic	821	2	Mylar	205

(Continued)

Table 22-2 (*Continued*)

PicoFarad (pF)	NanoFarad (nF)	MicroFarad (mF, µF or mfd)	Capacitance Code
1000	1 or 1n	0.001	102
1500	1.5 or 1n5	0.0015	152
2200	2.2 or 2n2	0.0022	222
3300	3.3 or 3n3	0.0033	332
4700	4.7 or 4n7	0.0047	472
6800	6.8 or 6n8	0.0068	682
10000	10 or 10n	0.01	103
15000	15 or 15n	0.015	153
22000	22 or 22n	0.022	223
33000	33 or 33n	0.033	333
47000	47 or 47n	0.047	473
68000	68 or 68n	0.068	683
100000	100 or 100n	0.1	104
150000	150 or 150n	0.15	154
220000	220 or 220n	0.22	224
330000	330 or 330n	0.33	334
470000	470 or 470n	0.47	474

remove the battery from the circuit. Now, you will have to troubleshoot the circuit to see if you made any errors in parts placement or wiring.

Take a look at the resistors and compare the color codes of each one against the color code chart to make sure that you have the right component in the correct location, because this is a common error. Next, take a look at the capacitors and make sure that the electrolytic capacitors have their polarity markings facing the right direction. Remember that electrolytic capacitors have polarity which must be observed in order for the circuit to work properly.

Next, you will want to re-check the transistor placement to make certain that you mounted the transistors correctly. Often it is a common error to place a transistor backwards in the circuit. Look at the transistor pin-out diagram and verify the lead locations; remember that the transistor's Collectors are connected to the transformers, and that the Emitter leads are connected to the resistors R4, R8 and R13, as seen on

the schematic. If you are confused on this, have an electronic enthusiast help you take a look a the circuit.

Finally, you will need to verify that the SCRs have been installed correctly. Look at the SCR pin-out diagram and verify the pin-out locations; remember that the anodes of the SCRs are connected to the lamps, and that the cathode leads are connected to the lamp common and the transformer secondary.

Re-check your lamp wiring at this point, and when you are sure that everything looks fine, you can reconnect your audio source, then connect the 9-volt power source and finally reconnect the 110-volt AC line to the lamps. Remember, this circuit is the only circuit in this book that connects directly to line voltage and you could get a serious burn or shock if you are not careful.

Your 3-Channel Color Light Organ is now complete and ready for your next party. This project is great for parties and gatherings. You could also use this circuit to control lamps at Halloween or Christmas if desired. Be sure, be safe, and have fun!

Chapter 23

8-Watt Power Audio Amplifier

Parts List

Parts Bin

R1 5 k ohm, 5%, ¼-watt resistor

R2 2 k ohm, ¼-watt resistor

R3, R4, R5 100k ohm, ¼-watt resistor

R6 1 ohm, ¼-watt resistor

P1 10 k ohm trim potentiometer

C1 2.2 µF, 35-volt electrolytic capacitor

C2 0.001 µF, 35-volt disk capacitor

C3 100 µF, 35-volt electrolytic capacitor

C4 22 µF, 35-volt electrolytic capacitor

C5, C6 0.1 µF, 35-volt disk capacitor

C7 470 µF, 35-volt electrolytic capacitor

C8 1000 µF, 35-volt electrolytic capacitor

U1 TDA2030 8-watt audio amplifier IC

D1, D2 1N4001 silicon diode

SP 8-ohm 3-inch mini speaker

S1 SPST power "on/off" switch

Misc PC board, wire, IC socket, 12-volt power supply, enclosure, etc.

The 8-Watt Power Audio Amplifier, in Figure 23-1 is an ideal kit for amplifying the audio signal outputs from any audio source. The 8-Watt audio amplifier would be a great addition to the Theremin project as well as for use with the voice changer and electronic organ projects in order to amplify the sound output from these devices. You could also use the audio amplifier with a microphone as a public address amplifier.

The 8-Watt Audio Amplifier, shown in Figure 23-2 is a pretty straightforward design. An audio signal is fed to the input of the 8-Watt Audio Amplifier via resistor R1. Capacitor C2 is used to limit input noise. Potentiometer P1 is used to adjust the input signal to the power amplifier circuit. The audio signal is then fed to capacitor C1. The signal leaving C1 is then fed to the audio amplifier IC module, a TDA2030. Resistor R5 sets up the overall gain of the circuit. The output of the audio amplifier module is next fed to capacitor C8,

Figure 23-1 *8-Watt Audio Amplifier*

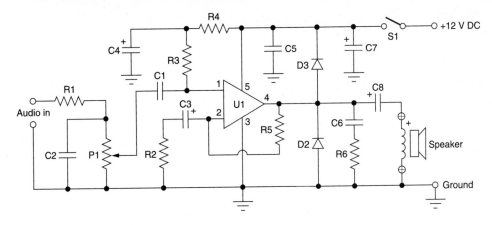

Figure 23-2   *8-Watt Audio Amplifier*

which couples the audio amplifier to the speaker. The resistor and capacitor network formed by R6 and C6 help to prevent the circuit from going into oscillation. Capacitor C7 is used to smooth the power supply voltage to the amplifier circuit. Capacitor C5 is a bypass to ground. The 8-watt power amplifier is powered from a 12-volt DC power source, which could be provided from a 12-volt battery, or eight "C" cells or from a 12-volt DC "wall-wart" power supply.

The 8-watt audio amplifier can be built on a 2¼ × 1¾ inch glass epoxy circuit board in about an hour or hour and a half. Before we begin constructing the circuit, take a few minutes to get organized. First, you will want to secure a large worktable or workbench, in a well-lit room. You want to make sure that you have enough room to spread all the components, diagram and tools in order to build the 8-watt amplifier circuit. Place all the project parts in front of you along with the schematic and pictorial diagram, also locate the

necessary charts and tables. You will need a 27 to 33-watt pencil tip soldering iron, a hank of #22 ga 60/40 rosin core solder, a wet sponge and a small tin of "Tip Tinner" from your local Radio Shack store. You will need a few tools for building the project; try to find a pair of small needle-nose pliers, a pair of small "flush-cut" wire-cutters, a pair of tweezers, a magnifying glass and a few small screwdrivers. Finally, try to locate an anti-static wrist band to help protect the audio amplifier module from static damage when handling it. Place the anti-static wrist strap around the wrist or the hand you will be using to handle the IC module and plug the end of the strap into the grounded outlet. Now, you can turn on your soldering iron and we will begin building the circuit.

First, take a look at the pictorial diagram shown in Figure 23-3. Next, you will want to place all of the project resistors in front of you. Refer to the resistor color chart in Table 23-1, which will help you to

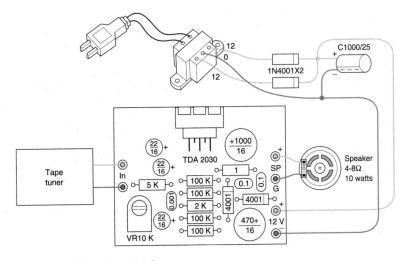

Figure 23-3   *8-Watt Audio Amplifier pictorial diagram*

Table 23-1

Resistor Color Code Chart

Color Band	1st Digit	2nd Digit	Multiplier	Tolerance
Black	0	0	1	
Brown	1	1	10	1%
Red	2	2	100	2%
Orange	3	3	1000 (K)	3%
Yellow	4	4	10000	4%
Green	5	5	100000	
Blue	6	6	1000000 (M)	
Violet	7	7	10000000	
Gray	8	8	100000000	
White	9	9	1000000000	
Gold			0.1	5%
Silver			0.01	10%
No color				20%

identify the resistors. Resistors will generally have three or four color bands printed on the body of the resistor. The first color band will start at one end of the resistor and will represent the first digit of the resistor value. The second color band will represent the second digit of the resistor value, while the third color band will denote the resistor's multiplier value. The fourth band will show the resistor's tolerance value. If there is no fourth band then the resistor will have a 20% tolerance, while a silver color band will indicate a 10% tolerance value. A fourth gold colored band will note a 5% tolerance value for the resistor.

Look through the resistors and try to locate a resistor which has a (green) (black) (red) set of color bands. The green band will represent the digit number (5), a black band will denote a (0) and a third red colored band will indicate a (00) or thousand multiplier, so the resistor value will be 5000 or 5k ohm. Identify the remaining resistors and install them on the circuit board. When the resistors have been installed, you can go ahead and solder them in place. When finished soldering the resistors on to the circuit board, you can flush-cut the excess component leads.

Now locate the adjustable gain control potentiometer P1. Potentiometers will generally have three leads either all in a row or two end leads with an offset center lead which is the adjustable wiper arm. Install the potentiometer on the circuit board and solder it in place. Cut the excess component leads.

Now we will move on to identifying and installing the capacitors. Refer to the capacitor code chart in Table 23-2, which illustrates the three digit code that is often used to identify capacitor values. The 8-Watt Audio Amplifier project has a number of capacitors. Look through the capacitors and you should see a few disk capacitors marked (104), refer to the capacitor code table and you will readily see that a (104) code corresponds to a 0.1 µF capacitor value, this value will be used for capacitors C5 and C6. Now look for a capacitor marked with (102), this capacitor will be a 0.001 µf capacitor located at C2. Solder these capacitors to the circuit board, then remove the excess leads.

The project also contains a number of electrolytic capacitors, which should have their actual value printed on the capacitor body. Electrolytic capacitors, as you will remember, have polarity, which must be observed in order for the circuit to work properly. On the electrolytic capacitor body, you should see either a black or white color band. Next to this color band, you

Table 23-2
Capacitance Codebreaker Information

This table is designed to provide the value of alphanumeric coded ceramic, mylar and mica capacitors in general. They come in many sizes, shapes, values and ratings; many different manufacturers worldwide produce them and not all play by the same rules. Most capacitors actually have the numeric values stamped on them; however, some are color coded and some have alphanumeric codes. The capacitor's first and second significant number IDs are the first and second values, followed by the multiplier number code, followed by the percentage tolerance letter code. Usually the first two digits of the code represent the significant part of the value, while the third digit, called the multiplier, corresponds to the number of zeros to be added to the first two digits.

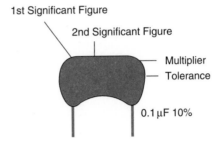

Value	Type	Code	Value	Type	Code
1.5 pF	Ceramic		1000 pF/0.001 µF	Ceramic/Mylar	102
3.3 pF	Ceramic		1500 pF/0.0015 µF	Ceramic/Mylar	152
10 pF	Ceramic		2000 pF/0.002 µF	Ceramic/Mylar	202
15 pF	Ceramic		2200 pF/0.0022 µF	Ceramic/Mylar	222
20 pF	Ceramic		4700 pF/0.0047 µF	Ceramic/Mylar	472
30 pF	Ceramic		5000 pF/0.005 µF	Ceramic/Mylar	502
33 pF	Ceramic		5600 pF/0.0056 µF	Ceramic/Mylar	562
47 pF	Ceramic		6800 pF/0.0068 µF	Ceramic/Mylar	682
56 pF	Ceramic		0.01	Ceramic/Mylar	103
68 pF	Ceramic		0.015	Mylar	
75 pF	Ceramic		0.02	Mylar	203
82 pF	Ceramic		0.022	Mylar	223
91 pF	Ceramic		0.033	Mylar	333
100 pF	Ceramic	101	0.047	Mylar	473
120 pF	Ceramic	121	0.05	Mylar	503
130 pF	Ceramic	131	0.056	Mylar	563
150 pF	Ceramic	151	0.068	Mylar	683
180 pF	Ceramic	181	0.1	Mylar	104
220 pF	Ceramic	221	0.2	Mylar	204
330 pF	Ceramic	331	0.22	Mylar	224
470 pF	Ceramic	471	0.33	Mylar	334
560 pF	Ceramic	561	0.47	Mylar	474
680 pF	Ceramic	681	0.56	Mylar	564
750 pF	Ceramic	751	1	Mylar	105
820 pF	Ceramic	821	2	Mylar	205

(Continued)

Table 23-2 (*Continued*)

PicoFarad (pF)	NanoFarad (nF)	MicroFarad (mF, µF or mfd)	Capacitance Code
1000	1 or 1n	0.001	102
1500	1.5 or 1n5	0.0015	152
2200	2.2 or 2n2	0.0022	222
3300	3.3 or 3n3	0.0033	332
4700	4.7 or 4n7	0.0047	472
6800	6.8 or 6n8	0.0068	682
10000	10 or 10n	0.01	103
15000	15 or 15n	0.015	153
22000	22 or 22n	0.022	223
33000	33 or 33n	0.033	333
47000	47 or 47n	0.047	473
68000	68 or 68n	0.068	683
100000	100 or 100n	0.1	104
150000	150 or 150n	0.15	154
220000	220 or 220n	0.22	224
330000	330 or 330n	0.33	334
470000	470 or 470n	0.47	474

should also see either a plus (+) or minus (−) marking. When installing these capacitors you will want to make sure that the polarity marking correspond to the polarity marking on the schematic or pictorial diagram. For example, on the schematic you will see that electrolytic capacitor C4 has its plus (+) side closest to resistor R4. So the (+) marking on the capacitor C4 should face resistor R4. Identify the remaining electrolytic capacitors and install them on the circuit board. Next solder the electrolytic capacitors to the circuit board, then remove the excess component leads.

The audio power amplifier circuit has two silicon diodes placed across the output of the IC module. Diodes, as you remember, have polarity and it must be observed for the circuit to work properly. At one of the diodes, you should find either a black or white colored band. This color band denotes the diode's cathode lead. Looking at the circuit schematic you will note that a diode's anode is an arrow or triangle pointing toward a line which is the cathode lead. When installing the diodes make sure that you observe the cathode leads carefully, this helps identify how the diode will be

mounted. In this audio amplifier circuit both diodes have their cathodes pointing toward the plus (+) 12-volt power bus. Install the two diodes, solder them in place, and follow up by removing the excess diode leads from the circuit board.

Now refer to the semiconductor pin-out diagram shown in Figure 23-4. Before installing the integrated circuit, it's a good idea to consider using integrated circuit sockets, in case there is ever a circuit failure at a later date. An IC socket is cheap insurance for an easy repair to the circuit. Very few people can unsolder integrated circuits from a circuit board without damaging the board. Locate the TDA2030 audio amplifier module and you should notice that the IC will have either a small indented circle, a notch or a cutout at one edge of the IC package. Just to the left of the notch or cutout will be pin 1. With your anti-static strap on, place the IC into the socket, making sure that you have aligned pin 1 of the IC with pin 1 of the socket. Look for reference on the schematic, note that pin 3 of U1 is connected to ground while pin 5 of the IC is connected to the power source; this should help orient

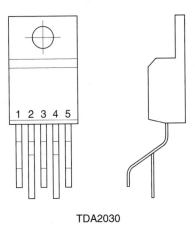

TDA2030

Figure 23-4 *Audio Amplifier IC*

the IC with the socket. If you are confused or can't seem to figure out which pin is which, then have a knowledgeable friend help you.

The 8-ohm speaker was mounted off-board, so you will have to prepare two 6 to 8 inch lengths of #22 ga stranded-insulated hookup wire to connect the speaker to the power amplifier circuit board. Solder each wire to the two speaker terminals, then solder the free ends of the two wires to the circuit board, where it's labeled speaker.

The last thing you will need to do is connect the 12-volt battery to the 8-watt power amplifier circuit board. The minus (−), negative or black battery clip will be connected to the ground bus of the circuit, which is pin 3 of U1, also to D2, P1 and the speaker, etc. The plus (+) or red battery wire will be placed in series with the optional power switch at S1. Your 8-watt power amplifier circuit is now ready to serve.

Now is an excellent time to take a short break, and when we return we will inspect the circuit for any possible "short" circuits or "cold" solder joints. Turn the circuit board over so that the foil side faces upwards toward you. Now, look over the foil side of the circuit board at all of the solder joints to see if they all look

clean, smooth and shiny. While inspecting the circuit board if you see any of the solder joints looking dull or "blobby", then unsolder, clean the solder pad and resolder the joint so it will look clean, smooth and shiny like the others. With that finished, look the circuit board over again; this time we will inspect the circuit board for possible "short" circuits. Often when removing excess component leads, they will sometimes stick to the circuit board from the rosin material in the solder. Make sure that there are no component leads between the circuit pads or PC lands. Once your inspection is complete, you can connect up you signal source to the amplifier. Connect your 12-volt power source and adjust the gain control at P1 on your amplifier. If all is well, you should begin hearing audio from the amplifier's speaker.

In the event that the circuit does not work properly, you will have to disconnect the battery and signal source and reinspect the circuit board for any possible errors. First you will want to make sure that you have placed the correct resistor in the right location on the circuit board. This is a common mistake and it could occur easily. When you have inspected the resistors and their placements, then you can move on to inspecting the capacitors to make certain that you placed the electrolytic capacitor on the board correctly by observing the polarity. Re-check the color band on the capacitors once again to make sure they are in the right location and mounted correctly. Check the diodes to make sure you have inserted them with the cathodes facing the plus side of the circuit.

Finally reconnect the audio source, the battery and speaker. Adjust the gain control on the source and on the circuit board and you should begin hearing the source material. Your 8-watt amplifier is now ready to perform for you. This amplifier can serve many purposes around your home as well as with projects listed in this book. Have fun and enjoy!

Chapter 24

Infrared Target Game

Parts List

Parts Bin

R1 10 k ohm, 5%,
¼-watt resistor

R2 680 k ohm, 5%,
¼-watt resistor

R3 100 ohm, 5%, ¼-watt
resistor

P1 2 k trimmer
potentiometer

C1 0.1 µF, 35-volt
mylar capacitor

C2, C4 0.01 µF, 35-volt
disk capacitor

C3, C5 10 µF, 35-volt
electrolytic capacitor

C6 27 pF, 35-volt disk
capacitor

C7 47 µF, 35-volt
electrolytic capacitor

D1 1N4148 silicon diode

U1 LM386 IC

IR infrared sensor—
Digi-Key PNZ323B-ND

L1, L2, L3, L4 red
LEDs

Q1 2N3904 transistor or
NTE-123AP

Q2 2N3906 transistor or
NTE159

Q3 TIP42 transistor PNP
power type or NTE197

T1 mini transformer 8
ohm to 100 k ohm

S1 SPST toggle or slide
switch (optional)

B1 9-volt transistor
radio battery

Misc PC board, wire, IC
socket, battery clip,
etc.

The Infrared Target Game illustrated in Figures 24-1 and 24-2, will permit you many hours of great fun for you and/or your children. This game is the same electronic circuit used in many commercial kids' target games, where you point the light gun at your opponent and when you hit their shield the lights all light up and you know you have been "hit." In fact this project will use any of the commercial toy light guns. If you do not have a light gun you can build the companion IR light transmitter.

The Infrared Target Game receiver circuit diagram shown in Figure 24-3 flashes a red diamond of LEDs every time its infrared sensor is "hit" by an infrared beam. An infrared beam consists of a track of light pulses, which causes the IR sensor to produce a train of electric pulses. These electric pulses are amplified, rectified and then used as a drive signal to light up the four high intensity LEDs. Amplification is accomplished by a one-transistor stage consisting of Q1, and integrated circuit at U1and associated components. The signal is then stepped up by a transformer T1, then

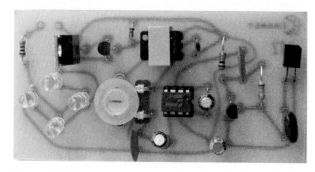

Figure 24-1 *IR Target Game receiver*

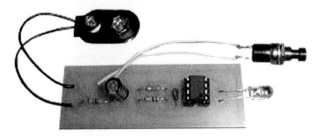

Figure 24-2   *IR Target game transmitter*

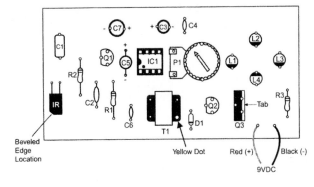

Figure 24-4   *IR Target receiver pictorial diagram*

rectified by a diode, filtered by it and used as a DC current to turn transistor Q2 and Q3 "on" thus providing current to display LEDs 1 through 4. Capacitor C2 is used to prevent "motorboating," which is a low frequency oscillation. Potentiometer P1 is used to set the intensity level and also prevent accidental triggering by infrared light from the nearby environment. The IR Target receiver circuit is built on a small circuit board and can be hid behind a cardboard "shield" or the circuit could be hung on a robe around your neck if desired. The IR Target Game receiver is powered by a conventional 9-volt transistor radio battery.

The IR Target Game receiver circuit was built on a 4 inch × 2 inch printed circuit board. Before we get started building the IR Target Game, let's take a few moments to secure a well-lit table or workbench, so we can spread out all the components, diagrams, charts and tools we will need for this project. First, you will need to locate some small tools, a pair of needle-nose pliers, an end-cutter or small diagonals, a pair of tweezers, and a magnifying glass. Next, you will want to locate a low wattage 27 to 33-watt soldering iron, some 60/40 rosin core solder, a wet sponge and a small jar of "Tip Tinner" from you local Radio Shack store. "Tip Tinner"

is used to clean and prepare your solder iron tip between solder joints. After about four or five solder joints, you will want to place the tip into the "Tip Tinner" and rotate the iron in the jar. Use the sponge when needed to keep the tip clean. You will also want to locate an anti-static wrist strap, which is used to prevent static damage to the integrated circuits when handling the integrated circuits. Often high voltage static charges can build up when moving around the worktable or getting in and out of your chair. Ground the wrist strap and use it when inserting the IC into the circuit or IC sockets.

The components for the receiver circuit board were laid out as shown in the pictorial diagram in Figure 24-4. The IR sensor was mounted at one end of the circuit board while the four display LEDs were mounted at the opposite end of the PC board. An IC socket was used to mount the integrated circuit. In the event of a circuit failure at some later date, it is much easier to repair the circuit if IC sockets are used. Most people cannot unsolder IC pins without destroying the printed circuit.

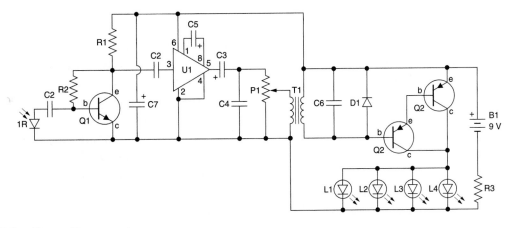

Figure 24-3   *Target Game receiver*

Table 24-1

Resistor Color Code Chart

Color Band	1st Digit	2nd Digit	Multiplier	Tolerance
Black	0	0	1	
Brown	1	1	10	1%
Red	2	2	100	2%
Orange	3	3	1000 (K)	3%
Yellow	4	4	10000	4%
Green	5	5	100000	
Blue	6	6	1000000 (M)	
Violet	7	7	10000000	
Gray	8	8	100000000	
White	9	9	1000000000	
Gold			0.1	5%
Silver			0.01	10%
No color				20%

When populating the circuit board or mounting the components, you will need to pay particular attention to considerations of polarity of the components.

Let's actually begin building the circuit. First you will need to locate all the resistors for the IR Target Game. Refer to Table 24-1, which illustrates the resistor color codes. Grab the resistors and look at them, now look at the color code chart. Note that each resistor will have three or four color bands. The bands start nearer to one edge. The first color band is the first digit, while the second color band is the second digit, and the third band is the multiplier value. Often there is also a fourth band which is the resistor tolerance value. If there is no band then the resistor has a 20% tolerance. If a resistor has a silver band then it will have a 10% tolerance value, and if the resistor has a gold band then the tolerance value is 5%. So, let's try to locate resistor R1, which has a 10 k ohm value. You will be looking for a resistor with the first color band having a brown band. The second color band should be black and the third band should be orange. Brown represents a first digit value of 1, a zero value for the second digit is represented as a black band, a third band with a 1000 multiplier would be orange. After you locate the 10 k ohm resistor, you can bend the leads and mount it on

the PC board. Next locate R2 and R3 and mount those resistors, and then solder all the resistors to the circuit board. Remember to cut the excess resistor leads from the circuit board using your end-cutter. Cut the excess leads flush to the edge of the circuit board.

Now locate the capacitors; there are five capacitors in the circuit. Note that capacitors C3 and C5 are electrolytic types, so you must observe the polarity markings when installing these components. You will see the component's value marked on the body of the capacitor and you should also see a white or black band at either end of the capacitor. Usually the band will have either a (–) minus or a (+) plus marked in or near the band. Refer to the schematic and pictorial diagram when installing the capacitors to ensure that you install the capacitors correctly by paying attention to the orientation of the polarity, with respect to the plus or ground bus, and to ensure that you do not damage the circuit when power is applied. There are also a few disk capacitors in this circuit, which may or may not be marked with their actual value. Often for small capacitors, their actual value may be too large to print, so the manufacturers will use a code. Refer to the chart used in Table 24-2, which illustrates this three-digit code. Take a look at the capacitors in your parts pile,

Table 24-2

Capacitance Codebreaker Information

This table is designed to provide the value of alphanumeric coded ceramic, mylar and mica capacitors in general. They come in many sizes, shapes, values and ratings; many different manufacturers worldwide produce them and not all play by the same rules. Most capacitors actually have the numeric values stamped on them; however, some are color coded and some have alphanumeric codes. The capacitor's first and second significant number IDs are the first and second values, followed by the multiplier number code, followed by the percentage tolerance letter code. Usually the first two digits of the code represent the significant part of the value, while the third digit, called the multiplier, corresponds to the number of zeros to be added to the first two digits.

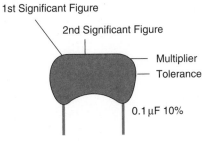

Value	Type	Code	Value	Type	Code
1.5 pF	Ceramic		1000 pF/0.001 μF	Ceramic/Mylar	102
3.3 pF	Ceramic		1500 pF/0.0015 μF	Ceramic/Mylar	152
10 pF	Ceramic		2000 pF/0.002 μF	Ceramic/Mylar	202
15 pF	Ceramic		2200 pF/0.0022 μF	Ceramic/Mylar	222
20 pF	Ceramic		4700 pF/0.0047 μF	Ceramic/Mylar	472
30 pF	Ceramic		5000 pF/0.005 μF	Ceramic/Mylar	502
33 pF	Ceramic		5600 pF/0.0056 μF	Ceramic/Mylar	562
47 pF	Ceramic		6800 pF/0.0068 μF	Ceramic/Mylar	682
56 pF	Ceramic		0.01	Ceramic/Mylar	103
68 pF	Ceramic		0.015	Mylar	
75 pF	Ceramic		0.02	Mylar	203
82 pF	Ceramic		0.022	Mylar	223
91 pF	Ceramic		0.033	Mylar	333
100 pF	Ceramic	101	0.047	Mylar	473
120 pF	Ceramic	121	0.05	Mylar	503
130 pF	Ceramic	131	0.056	Mylar	563
150 pF	Ceramic	151	0.068	Mylar	683
180 pF	Ceramic	181	0.1	Mylar	104
220 pF	Ceramic	221	0.2	Mylar	204
330 pF	Ceramic	331	0.22	Mylar	224
470 pF	Ceramic	471	0.33	Mylar	334
560 pF	Ceramic	561	0.47	Mylar	474
680 pF	Ceramic	681	0.56	Mylar	564
750 pF	Ceramic	751	1	Mylar	105
820 pF	Ceramic	821	2	Mylar	205

(Continued)

Table 24-2 (*Continued*)

PicoFarad (pF)	NanoFarad (nF)	MicroFarad (mF, µF or mfd)	Capacitance Code
1000	1 or 1n	0.001	102
1500	1.5 or 1n5	0.0015	152
2200	2.2 or 2n2	0.0022	222
3300	3.3 or 3n3	0.0033	332
4700	4.7 or 4n7	0.0047	472
6800	6.8 or 6n8	0.0068	682
10000	10 or 10n	0.01	103
15000	15 or 15n	0.015	153
22000	22 or 22n	0.022	223
33000	33 or 33n	0.033	333
47000	47 or 47n	0.047	473
68000	68 or 68n	0.068	683
100000	100 or 100n	0.1	104
150000	150 or 150n	0.15	154
220000	220 or 220n	0.22	224
330000	330 or 330n	0.33	334
470000	470 or 470n	0.47	474

and you should see one marked with a (104); this will represent a capacitor with a value of 0.1 µF, which is capacitor C1. Capacitor C1 does not have any polarity, so it can be installed in either direction. Go ahead and install and solder the capacitors on the circuit board. Next follow up by cutting the excess leads from the PC board.

The IR Target Game receiver has one silicon diode which is connected across capacitor C6. Diodes have polarity markings and they must be observed in order to correctly install them. One end of the diode will usually have a black band and this denotes the diode's cathode or minus (–) lead. The anode of the diode is usually denoted as an arrow or triangle pointing to the cathode, which is shown as a line. Refer to the schematic diagram and pictorial diagrams to ensure that you can identify that the cathode is connected to one end of T1 and the Emitter of Q3. Install the diode and solder it in place, then remove the excess component leads.

The IR detector, marked (IR) was mounted at one end of the IR receiver circuit board.

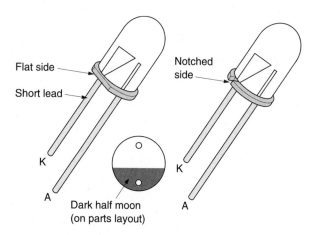

Figure 24-5   *LED identification*

Now you should try to locate the four red LEDs. Refer to the diagram at Figure 24-5, which illustrates the mounting configuration of the LEDs. The LEDs will usually have a flat edge at one side of the LED package. The flat edge usually corresponds to the shorter lead, which is the cathode or minus (–) lead. Note that all the

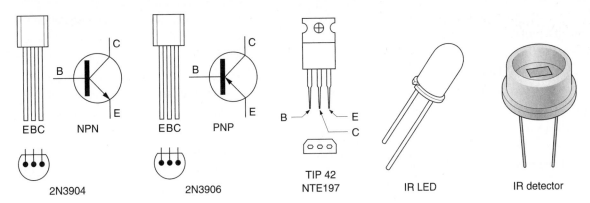

Figure 24-6 *IR Target semiconductors*

LEDs have their cathode leads connected together through R3 which is set to the minus (–) battery bus lead.

Locate the trimmer potentiometer and observe the pin-outs on the bottom of the device. There will be three leads. The center lead is usually the wiper arm of the device and in this circuit the wiper arm connects to T1 as shown. Sometimes all three leads will be in-line with the wiper in the center, or other potentiometers may have the center lead offset or staggered.

Locate the three transistors and place them in front of you. Look at the diagram shown in Figure 24-6, which shows the pin-outs and package layouts for the transistors. Q1 and Q2 are a complementary pair, i.e. an NPN and a PNP transistor. Both of these device packages will appear the same. Each of these transistors will have a Base, Collector and an Emitter lead, as shown in the diagram. Q1 and Q2 are TO92 packages. As you look at the top view in the diagram you will note that there is a flat edge on the transistor. At the top of the layout diagram you will see the Emitter pin with the Base pin in the center and the Collector lead at the other end. Make sure you pay particular attention to orientation when placing the transistor into the circuit board. It's very easy to make a mistake on visual orientation. Transistor Q3 is a power transistor which uses a large package called a TO220 style. Some power transistors also have TO5 cases. Refer to the diagram when installing Q3. Note that if the transistor package looks like a TO220 style then the Base lead will be at one end while the Emitter is at the other. If the package style is a TO5 style then the center or offset pin is the Base lead. The Emitter lead will be below the tab on the metal case. Be very careful when installing the

transistors to ensure that you have placed the correct lead into the correct PC board hole. Refer back and forth between looking at the actual component and the schematic and or pictorial diagram for accurate results. Transistor reference diagrams usually show the leads from the bottom of the device.

Locate the integrated circuit and observe that it will have either a small indented circle or a cutout notch at one end of the IC package. Note that pin 1 of the IC will be just to the left of the indented circle or the cutout. Refer to the schematic and the pictorial diagram and identify and make note of where pin 1 of the IC socket is, so you can mate pin 1 of the IC to pin 1 of the IC socket. Note that pin 1 of the IC is connected to the minus (–) side of capacitor C5, which is different from most integrated circuits. Be sure to carefully check your work over again to avoid mistakes that can destroy the circuit when power is applied. If you have difficulties identifying the transistor or integrated circuit pin-out, then ask a knowledgeable electronics enthusiast for help.

Your IR Target receiver is almost completed. Locate the 9-volt battery clip and solder it to the circuit board. The red battery clip lead is soldered to the plus (+) voltage bus on the circuit board. Refer to the schematic and observe that the plus (+) bus is connected to transistor Q3 diode D1 C6 and transformer T1. The black or minus (–) battery clip lead is connected to the free end of resistor R3. Remember R3 connects to the four LED cathodes. Note that you can connect an "on/off" switch in series with the plus (+) or red battery clip lead.

The IR Target Game receiver is now ready. But before you apply power, you should take a quick break and then inspect the circuit for both "cold" solder joints and "stray" component leads which may have stuck to the circuit board during construction. First look the PC board over and check for "cold" solder joints. All the solder joints should be clean, smooth and shiny. If any of the solder joints do not look clean, smooth and shiny then you will need to unsolder and clean the solder joint, perhaps first using a solder-sucker to remove the original solder. Once the old solder is removed then just go ahead and re-solder the solder joint, so that it looks clean, smooth and shiny. Inspect all the solder joints and re-solder them as needed. Now re-inspect the circuit board and look for any "stray" component leads which may have "stuck" to the circuit board foil. Any "stray" leads may "short" out the circuit when power is first applied to the circuit.

Now you are ready to apply power to the circuit and test it out. Locate a 9-volt battery and connect it to the battery clip on the IR receiver unit.

In a semi-dark room adjust the trimmer potentiometer on the IR receiver counter-clockwise until the LEDs just light up; then turn the trimmer clockwise until the LEDs just go out. Now activate the transmitter with the pushbutton and the target receiver should respond and the four LEDs on the IR Target Game should all light up. If the diamond pattern of LEDs flashes, then you will know that the receiver is working OK. Now you can mount the receiver in a small project enclosure behind a cardboard, plastic or metal shield, so the LEDs can be seen, and so that the detector can "look" out for a signal from the transmitter unit. Line-up the IR transmitter with the IR receiver, so that the LED in the transmitter is directly pointed to the IR receiver's IR detector (IR).

In the event that the IR receiver does not appear to work, then you will have to remove the battery and re-check the circuit for any possible errors. Errors do happen! If one of the LEDs does not light-up, then it is a simple matter to unsolder the joint, remove it from the joint and turn the LED around and re-solder it back in place. If the circuit does not light any of the LEDs then you have a more serious problem. Next, you will have to check to make sure that each resistor is in its proper location, check the color codes carefully or use an ohmmeter to check its value. Note that you may have to "lift" one resistor lead out of the circuit to test it. Now, check the electrolytic capacitors to be sure that they have been installed correctly, paying attention to

the polarity markings. Re-check the installation of the silicon diode, carefully observe the polarity. Next, check the orientation of the transistors. Refer to the schematic, pictorial diagram and the parts pin-out diagram to make certain you have installed the transistors correctly. Finally make sure that the IC is installed right. Power is applied to pin 6 and ground is at pin 2. One final note: if the detector was installed backwards the circuit will never detect IR light from the transmitter, so as a last resort make sure that it is installed correctly.

Once you are certain that you have corrected the problem, re-connect the battery clip to the battery and aim the receiver at the transmitter to see if the system works.

Companion Infrared Transmitter

Parts Bin

R1, R3 4.7k ohm, 5%, ¼-watt resistor

R2 100k ohm, 5%, ¼-watt resistor

C1 100µF, 35-volt electrolytic capacitor

C2 0.01µf, 35-volt disk capacitor

L1 infrared LED— Digi-Key-LN64PA-ND

U1 LM555 Timer IC

S1 momentary pushbutton switch (Normally open type)

B1 9-volt transistor radio battery

Misc PC circuit board, IC socket, wire, battery clip, enclosure, etc.

The companion IR transmitter shown in Figure 24-2 is matched to the Infrared Target Game receiver. The IR transmitter circuit depicted in schematic diagram

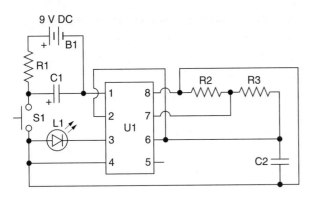

Figure 24-7   *IR Target Game. Courtesy Chaney (CH)*

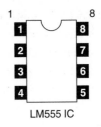

LM555 IC

Figure 24-8   *IR Transmitter IC diagram*

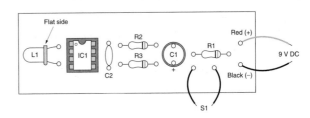

Figure 24-9   *Target Transmitter pictorial diagram*

Figure 24-7 is built around an LM555 timer IC. The LM555 at U1 is used to create a stream of pulses which are sent to the IR LED. The integrated circuit at U1 is configured as an astable multivibrator or oscillator that has a frequency determined by resistor R2, R2 and capacitor C2. The battery charges capacitor C1 through resistor R1. When pushbutton S1 is depressed, the DC or direct current charge on C1 is applied to the (+) input leads of U1 causing it to pulse the infrared LED for about one second. The pulse is actually a series of light pulses occurring in the one second time interval. The frequency of the pulses is determined by components R2, R3 and capacitor C2. The frequency is not determined by C1 and S1; these components are used just to limit the time or burst of pulses produced. The output of U1 from pin 3 is applied directly to the special infrared LED.

The IR transmitter circuit is pretty straightforward to construct and can be built in about an hour or so. The original IR transmitter was constructed on a small printed circuit board measuring 2¾ inch × 1 inch and can be housed in a small plastic enclosure. Refer to the layout diagram for component locations. The first thing you will want to do is find an eight pin integrated circuit socket for the LM555 timer chip. Use of an IC socket will greatly help you in the event of a circuit failure which could occur at a later date. Unsoldering integrated circuits are very difficult to do without damaging the PC board. First locate the three resistors, then refer to Table 24-1 which depicts the resistor color codes as mentioned in the IR Target receiver text above. Bend the resistor leads to fit the PC board and then install the resistors on the PC board, solder the resistors to the circuit board and then remove the excess component leads. Next, locate the capacitors; C1 is an electrolytic capacitor and it will have polarity marking

which will designate either a plus (+) or minus (−) marking at one edge of the capacitor body. It is important that you observe the polarity marking and install the capacitor with respect to its polarity to avoid damage to the circuit when power is first applied. Capacitor C2 is a disk capacitor and it will not have any polarity marking but may be marked with its value or the designation (103). Refer to the capacitor identification chart shown in Table 24-2. Install the capacitors, solder them to the circuit board and then cut the excess capacitor leads flush to the edge of the circuit board.

Now, install the integrated circuit, refer to IC pin-out diagram in Figure 24-8 and the IR transmitter layout diagram depicted in Figure 24-9. Most ICs have some sort of orientation markings on them to help install them. Look for a small indented circle, a small notch or cutout at one end of the plastic IC package. Then look just to left of the notch or cutout and you will find pin 1. Make sure that you line-up pin 1 of the IC with pin 1 of the socket before pushing the IC into the socket. Note that pin 1 of the IC is connected to the minus (−) side of capacitor C1, and the minus (−) terminal of the battery. Be sure to use your anti-static wrist strap when handling the integrated circuit. Pin 8 of the LM555 timer chip at U1 is connected to R2 and the junction of S1, C2 and LED L1. Using your anti-static wrist band bend all of the IC leads at one time against the table and insert the IC into the socket.

Next, locate the red LED. You will need to identify the flat edge on the LED; this lead will be the cathode lead (see LED pin-out diagram in Figure 24-5). The cathode or minus (–) lead of the LED may also be identified by the shorter LED lead. Prepare a 6 inch length of #22 ga stranded-insulated hookup wire to connect the pushbutton switch to the circuit board. Solder one end of each wire to the terminals on the pushbutton. Now, solder the two free ends of the wires to the circuit board. One switch lead will connect to pin 4 of the U1, while the other pushbutton lead will connect to the junction of R1 and C1. Finally locate the 9-volt battery clip and solder it to the circuit board. The red or plus (+) lead from the battery clip will be connected to the free end of resistor R1. The black or minus (–) battery lead will connect to pin 1 of the integrated circuit. You could elect to use an "on/off" switch to power the circuit if desired. The power switch can be wired in series with the red or positive battery clip lead.

Take a short break and then we will proceed to inspect the IR transmitter PC board for any "short" circuits and "cold" solder joints. Turn over the circuit board, so the copper foil side is facing up toward you, and so that you can inspect the board. Look for all the solder joints to be smooth, clean and shiny. If all the solder joints look good then you can move on. If any of the solder joints look dull or "blobby" then you will want to de-solder the connection and re-do the solder connection until it looks clean, smooth and shiny; refer to the earlier chapter on soldering if necessary. Next you will re-inspect the circuit board for any "shorts" caused by "stray" component leads which may have stuck to the circuit board after cutting the component leads. Often "cut" component leads may adhere to the PC board and if these bare wires are left on they may cause the board to "short-out" when power is first applied to the circuit. Now that the inspection process has been completed, you can move on to testing the IR transmitter board.

Connect a fresh 9-volt battery to the battery clip on the IR transmitter, and install a 9-volt battery in the IR target receiver, then aim the IR transmitter toward the IR receiver in a darkened room. Press the pushbutton on the IR transmitter, carefully adjust the target sensitivity control on the IR target receiver to detect the IR

transmitter. Make sure that you carefully aim the transmitter in-line with the IR target receiver. Position the IR transmitter about three feet or so from the receiver for the initial test. Once set up properly you should be able to detect the IR transmitter up to 50 feet away. If you desire longer range then consider placing a magnifying lens about 6 inches in front of the IR LED in the IR transmitter.

If everything checks out OK, then you can mount the IR transmitter circuit in a phasor gun, pistol or rifle housing of your choice. Pushbutton S1 is your trigger which causes short bursts of IR energy to be targeted toward you opponent. Note, you must release the pushbutton control quickly in order to charge capacitor C1 before the next "shot" can be "fired" again. Also note that at a separation distance of less than two feet between the IR transmitter and IR receiver the transmitter may emit enough IR energy to continuously trigger the target. To prevent battery drain when not in use, be sure to use either an on-off power switch to remove the battery from the circuit or disconnect the battery clip from the battery.

If your IR transmitter does not appear to work correctly, you will have to remove the battery and troubleshoot the circuit for any possible errors. First check to make sure the resistors have the right color code or value for their respective locations. Also make sure that the capacitor at C1 is installed correctly. Now, you must make certain that you installed the LED correctly. If the LED is not wired correctly, it will never send a signal to the IR receiver. Finally, make sure that the LM555 timer is installed correctly, with pin 1 connected to the minus (–) side of the battery and that pin 8 is connected to S1, C2 and L1, which has its path to the plus side of the battery.

Your IR receiver unit and the IR Target Game system is now ready for use and some really fun hunting or target practice, or simulated "paint ball" type fun!

This game will provide many hours of fun for gamers of all ages. You can build a few for you friends, or you and your friends can have a building competition to see how nicely you can customize your IR guns or IR target receivers!

Chapter 25

Electronic Hourglass

Parts List

Parts Bin

R1, R3, R10, R17   10 k ohm, 5%, ¼-watt resistor

R2   1 k ohm, 5%, ¼-watt resistor

R4, R5, R6   5 k ohm, 5%, ¼-watt resistor

R7, R8, R9   5 k ohm, 5%, ¼-watt resistor

R11, R13, R18, R21   330 ohm, 5%, ¼-watt resistor

R12, R14, R19, R20   300 ohm, 5%, ¼-watt resistor

R15, R16,   820 ohm, 5%, ¼-watt resistor

P1   100 k ohm potentiometer

C1   47 µF, 35-volt electrolytic capacitor

C2   10 µF, 35-volt electrolytic capacitor

C3   0.01 µF, 35-volt disk capacitor (103)

D1   1N4004 silicon diode

D2, D3, D4, D5, D6   1N4148 silicon diode

D7, D8, D9, D10   1N4148 silicon diode

D11, D12, D13   IN4148 silicon diode

D14, D15, D16   1N4148 silicon diode

Q1, Q2, Q3, Q4   C458 NPN transistor—NTE85 replacement

Q5, Q6, Q7   C458 NPN transistor—NTE85 replacement

U1   LM555 timer IC (National Semiconductor)

U2   CD4017 decade counter IC (National Semiconductor)

L1, L2, L3, L4, L5   red LEDs

L6, L7, L8, L9, L10, L11   red LEDs

L12, L13, L14, L15, L16   red LEDs

L17, L18, L19, L20, L21, L22   red LEDs

S1   momentary pushbutton switch (normally open)

S2   optional "on/off" switch

B1   9-volt transistor radio battery

Misc   PC board, IC sockets, battery clip, wire, hardware, case, etc.

The Electronic Hourglass is a fun, yet useful project sure to please the young as well as the old, and it's fun to watch. The Electronic Hourglass is highlighted in Figure 25-1; it is modeled after the old-fashioned sand

Figure 25-1 *Electronic Hourglass*

hourglass which is a memory of the past. The Electronic Hourglass can be used for any sort of game, from board games to charades. The project features adjustable speed, bright LEDs and safe 9-volt operation.

The Electronic Hourglass circuit, shown in the schematic at Figure 25-2, illustrates the circuit which can be divided into two main parts or sections. The first part of the circuit is the frequency oscillator formed around the integrated circuit at U1. Integrated circuit U1 is an LM555 timer IC. The support components for the timer circuit comprise R1, R2, C2, C3 and potentiometer P1, which controls the speed of the display circuit. The output of the timer IC at U1 is used to "clock" the decade counter in part two of the Electronic Hourglass circuit. The decade counter at U2 has ten outputs but only six

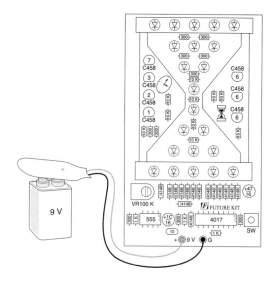

Figure 25-3 *Hourglass pictorial diagram*

outputs are used in this project. The output of the CD4017 decade counter is connected to the driver transistors Q1 through Q7. The decade counter will automatically reset itself at the end of the count, but if you jump J1, then the reset switch is used to reset the counting circuit. Pin 1 of U2 is used to reset the timer circuit through the RESET switch at S1. The Electronic Hourglass display consists of 22 LEDs, 11 of the LEDs are used in the top part of the hourglass and 11 more LEDs are used in the bottom of the hourglass, as depicted in the pictorial diagram Figure 25-3.

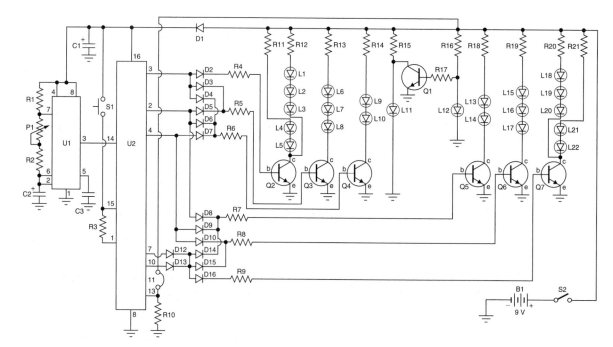

Figure 25-2 *Electronic Hourglass*

Table 25-1

Resistor Color Code Chart

Color Band	1st Digit	2nd Digit	Multiplier	Tolerance
Black	0	0	1	
Brown	1	1	10	1%
Red	2	2	100	2%
Orange	3	3	1000 (K)	3%
Yellow	4	4	10000	4%
Green	5	5	100000	
Blue	6	6	1000000 (M)	
Violet	7	7	10000000	
Gray	8	8	100000000	
White	9	9	1000000000	
Gold			0.1	5%
Silver			0.01	10%
No color				20%

Before we begin constructing the Electronic Hourglass, let's take a few minutes to secure a well-lit table or workbench for building the circuit. You will want to locate a 27 to 33-watt pencil type soldering iron with either a sharp pointed tip or a small flat wedge type tip. You will need to locate a small wet sponge to clean the soldering tip as well as some #22 ga 60/40 rosin core solder. Also try to locate some "Tip Tinner" from your local Radio Shack store. "Tip Tinner" is used to clean and dress the tip of a soldering iron.

As we will be working with integrated circuits, it would be a wise move to locate an anti-static wrist band to prevent damage to the integrated circuits when handling them. Moving across a carpet or getting up and down from your chair will create static which could destroy the integrated circuits, so the grounded anti-static wrist strap is a must. Now find a small wire-cutter, a needle-nose pliers, a magnifying glass, and a pair of tweezers.

Locate the schematic and the layout diagrams as well as all of the charts or tables and we will begin constructing the project. Heat up your soldering iron and we will get ready. Place the project parts in front of you and begin looking for the project resistors. When you find all the resistors, you can refer to the resistor colors found in Table 25-1. Each resistor will have three

or four color bands on the body of the resistor. The colors will start from one end of the resistor body. This first color code will represent the first digit of the code, while the second color code represents the second digit of the code. The third color code denotes the resistor's multiplier digit and if the resistor has a fourth color band then that color is noted as the tolerance of the resistor. If the fourth color band is silver then the resistor has a 20% tolerance and if the fourth band is gold then the resistor has a 5% tolerance value. Look through the resistors and see if you can find a resistor with color bands as follows; a (brown) (black) (orange) band. This resistor will have a 10 k ohm value, and is resistor R1 between pins 4 and 7 of U1 (see Table 25-1). Identify all of the resistors using the color code chart. Once you have identified all the resistors, you can go ahead and install them in their respective locations on the circuit board. When the resistors have been placed on the circuit board, you can solder them in place. Repeat the process for all of the resistors. When they are all soldered on the board, then you can use your wire-cutter to cut the excess component leads flush to the edge of the board.

Now we will move on to locating and identifying the capacitors in the Electronic Hourglass project. This project has only three capacitors and two of them are electrolytic types. Electrolytic capacitors are polarity sensitive and must be installed based on their polarity marking in order for the circuit to work properly. Look on the body of the electrolytic capacitor and you will find a black or white colored band on the side of the capacitor body. On or near the colored band, you will see a plus (+) or minus (–) marking to denote the component's polarity. When installing these capacitors pay close attention to the polarity markings and refer to the schematic, for example, capacitor C1 has its plus (+) marking facing the 9-volt power source. Often other types of capacitors will not have their values on the body of the capacitor as do the electrolytic capacitors. Sometimes capacitors will have a three-digit code printed instead. The capacitor at C3, a disk capacitor, will have a three-digit code printed on one side of the capacitor body. Refer to the chart in Table 25-2, which lists the most common code values found on capacitors. Capacitor C3 should have (103) marked on the body of the capacitor, which says its value is 0.01 μF. Go ahead and install the capacitors on the circuit board and solder them in place. Remove the excess component leads after soldering in the capacitors.

Now identify the 100 k trimmer potentiometer located at P1, which is in series with R2 across pins 6 and 7 of U1. Potentiometers have three leads which are usually in-line all in a row, with the adjustable wiper lead in the center. Sometimes the center lead is offset from the end leads. Mount the potentiometer on the circuit board and solder it in place. Potentiometer P1 is used to adjust the speed of the timing of the hourglass circuit. If you wish to make the timing longer than the control allows, then you may have to increase the resistance of R2 or increase the value of capacitor C2.

The Electronic Hourglass has 15 silicon diodes at the outputs of the decade counter. Diodes are polarity sensitive and must be mounted in a certain way to prevent circuit damage. Diodes have an anode and a cathode. Schematically the anode of the diode is shown as a triangle or arrow which points to the cathode, which is shown as a line. Diodes will have polarity markings on them in the form of a colored band, which will be either white or black. The cathode lead is the minus (–) lead associated with the colored band. Take a look at the schematic and you will see that diode D1 will have its cathode facing toward pin 16 on U2, so the colored band will face or point to pin 16 on U2. Go ahead and install the diodes on the circuit board, and

solder them in place. Remember to cut the excess component leads from the diodes.

Next, locate all of the transistors for the project and place them in front of you. Note that all the transistors are NPN types. Transistors will generally have three leads extending from the bottom of the transistor case or package. Refer to the diagram shown in Figure 25-4, which illustrates the pin-outs of the transistors and integrated circuits. The transistors will be of the TO-92 package or case style, which will have the Emitter lead at one end, a Base lead in the center and a Collector lead opposite the Emitter at the other end of the transistor package. The reference for the transistor is generally shown from the bottom of the plastic transistor package. Make sure that you can identify all of the leads, then refer to the schematic to see where each of the leads will go. Transistor Q1 will have its Collector pointing or connected to the ground bus. The Base lead of Q1 connects to resistor R17, while the Collector of Q1 is connected to resistor R15. If you have problems identifying the transistor leads or deciding where a particular lead is placed, check with a knowledgeable electronics friend to help you. You do not want to install the transistors incorrectly, as installing incorrectly can damage the circuit. Once you have identified all of the transistors and where they should be placed, you can go ahead and install them on the circuit board. Solder all the transistors to the circuit board and, when finished, cut the excess component leads.

Next, we will locate and install the integrated circuits. The first thing you should do is install integrated circuit sockets, they are a cheap insurance against a circuit failure at a later date. Sometimes integrated circuits will go bad, and if one does it is much easier to unplug the IC and install another one rather than trying to unsolder an IC from the circuit board. Most people cannot unsolder an integrated circuit without damaging the circuit board, so it is much easier to just install the sockets when building a circuit. Integrated circuits have to be installed in a certain way in order to make the circuit work properly and to not destroy the circuit. Each IC will have either a small indented circle, or a small notch or cutout at one end of the IC package. Pin 1 of the IC will be just to the left of the circle or notch. After you identify the two integrated circuits, you will have to match up pin 1 of the IC with pin 1 on the socket. Look at the IC pin-out diagram in Figure 25-4, from that you will see that pin 1 of the IC will be at the top left side of

Table 25-2

Capacitance Codebreaker Information

This table is designed to provide the value of alphanumeric coded ceramic, mylar and mica capacitors in general. They come in many sizes, shapes, values and ratings; many different manufacturers worldwide produce them and not all play by the same rules. Most capacitors actually have the numeric values stamped on them; however, some are color coded and some have alphanumeric codes. The capacitor's first and second significant number IDs are the first and second values, followed by the multiplier number code, followed by the percentage tolerance letter code. Usually the first two digits of the code represent the significant part of the value, while the third digit, called the multiplier, corresponds to the number of zeros to be added to the first two digits.

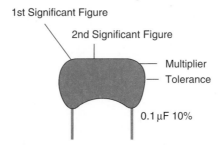

Value	Type	Code	Value	Type	Code
1.5 pF	Ceramic		1000 pF/0.001 µF	Ceramic/Mylar	102
3.3 pF	Ceramic		1500 pF/0.0015 µF	Ceramic/Mylar	152
10 pF	Ceramic		2000 pF/0.002 µF	Ceramic/Mylar	202
15 pF	Ceramic		2200 pF/0.0022 µF	Ceramic/Mylar	222
20 pF	Ceramic		4700 pF/0.0047 µF	Ceramic/Mylar	472
30 pF	Ceramic		5000 pF/0.005 µF	Ceramic/Mylar	502
33 pF	Ceramic		5600 pF/0.0056 µF	Ceramic/Mylar	562
47 pF	Ceramic		6800 pF/0.0068 µF	Ceramic/Mylar	682
56 pF	Ceramic		0.01	Ceramic/Mylar	103
68 pF	Ceramic		0.015	Mylar	
75 pF	Ceramic		0.02	Mylar	203
82 pF	Ceramic		0.022	Mylar	223
91 pF	Ceramic		0.033	Mylar	333
100 pF	Ceramic	101	0.047	Mylar	473
120 pF	Ceramic	121	0.05	Mylar	503
130 pF	Ceramic	131	0.056	Mylar	563
150 pF	Ceramic	151	0.068	Mylar	683
180 pF	Ceramic	181	0.1	Mylar	104
220 pF	Ceramic	221	0.2	Mylar	204
330 pF	Ceramic	331	0.22	Mylar	224
470 pF	Ceramic	471	0.33	Mylar	334
560 pF	Ceramic	561	0.47	Mylar	474
680 pF	Ceramic	681	0.56	Mylar	564
750 pF	Ceramic	751	1	Mylar	105
820 pF	Ceramic	821	2	Mylar	205

(Continued)

Table 25-2 (*Continued*)

PicoFarad (pF)	NanoFarad (nF)	MicroFarad (mF, μF or mfd)	Capacitance Code
1000	1 or 1n	0.001	102
1500	1.5 or 1n5	0.0015	152
2200	2.2 or 2n2	0.0022	222
3300	3.3 or 3n3	0.0033	332
4700	4.7 or 4n7	0.0047	472
6800	6.8 or 6n8	0.0068	682
10000	10 or 10n	0.01	103
15000	15 or 15n	0.015	153
22000	22 or 22n	0.022	223
33000	33 or 33n	0.033	333
47000	47 or 47n	0.047	473
68000	68 or 68n	0.068	683
100000	100 or 100n	0.1	104
150000	150 or 150n	0.15	154
220000	220 or 220n	0.22	224
330000	330 or 330n	0.33	334
470000	470 or 470n	0.47	474

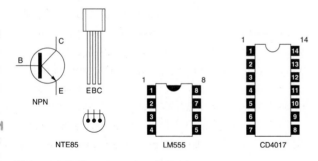

Figure 25-4   *Electronic Hourglass semiconductors*

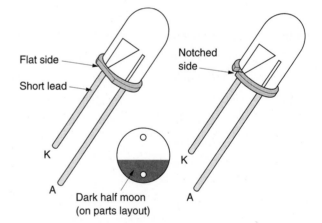

Figure 25-5   *LED identification*

the package. Pin 1 of U1 is connected to ground as is pin 8 of U2. Note that pins 4 and 8 of U1, and pin 16 of U2 are connected to the plus (+) 9-volt supply. Have a friend who is an electronic enthusiast help you if you have difficulties installing the integrated circuits.

Now, we will locate the LEDs and install them on the circuit board. Refer to the diagram in Figure 25-5, which illustrates the LED pin-outs. Note that the LED's cathode is the shorter lead which is on the side of the

LED with the flat edge. A dark half moon on the layout diagram also denotes the cathode lead of the LED. Remember that an LED and diode both have an arrow or triangle pointing to a line which is the cathode of the devices. Once you have identified all of the LEDs check where they should be placed on the circuit board. There are 22 LEDs in this project, 11 of them on one side of the hourglass and the other 11 are on the other side of the hourglass, as shown in the pictorial or layout diagram. Make sure that you refer to the schematic and layout diagram when installing the LEDs. You will note that LEDs 1 through 11 will be mounted on one end of the hourglass and LEDs 12 through 22 are mounted on the other side of the hourglass.

If you chose to use a printed circuit momentary pushbutton switch, then you can go ahead and mount the switch to the circuit board. If you elected to use a chassis mounted reset switch, then you will have to prepare two 6 inch lengths of #22 ga stranded-insulated wire. Solder the free ends to your normally open pushbutton switch. The remaining ends of the two wires will be soldered to the circuit board between pin 15 of U2 and the plus 9-volt line at pin 16 of U2. If you elect to use an "on/off" switch at S2 then you can wire it in series with the plus 9-volt line to the circuit at pin 16 of U2 and the plus + 9-volt bus line. Solder the 9-volt battery to the circuit board, the black wire on the battery clip is the minus (–) lead and is connected to the circuit ground. The plus (+) lead from the battery clip goes to the switch at S2 or to the plus (+) line of the circuit.

Take a short, well-deserved break and when we return we will inspect the circuit board for "short" circuits and "cold" solder joints. First, turn the circuit board, so the foil side faces up toward you. Take a look at the circuit board and examine the solder joints. The solder joints should all look clean, smooth and shiny. If you find a solder joint that looks dull or "blobby", then you should unsolder the joint and remove the solder. After removing the solder from the "bad" solder joint, you can re-solder the joint so that it looks clean, smooth and shiny.

Now once again take a look at the circuit board; this time we are going to check for "short" circuits. Sometimes after constructing a circuit, you may find that "stray" component leads have remained on the circuit board after you cut the excess leads. Often the rosin core from the solder will leave sticky residue which will make leads adhere to the circuit board. These "stray" wire or component leads can connect between circuit traces and "short" out the circuit. If power is applied to the circuit board and a "short "exists, it can cause the circuit to fail,

so you want to make sure that there are no "short" circuits when applying power to the circuit.

Now your Electronic Hourglass is complete and ready to test. Connect a 9-volt battery to the battery clip lead, turn on the power switch and the LEDs should begin lighting up. Remember the jumper at J1, if you want to manually reset the circuit.

If the lights do not start up, you will have to disconnect the battery and re-inspect the circuit board. If one of the LEDs does not light up then it is likely that you installed an LED backwards. It is easy to solve this problem: first unsolder the LED and remove the solder from the connections. Next turn the LED around and re-install it correctly. If the circuit does not light up at all, then the problem is more serious.

You will have to look and make sure that all the resistors are in their correct place and that their color codes are correct for each particular location. Next, move on to the electrolytic capacitors to make sure they have been installed correctly, observing the polarity markings.

Now move to inspecting the diodes, to be sure that they have been installed correctly, with the cathodes facing the right direction. Then you will have to make sure that the transistors are installed correctly. Remember the Base lead is the center lead, and that the Emitter's leads are connected to the circuit ground and that the Collector leads all connect to the LEDs. Finally, check that the integrated circuits have been placed in the sockets properly, observe the schematic carefully and possibly have a knowledgeable friend help you identify the pin-outs and where they connect to on the circuit board. Once you have carefully inspected the circuit board, you can reconnect the battery and re-test the Electronic Hourglass. Hopefully this time-around the circuit will work perfectly. Remember to adjust the potentiometer if you wish the circuit to time faster/slower. You can use this circuit with board games, card games, etc. Note, you can adjust the time of the hourglass using P1 (see text). Once the display has "spilled" its sand from one section to the other, you will have to use the pushbutton to reset and start the timing all over again for the next cycle. Have fun with your new Electronic Hourglass!

Chapter 26

Electronic Windmill/Propeller

Parts List

Parts Bin

R1, R7, R12 10 k ohm, 5%, ¼-watt resistor

R2, R8, R13 1 k ohm, 5%, ¼-watt resistor

R3, R9, R14 5 k ohm, 5%, ¼-watt resistor

R4 1.2 k ohm, 5%, ¼-watt resistor

R5, R6, R10, R11 220 ohm, 5%, ¼-watt resistor

R11, R15, R16 220 ohm, 5%, ¼-watt resistor

P1 10 k ohm potentiometer (trimmer)

C1 1 µF, 35-volt electrolytic capacitor

C2, C3, C4 10 µF, 35-volt electrolytic capacitor

Q1, Q2, Q3 C458 NPN transistor—NTE85 replacement

Q4, Q5, Q6 C458 NPN transistor—NTE85 replacement

L1, L2, L3, L4, L5, L6 red LEDs

L7, L8, L9, L10, L11 red LEDs

L12, L13, L14, L15 red LEDs

L16, L17, L18, L19 red LEDs

L20, L21, L22, L23 red LEDs

L24, L25 red LEDs

S1 SPST toggle or slide switch

B1 9-volt transistor radio battery

Misc printed circuit board, battery clip, wire, etc.

The Electronic Windmill is a fun project which lights up 25 LEDs and creates what appears to be a spinning propeller or windmill effect to your eyes, see Figure 26-1. This project is a great attention getter, which will attract attention to any display or object. It's fun to watch and will delight any observer for many hours. The circuit is easy to build and operates from a single 9-volt battery and it can be built in about an hour or so.

Figure 26-1 *Electronic Windmill*

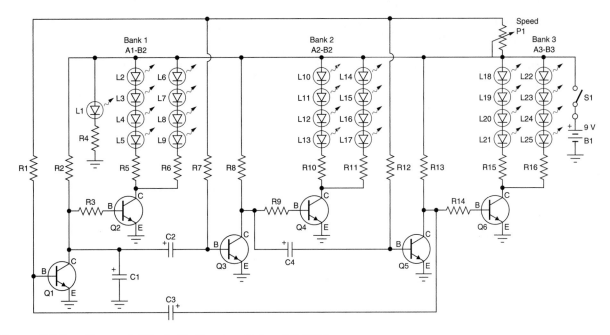

Figure 26-2 *Electronic Windmill. Courtesy Future Kit (FK)*

The Electronic Windmill shown in the schematic at Figure 26-2 is based on an oscillator formed by transistors Q1, Q3 and Q5. The frequency of the master oscillator is formed by resistor R1 and potentiometer P1 and 10 μF capacitors at C2 and C3. The speed of the display circuit can be controlled by P1, a 10 k ohm potentiometer. Transistors Q2, Q4 and Q6 are used to invert the output of the oscillator and to drive the LEDs bands in sequence. Transistor Q2 will drive LEDs in bank one, while Q4 drives LEDs in bank two, and transistor Q6 drives LEDs bank three. The 220 ohm resistors in series with the LED's strings are used to limit the current of each string of LEDs. The Electronic Windmill circuit is powered by a 9-volt transistor radio battery at B1; power is applied to the circuit via the on/off switch at S1.

The Electronic Windmill circuit layout diagram is shown in Figure 26-3, which illustrates the placement of the components on the circuit board. The circuit is pretty straightforward and can be constructed in less than two hours. The key to the circuit display is the arrangement of the LEDs in the circuit. The Electronic Windmill circuit can be constructed on a 2¾ inch × 2½ inch circuit board. The placement of the LEDs actually creates the illustration of the spinning windmill or propeller effect.

Before we begin building the Electronic Windmill, you should prepare a large, clean, well-lit worktable or workbench so you can spread out all of the components for the project as well as all of the necessary diagrams. You will also want to locate some small tools such as a small flat-blade screwdriver, a small Phillips head screwdriver, a small pair of end-cutters, a small pair of needle-nose pliers, as well as a 27 to 33-watt pencil tip soldering iron. Obtain a small spool of 60/40 rosin core solder, a small tin of "Tip Tinner" and a wet sponge to clean the soldering iron tip. Locate the schematic, pictorial diagram, and the various charts and tables for the project as well as the parts and circuit board and we will begin. Heat up your soldering iron!

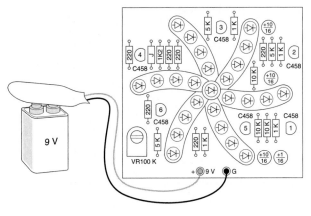

Figure 26-3 *Electronic Windmill pictorial diagram*

Table 26-1

Resistor Color Code Chart

Color Band	1st Digit	2nd Digit	Multiplier	Tolerance
Black	0	0	1	
Brown	1	1	10	1%
Red	2	2	100	2%
Orange	3	3	1000 (K)	3%
Yellow	4	4	10000	4%
Green	5	5	100000	
Blue	6	6	1000000 (M)	
Violet	7	7	10000000	
Gray	8	8	100000000	
White	9	9	1000000000	
Gold			0.1	5%
Silver			0.01	10%
No color				20%

Let's begin by first locating all the resistors for the project. Now refer to the chart in Table 26-1, which illustrates the resistor color codes. Each resistor will have three or four color bands beginning at one end of the resistor body. The first color represents the first digit of the resistor value. The numbers range from 0 through 9. The second color code represents the resistor's second digit which ranges from 0 to 9. The third color band represents the multiplier value for the resistor. If there is a fourth band this indicates the tolerance of the resistor. If the fourth color band is silver then the tolerance value is 10%, if the fourth band is gold, then the resistor tolerance is 5% and so on. Usually if there is no color band then the resistor will have a 20% tolerance value. So, let's look through the resistor pile and try to locate resistor R1, which has a 10 ohm value. Look for a resistor with (brown) (black) (orange), this will be a 10 k ohm resistor. From the chart, brown is the digit 1, while black is zero for the second digit, and an orange color band says that you multiply by 1000, so the value is 10000 ohm. Locate the remaining resistors and identify them. When you have identified all of the resistors, you can install them on the circuit board. After all the resistors have been placed on the circuit board, you can go ahead and solder them in place, then remove the excess component leads with the small end-cutters. Cut the excess leads flush to the circuit board.

Locate the PC board trimmer potentiometer from the parts pile. Notice that the potentiometer will have three leads on it. There will be two outside leads and a center lead. Sometimes all three leads are in a row or in-line, or sometimes the center lead will be offset from the two other leads. The center lead is the potentiometer's wiper or adjustable lead. In the Electronic Windmill project the center lead and one of the end leads are tied together and connected to the plus (+) 9-volt power bus. Install the potentiometer on the lower left side of the PC board as shown, and then solder it in place on the circuit board. After the potentiometer has been soldered to the circuit board, then you can remove the excess leads by flush-cutting them.

Next, locate the capacitors for the project. All of the capacitors in this project are electrolytic types. When installing electrolytic capacitors you will have to observe polarity of the capacitor as well as the placement of it on the circuit board. All electrolytic capacitors will have a white or black color band on the body of the resistor. Located inside or near the band you will see either a plus (+) or minus (−) marking, this is

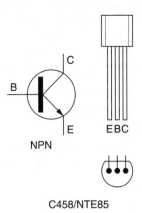

Figure 26-4 *Electronic Windmill semiconductors*

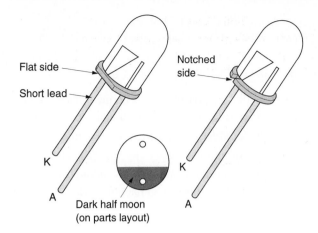

Figure 26-5 *LED identification*

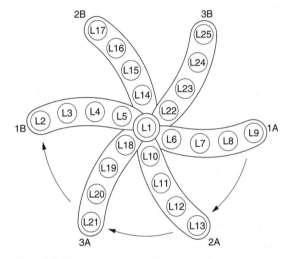

Figure 26-6 *Windmill LED orientation*

the polarity designation. Now look at the schematic diagram and observe how and where the capacitor goes into the circuit. On the schematic an electrolytic capacitor will have a plus (+) marking to tell you how to orient the capacitor on the circuit board. Observe this marking when installing the electrolytic capacitors. Locate and install all of the electrolytic capacitors on to the circuit board. Finally solder each of the capacitors to the circuit board, and when you are certain you have installed them correctly remove the excess capacitor leads from the circuit board.

Now, go ahead and locate the six NPN transistors used in the project and we will install them on the circuit board. Refer to the diagram shown in Figure 26-4, which illustrates the pin-out locations of the transistor. Transistors generally have three leads, an Emitter, a Base and a Collector lead as shown. Refer to the schematic for the transistor symbol and you will notice that the transistor's Base lead is represented by a flat line with the Collector and Emitter lead connected to it. The Emitter lead will have an arrow on it, and it will point toward or away from the Base lead. If the arrow points toward the Base lead then the transistor is a PNP type, and if the arrow points away from the Base lead then the transistor is an NPN type. Install the transistors, solder them in place and then cut the excess leads from the PC board.

Now we only have to install the LEDs on the circuit board and the Electronic Windmill will be almost finished. There are 25 LEDs in this project. LED 1, or L1, is at the center of the pinwheel. Note that LEDs have two leads, an anode and a cathode. In Figure 26-5, you will notice the pin-out for the LEDs shown. The cathode of the LED is shown in the schematic as a line

with an arrow pointing to it. Opposite the cathode lead is the anode, which is the arrow. Note that the anode, or (A) lead is the longer of the two LED leads and the (K) or cathode lead is the shorter lead. Take a look at the diagram depicted in Figure 26-6, which illustrates where the LEDs are installed on the PC board. The secret of the spinning effect is in the placement of the LEDs. Note that Blade-1 of the windmill is the LED bank 1. LEDs L6 through L9 are represented as A1, while L2 through L5 are represented as 1B or Blade-1 or Arm-1 of the windmill. Now, observe that Blade-2 is made up of 2A and 2B. LEDs L10 through L13 are designated as 2A, while LEDs L14 through L17 are shown as 2B. Finally Blade-3 or Arm-3 of the windmill is represented as 3A and 3B. LEDs L18 through L21 are section 3A and LEDs L22 through L25 are shown as section 3B. Once powered the circuit will have three "blades" or "arms" which appear to rotate clockwise and the speed of the rotation is controlled by potentiometer P1. Install all of the LEDs onto the PC board. Next, solder the LEDs to the circuit board and follow-up by removing the excess leads.

With all the components placed on the printed circuit board, we are ready to connect the 9-volt battery clip to the circuit. An SPST power switch was inserted in series with the red or plus (+) battery clip lead. The black battery clip lead is the minus (–) lead and it is connected to the circuit ground bus connection on the circuit board.

Before we apply power to the circuit, we will inspect the circuit board for any possible "short" circuits or "cold" solder joints. Take a short, well-deserved break and when we return, we will inspect the PC board. Turn the circuit board over so that the foil side of the PC board is facing upwards toward you. Look the PC board over carefully, for any possible "cold" solder joints. All of the solder joints should look clean, shiny and smooth. While inspecting the PC board, if you should see any of the solder joints that look dull or "blobby" dirty, then you should remove the solder from the particular solder joint and re-solder the joint, so that it looks clean, shiny and smooth. Next we will inspect the circuit board to see if there are any "short" circuits caused by "stray" component leads, solder or wires. Often excess component leads will "short-out" or "bridge" some solder pads, causing a "short-circuit." A "short-circuit" can cause damage to the circuit when first powered up, so we want to make sure that these conditions do not exist. Rosin core solder residue will often cause component leads, wires, metal solder balls to stick to the underside of the circuit board. Pay particular attention while inspecting the circuit board for these types of problems.

Once the board has been inspected carefully, you can go ahead and connect up a 9-volt battery to the battery clip. When power is connected, you should begin to see the "blades" or windmill "arms" appear to move clockwise. If the windmill "blades" begin to rotate, then all is well and the circuit is working fine. You can adjust the potentiometer to change the speed of the rotation of the windmill "arms."

In the event that the circuit does not work at all or if some LEDs do not light up, you will need to remove the 9-volt battery and you will have to troubleshoot the circuit board. If an LED or a few LEDs do not light up, then the problem is almost obvious, in that you may have installed an LED backwards or perhaps an LED is defective. If the circuit does not work at all, check your battery to make sure that it is good. Next, you will need to make sure that each of the resistors is the correct value and in the correct location. Once you are sure that each resistor has the right value for its location, then you can move on. Now, you will need to make sure that the capacitors are in their correct locations and that you have installed them correctly. Remember that the electrolytic capacitors have polarity and you will need to ensure that you have inserted them with respect to their polarity markings. It is a common mistake to install capacitors backwards.

Finally, you will have to make sure all six transistors have been installed correctly. Re-examine the schematic diagram, the pictorial diagram and the transistor pin-out diagram and compare this information against the circuit board in front of you to see if the transistors have been installed correctly. Commonly people orient the transistor incorrectly when installing them. Make certain that you can identify which end of the transistor is the Emitter. Most plastic TO-92 transistor cases are arranged with the Emitter at one end, with the Base lead in the center and the Collector at the opposite end from the Emitter. Most layout or pin-out diagrams show the transistor package from the bottom, so make sure that you know the reference point of the diagram when trying to identify the transistor. If you are still having difficulty identifying the pin-outs of the transistor or are unsure of yourself, then contact a knowledgeable electronics enthusiast or a parent who may be able to help you. Once you are sure of the transistor pin-outs and you are satisfied with your inspection, you can reconnect the 9-volt transistor radio battery. Turn on the switch at S1, and you should see the windmill "blades" turning round and round in quick succession. You can adjust the speed of the rotation by turning the control at P1. Your Electronic Windmill is now ready to entertain and amuse both you and your friends. You can use the Electronic Windmill to highlight a project, a product, or a display. Have fun with your new Electronic Windmill project!

Chapter 27

Electronic Roulette Game

Parts List

Parts Bin

R1, R2, R3, R4 1.2 k ohm, 5%, ¼-watt resistor

R5, R6, R7, R8, R9 10 k ohm, 5%, ¼-watt resistor

R10 820 ohm, 5%, ¼-watt resistor

R11, R20 100k ohm, 5%, ¼-watt resistor

R12 2.2 megohm, 5%, ¼-watt resistor

R13 47 k ohm, 5%, ¼-watt resistor

R14 330 k ohm, 5%, ¼-watt resistor

R15, R16 20 k ohm, 5%, ¼-watt resistor

R17 56 k ohm, 5%, ¼-watt resistor

R18 3.3 megohm, 5%, ¼-watt resistor

R19 1.5 k ohms, 5%, ¼-watt resistor

R21 4.7 megohm, 5%, ¼-watt resistor

R22 1 k ohm, 5%, ¼-watt resistor

R23 1.8 megohm, 5%, ¼-watt resistor

R24 270 k ohm, 5%, ¼-watt resistor

C1 0.022 µF, 35-volt disk capacitor (223 or 203)

C2 0.0033 µF, 35-volt mylar capacitor (322)

C3, C6 1 µF, 35-volt electrolytic capacitor

C4 0.001 µF, 35-volt disk capacitor (102)

C5 0.47 µF, 35-volt electrolytic capacitor

C7, C8 100 µF, 35-volt electrolytic capacitor

D1–D36 Red LEDs

D37, D38 Green LEDs

D39, D40, D42 1N4148 silicon diodes

D41, D43 1N4001 silicon diodes

Q1, Q2, Q3, Q4 2N3904 NPN transistor

Q7, Q8, Q9 2N3904 NPN transistor

Q5, Q6 2N3906 PNP transistor

U1, U3 CD4017 CMOS Counter

U2 CD4069 CMOS Inverter

S1 momentary pushbutton switch (normally open)

B1 9-volt transistor radio battery

BZ1 piezo sounder

Misc PC board, IC sockets, wire, screws, nuts, spacers, case, etc.

Figure 27-1   *Roulette Game*

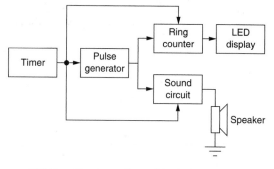

Figure 27-2   *Roulette Game block diagram*

You can pretend that you are at the gaming tables of your favorite casino any time you desire with the Electronic Roulette Game. In this electronic version of the Roulette Game, the Electronic Roulette circuit replaces the ivory ball with a circuit of flashing LEDs. The red LEDs are arranged in a circle next to a black or red number and two green LEDs are positioned next to "0" and "00," as shown in the Figure 27-1. When the pushbutton switch is pushed, the LEDs light one after the other, in a sequence that represents the movement of the ivory ball. The number next to the lit LED when movement stops is the winning number. During movement, the sound of a bouncing ball is generated. If the switch is not pressed again, the circuits will automatically turn off, to conserve the battery. A constant tone will alert you to check your number before automatic shutdown.

The block diagram shown in Figure 27-2 illustrates the main functional parts of the Roulette Game circuit. In the block diagram the timer circuit is used to turn all the other circuits on and off. The pulse generator makes pulses that create the sound and force the ring counter to move the position of the lit LED. The sound circuit generates the sound of a bouncing ivory ball, and a warning tone a few seconds before the circuit powers down. The ring counter lights each LED in a sequence. The LEDs represent the position of the ivory ball.

The timer circuit, shown in the main schematic diagram in Figure 27-3, begins when pushbutton switch S1, is first pressed; capacitor C7 is charged to the battery voltage. This is similar to flipping a "Timer Glass." Just as sand runs downwards in an "hourglass," capacitor C7 charges and forces transistors Q6, Q8 and

Q9 to turn "on," which begins the timer circuit. The output from the initial timer is next fed to the pulse generator via R24. The pulse generator is formed by U2:a, U2:b and U2:c. The pulses from U2 are next fed to the ring counter at U1. Each output from U1 is fed to a bank of LEDs. For example, the output of Q0 at pin 3 is used to activate D4, D11, D15 and D29. The CD4017 counter at U2:F is "clocked" by the CO pin at pin 12 of U1. This "clocked" output from pin 12 of U2:f, is a stream of pulses which is split into two branches. One leg of the output is sent to the input of U3 at the clock input pin 14. The clock pulses sent to U3 are sequenced through outputs Q0 to Q9 of U3. The outputs of U3 are used to "drive" the LEDs through transistors Q1, Q2, Q3 and Q4. The other branch from pin 12 of U1, the "clocked" output pulses, also goes to the inverters at U2:d and U2:e. These inverters are used to drive the piezo sounder at BZ1, which simulates the ball hitting the sides of the rotating wheel in an actual roulette wheel.

The Electronic Roulette Wheel Game was constructed on a 4¾ inch × 4¾ inch circuit board which contains all the circuitry and the LEDs arranged in a circle. The three pictorial diagrams of Figures 27-4, 27-5, and 27-6, show the placement of the electronic components. Each of the diagrams presents a portion of the total component count.

Let's begin building the Electronic Roulette Game. First, you will need to locate a well-lit work area, and a large, clean, worktable or workbench, so that you can lay out all of the project parts, and all the diagrams and tools. Locate a 27 to 33-watt pencil type soldering iron. You will also want to obtain a roll of #22 ga 60/40 rosin core solder, a wet sponge and a canister of "Tip Tinner" which is used to clean and dress the soldering iron tip. You will also want to obtain a few small tools, such as pair of needle-nose pliers, a wire-cutter, a magnifying

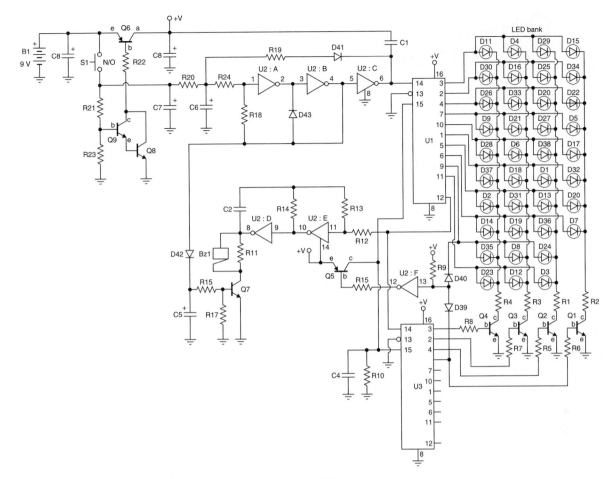

Figure 27-3 *Roulette Game schematic. Courtesy Elenco (EE)*

glass and a pair of tweezers. A small flat-blade and Phillips screwdriver would round out the tools. Plug the solder iron into the wall outlet and we will begin building the project.

First, locate the project resistors and place them in front of you. Now refer to the chart in Table 27-1, which will help you to identify the resistor codes. Resistors use color codes to identify the resistor's value. The color bands will start at one end of the resistor body. Take a look at the color code chart and you will see three columns. The first color band of the resistor will be the first digit of the code. The second color band will represent the second code value and the third color band represents the resistor's multiplier value. Many resistors will also have an additional fourth color band which denotes the resistor's tolerance value. If there is no fourth color band then the resistor has a 20% tolerance value. If the fourth color band is silver, then the resistor will have a 10% tolerance. If the fourth

color band is gold then the resistors will have a 5% tolerance value. Now take a look at the resistors in front of you; look for a resistor which has color code beginning with brown-red-red-gold. Note that you will find four resistors that have the same color codes, these resistors are R1 through R4, which have a 1.2 k ohm value. Once you have located these four resistors, you can identify where they should be placed on the circuit board. When you have determined where the resistors should be placed, you can install all four resistors in their respective locations, solder the resistors to the PC board and then cut the excess lead. Once you have soldered these four resistors in place, you can proceed to identify and install the rest of the project resistors.

Now let's move on to installing the project capacitors. The Electronic Roulette Game has eight capacitors, three of which are electrolytic types. Electrolytic capacitors will usually have their actual value printed on the body of the component. Electrolytic capacitors must be installed with

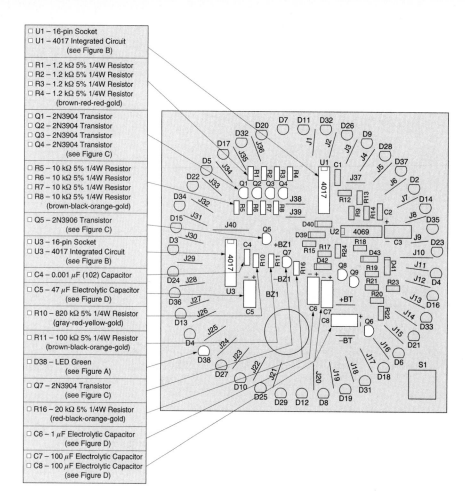

Figure 27-4 *Roulette Game pictorial diagram #1*

respect to their polarity markings, and they will have either a black or white band at one end of the capacitor body. You will have to identify this band and install the capacitor while paying attention to the colored band on the capacitor and to where the plus (+) or minus (−) markings are on the schematic diagram. Electrolytic capacitors are sensitive to polarity and must be installed correctly for the circuit to work. Depending on the particular circuit, if the capacitor is installed backwards damage may result and the entire circuit ruined, so pay careful attention to polarity issues when installing these parts.

The Roulette project also has some other types of capacitors in the circuit; these mylar and disk capacitors may or may not have their actual values printed on the body of the component. Often "codes" are used to identify these types of capacitors. Refer to the chart in Table 27-2 which illustrates some of the code values and how they correspond to the actual capacitor values. Take a look at the chart and you will notice that one of

the capacitors will have a code marking of (102); this corresponds to an actual value of 0.001 μF which corresponds to capacitor C4. You can identify the remaining capacitors, place them on the circuit board in their respective locations, and solder them in place. You can then cut the excess capacitors component leads from the PC board.

Next, locate the silicon rectifier diodes, i.e. D39, D40, D41, D42 and D43. These diodes must be installed with concern for polarity much the same as the capacitors. Silicon diodes will have either a black or white colored band at one end of the diode body. The anode of the diode is the triangle or arrow symbol on the electronic schematic, while the cathode of the diode is the straight line on the symbol. The cathode is the minus (−) designation, while the anode is the plus (+) designation for the diode. The black or white band denotes the diode's cathode lead. Once you have determined the polarity of the diode, you must look on

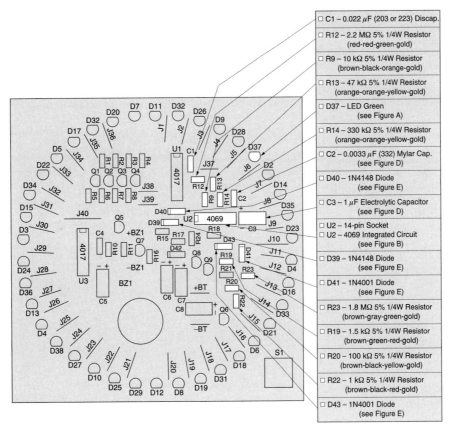

List items pointing to the board:

- C1 – 0.022 μF (203 or 223) Discap.
- R12 – 2.2 MΩ 5% 1/4W Resistor (red-red-green-gold)
- R9 – 10 kΩ 5% 1/4W Resistor (brown-black-orange-gold)
- R13 – 47 kΩ 5% 1/4W Resistor (orange-orange-yellow-gold)
- D37 – LED Green (see Figure A)
- R14 – 330 kΩ 5% 1/4W Resistor (orange-orange-yellow-gold)
- C2 – 0.0033 μF (332) Mylar Cap. (see Figure D)
- D40 – 1N4148 Diode (see Figure E)
- C3 – 1 μF Electrolytic Capacitor (see Figure D)
- U2 – 14-pin Socket
- U2 – 4069 Integrated Circuit (see Figure B)
- D39 – 1N4148 Diode (see Figure E)
- D41 – 1N4001 Diode (see Figure E)
- R23 – 1.8 MΩ 5% 1/4W Resistor (brown-gray-green-gold)
- R19 – 1.5 kΩ 5% 1/4W Resistor (brown-green-red-gold)
- R20 – 100 kΩ 5% 1/4W Resistor (brown-black-yellow-gold)
- R22 – 1 kΩ 5% 1/4W Resistor (brown-black-red-gold)
- D43 – 1N4001 Diode (see Figure E)

Figure 27-5 *Roulette Game pictorial diagram #2*

the schematic and identify which lead is the cathode and which is the anode when placing the diode on the circuit board. When you have identified the diodes—there are two different types in this project—you can go ahead and install the diodes, solder the diodes in place and then cut the excess leads.

Next, you will need to locate the transistors. There are two types of transistors in this project, the 2N3904, an NPN type, and the 2N3906, the PNP type transistor. Refer now to the diagram shown in Figure 27-7 which depicts the package type and pin-outs for the two different transistors, and the integrated circuits. Transistors will generally have three leads protruding from the case or package. Transistors for this project will most likely be packaged in the small plastic TO-92 package, as shown. The transistor's three leads are represented by a Base lead, a Collector lead and an Emitter lead. The TO-92 package will have the pin-outs beginning with the Emitter at one end with the Base lead in the center and the Collector lead opposite the Emitter lead. Transistors must be installed correctly in order for the circuit to work correctly. Remember that

the pin-out diagram usually views the transistor package as the bottom view for reference. Damage may occur if they are installed incorrectly, so pay close attention when installing them on the circuit board to avoid damage to the component or to the circuit itself. Take your time and do it right, the first time!

The use of integrated circuit sockets is highly recommended when building the circuit; they will greatly aid you in the event of a circuit malfunction at some later point in time. It is much easier to unplug the IC from the socket, rather than trying to unsolder the IC from the circuit board. Most people have great difficulty unsoldering integrated circuits and they often damage the circuit board in the process. The IC sockets should have some form of identifier on them, such as a notch or cutout at one of the sockets. Just to the left of this notch or cutout you will find pin 1 of the IC socket. Pin 1 of U1 is decoded output #5, while pin 16 of U1 is the plus (+) voltage supply pin. Pin 1 of U2 is input of inverter (A), while pin 14 is the power supply line. Pin 1 of U3 has the same pin-outs as U1. Solder all the IC sockets in place on the PC board. The integrated circuits

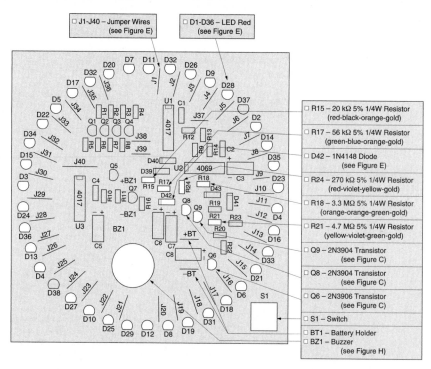

Figure 27-6 *Roulette Game pictorial diagram #3*

will have some form of identification or marking on them to aid in installing them. Look at the integrated circuit package and you will see a small indented circle, a small cutout or notch at one end of the IC package. Pin 1 of the integrated circuit will be just to the left of the indented circuit or notch. When installing the IC, you will also have to identify which pin on the IC socket corresponds to pin 1 in order to line-up pin 1 of the IC with pin 1 of the socket. Refer to the schematic and layout diagram to ensure that you can readily identify pin 1 of the IC before inserting the IC into the socket. Integrated circuits can be damaged quite quickly if they are installed incorrectly. If you have difficulty in identifying where pin 1 of a particular integrated circuit is, have a knowledgeable friend or parent help you out.

Finally it's time to locate the LEDs for this project; there are many red LEDs and only two green LEDs. Refer to the diagram shown in Figure 27-8, which illustrates the LED package and pin-outs. You will notice that the LED will have a flat edge on one side of it; this corresponds to the minus (–) or negative lead which is the shorter lead. Also note that the symbol for an LED is very similar to a diode with an anode and a cathode. LEDs like diodes are polarity sensitive and

must be installed so that the cathode, whose symbol is depicted by a single flat line, with the anode is a triangle pointing toward the cathode. In the symbol diagram the negative or cathode lead is represented as the dark half moon. When installing the LEDs make sure that you can identify the location of the cathode of a particular LED on the schematic before installing the LED. Remember that there are only two green LEDs which are placed at D37 and D38, i.e. "0" and "00."

Now you will need to install S1, a PC board type pushbutton between the junction of Q6 and C8 and R21/R20. Insert the pushbutton into the circuit board and solder it in place. Finally, locate the piezo sounder BZ1, it is connected between Q7 and pin 8 of U2:d. Note that a 100 k ohm resistor is soldered across the piezo sounder.

Now is a good time to take a short break, and when we return we will inspect the circuit board for any possible "cold" solder joints and "short" circuits. Turn the circuit board over, so that the copper foil side faces up toward you. Look closely at all the solder joints; they should all be clean, smooth and shiny. You will want to eliminate any possible "cold" solder joints as they can cause the circuit to work intermittently or to fail prematurely. If any solder joints look dull or "blobby" then remove the solder

Table 27-1

Resistor Color Code Chart

Color Band	1st Digit	2nd Digit	Multiplier	Tolerance
Black	0	0	1	
Brown	1	1	10	1%
Red	2	2	100	2%
Orange	3	3	1000 (K)	3%
Yellow	4	4	10000	4%
Green	5	5	100000	
Blue	6	6	1000000 (M)	
Violet	7	7	10000000	
Gray	8	8	100000000	
White	9	9	1000000000	
Gold			0.1	5%
Silver			0.01	10%
No color				20%

from the joint and re-solder it, so that the solder joint looks clean, smooth and shiny. Now re-inspect the circuit board; we will be looking for any "short" circuits which may be caused from "cut" component leads when you were trimming the component leads in preparing the circuit board. "Stray" component leads can often cling or stick to the underside of the PC forming a "bridge" or "short" which can cause the circuit to malfunction and possibly destroy a component. Take your time and inspect the circuit carefully. Once you are satisfied that you have inspected the circuit board then you can prepare to solder the battery clip and test the Roulette Game circuit.

Locate the 9-volt battery clip; the red or positive (+) lead will be soldered to the plus ties points which include pin 16 of the CD4017s, and pin 14 of the CD4069 and the junction of S1/C8 and Q6. The minus or negative (–) battery clip lead is soldered to the ground ties point throughout the circuit.

Once the battery clip is soldered to the circuit, you are ready to test the Roulette Game. Insert a 9-volt battery into the clip and press pushbutton S1. When S1 is pressed the LEDs will light up one after the other, in a sequence that represents the movement of the

conventional ivory ball. After releasing the pushbutton, the number next to the lit LED when the movement stops is the winning number. During the movement of the LEDs around the circle, the sound of the bouncing ball will be heard. If the S1 is not pressed again the circuit will go to sleep and shut down.

In the event that the LEDs do not light up in sequence when S1 is pressed, or if certain LEDs light up and not others, or if you heard no sound from the sounder, then you will have to remove the battery and inspect the circuit board. After removing the 9-volt battery, you will have to troubleshoot the circuit. If one of the LEDs does not appear to light up, it will be simple to unsolder the LED, remove the solder from the joint, rotate the LED 180 degrees, and then re-solder the LED back into the circuit. Connect the battery back to the circuit and you should be ready to go!

If your Roulette Game still does not work properly or appears to be "dead," then you will have to delve further into the circuit. First, you will have to check the circuit to make sure that you have placed all the resistors into their correct locations. It is a common mistake to put the wrong resistor into a particular location. Re-check all the resistors to make sure that each of the color

Table 27-2

Capacitance Codebreaker Information

This table is designed to provide the value of alphanumeric coded ceramic, mylar and mica capacitors in general. They come in many sizes, shapes, values and ratings; many different manufacturers worldwide produce them and not all play by the same rules. Most capacitors actually have the numeric values stamped on them; however, some are color coded and some have alphanumeric codes. The capacitor's first and second significant number IDs are the first and second values, followed by the multiplier number code, followed by the percentage tolerance letter code. Usually the first two digits of the code represent the significant part of the value, while the third digit, called the multiplier, corresponds to the number of zeros to be added to the first two digits.

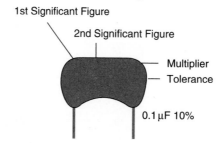

Value	Type	Code	Value	Type	Code
1.5 pF	Ceramic		1000 pF/0.001 µF	Ceramic/Mylar	102
3.3 pF	Ceramic		1500 pF/0.0015 µF	Ceramic/Mylar	152
10 pF	Ceramic		2000 pF/0.002 µF	Ceramic/Mylar	202
15 pF	Ceramic		2200 pF/0.0022 µF	Ceramic/Mylar	222
20 pF	Ceramic		4700 pF/0.0047 µF	Ceramic/Mylar	472
30 pF	Ceramic		5000 pF/0.005 µF	Ceramic/Mylar	502
33 pF	Ceramic		5600 pF/0.0056 µF	Ceramic/Mylar	562
47 pF	Ceramic		6800 pF/0.0068 µF	Ceramic/Mylar	682
56 pF	Ceramic		0.01	Ceramic/Mylar	103
68 pF	Ceramic		0.015	Mylar	
75 pF	Ceramic		0.02	Mylar	203
82 pF	Ceramic		0.022	Mylar	223
91 pF	Ceramic		0.033	Mylar	333
100 pF	Ceramic	101	0.047	Mylar	473
120 pF	Ceramic	121	0.05	Mylar	503
130 pF	Ceramic	131	0.056	Mylar	563
150 pF	Ceramic	151	0.068	Mylar	683
180 pF	Ceramic	181	0.1	Mylar	104
220 pF	Ceramic	221	0.2	Mylar	204
330 pF	Ceramic	331	0.22	Mylar	224
470 pF	Ceramic	471	0.33	Mylar	334
560 pF	Ceramic	561	0.47	Mylar	474
680 pF	Ceramic	681	0.56	Mylar	564
750 pF	Ceramic	751	1	Mylar	105
820 pF	Ceramic	821	2	Mylar	205

(Continued)

Table 27-2 (*Continued*)

PicoFarad (pF)	NanoFarad (nF)	MicroFarad (mF, µF or mfd)	Capacitance Code
1000	1 or 1n	0.001	102
1500	1.5 or 1n5	0.0015	152
2200	2.2 or 2n2	0.0022	222
3300	3.3 or 3n3	0.0033	332
4700	4.7 or 4n7	0.0047	472
6800	6.8 or 6n8	0.0068	682
10000	10 or 10n	0.01	103
15000	15 or 15n	0.015	153
22000	22 or 22n	0.022	223
33000	33 or 33n	0.033	333
47000	47 or 47n	0.047	473
68000	68 or 68n	0.068	683
100000	100 or 100n	0.1	104
150000	150 or 150n	0.15	154
220000	220 or 220n	0.22	224
330000	330 or 330n	0.33	334
470000	470 or 470n	0.47	474

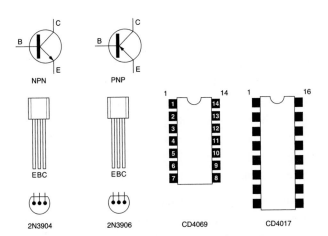

Figure 27-7 *Roulette Game semiconductor pin-out diagram*

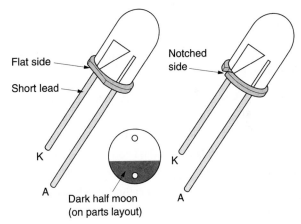

Figure 27-8 *LED identification*

codes is correct for that particular location. After checking the resistors, make sure that the electrolytic capacitors have been installed correctly. Again observe the polarity marking to be sure that each capacitor is installed correctly and that it is in the correct location; refer to the schematic and layout diagrams for help.

Next, check the polarity of all the diodes. Remember that the cathode is on the side of the color band on the diodes, so check the schematic to make sure that you are installing the diodes correctly. Now, check all the transistors and make sure that you can identify each of the three leads and where they belong in the circuit. If you are having trouble identifying the transistor leads and their correct placement, have an adult or knowledgeable friend help you. Remember that the

Table 27-3

Strategies	Explanation	Payoff
A) Single Straight	chips on a number from 1–36, incl 0 and 00	36 times
B) Split	chips on 2-numbers vert or horiz next to one another	18 times
C) Street	chips on 3-numbers horizontal on 1-line	12 times
D) Corner	chips on 4-numbers vert & horiz next to one another	9 times
E) Line	chips on 6-numbers in 2-horiz lines next to one another	6 times
F) Column	chips on 12-numbers in 1-vertical line	3 times
G) 1st Dozen	chips on 12 numbers in 1st twelve, 2nd twelve, or 3rd twelve	3 times
2nd Dozen		
3rd Dozen		
H) Low or High	chips on 18-numbers either from 1–18 or from 19–36	2 times
I) Red or Black	chips on "red" or "black," betting on all numbers which are red or black	2 times
J) Odd or Even	chips on "odd" or "even," betting on all numbers which are either odd or even	2 times

Chip Values

GOLD	$100.00
GREEN	$25.00
RED	$5.00
WHITE	$1.00

transistor pin-out diagram illustrates the transistor from the bottom view.

Now, check to make sure that the integrated circuits have been inserted into the sockets with correct orientation. Finally, look at all of the LEDs to make sure that they have been installed correctly. Make sure that you can identify which LED is the cathode or minus (–) lead. All of the LED cathode leads should point to the Collectors of transistors Q1 through Q4. Once you are satisfied that your inspection is complete, you can re-connect the 9-volt battery and test the circuit once again. Hopefully you have corrected your mistake and the circuit works perfectly now. Have fun!

Playing the Roulette Game

The object of playing the Roulette Game is to increase the value of your chips more than any other player.

Chips with gold centers are worth $100.00, green centers = $25.00. Red centers = $5.00 and white centers are worth $1.00. Each player starts with 1 green, 2 red, and 5 white chips, which equal $40.00. All the rest of the unused chips belong to the house. Determine how long the roulette table will be open, one hour for example. One person must act as the Croupier.

The Croupier is the attendant who collects and plays the stakes using the house's money. Since there is no way to predict the outcome of each spin, the Croupier may also be a player. It is possible for a person to play roulette and try to beat the house by increasing his total chip value.

The very first action in roulette is to place your wager on the game table. The types of bets and their rates of return are listed in the chart at Table 27-3. The method for placing a wager is shown in the chart at Table 27-4. Placing wagers starts when the Croupier announces "Place your Wagers." All wagers must be in place when the Croupier announces "no more wagers!" After all

Table 27-4

Gaming Chart

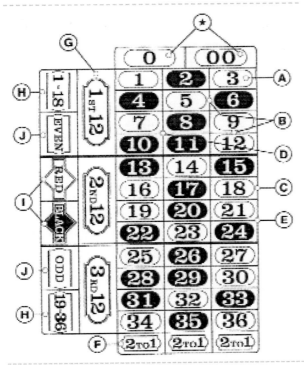

wagers have been placed, the start button is pressed by the Croupier and the lit LED that represents the ivory ball races around the circle adding excitement and anticipation to the game. The number next to the lit LED, when the motion stops, is the winning number. All wagers are paid by the Croupier according to the rates of return listed in Table 27-3.

The game ends when the house runs out of chips or the predetermined time period expires. To prevent a person from doubling his wager until he wins, a maximum limit of $100.00 should be placed on each wager. When a player loses all of their chips, they may borrow from other players at whatever interest rate that player demands. At no time may a player borrow more than $40.00. Once a player owes $40.00 and has lost all their chips, they are bankrupt and can no longer place wagers. A bankrupt player may assume the position of Croupier and earn $1.00 from the house for every 10 spins to remain in the game. A Croupier who is not bankrupt is paid no salary by the house.

Probability—If among (F+U) equi-probable and mutually exclusive events, F is regarded as favorable and U as unfavorable, then for a single event, the probability of favorable outcome is F/ F+U. The probability of an unfavorable outcome is 1 minus the probability of a favorable outcome. In other words, since there is the same chance that any number may win on any spin (mutually exclusive events), the chances of winning equals the number of winning numbers divided by the total number of possible numbers. Roulette has 38 possible numbers that may win. Therefore F+U is always equal to 38. If you wager on a single number, the chances of winning are 1 divided by 38, or approximately 97.37%. If you win, the house pays you 36 times your wager. Multiplying your chances of winning times your paycheck shows the advantage for the house. In this case, the number is 94.74% which means the house has 5.26% advantage over the players wagering on a single number.

If a wager is placed on black or red, the probability of winning is 18 divided by 38 because the number of black numbers and the number of red numbers is 18. The probability of a favorable outcome if one color is wagered equals 47.4%. The payout if you win is 2 to 1. This yields an advantage for the house of $1- (0.474 \times 2)$ or approximately 5.26%. As you can see, the house always has a 5.3% advantage.

Chapter 28

Hangman Game

Parts List

Parts Bin

R1, R2 390 k ohm,
  ¼-watt, 5% resistors

R3, R12, R13, R14, R15
  47 k ohm, ¼-watt,
  5% resistors

R4 330 k ohm

R5, R6, R7 4.7 k ohm,
  ¼-watt,
  5% resistors

R8, R9 10 k ohm,
  ¼-watt, 5% resistors

R10 22 k ohm, ¼-watt,
  5% resistors

R11 33 k ohm, ¼-watt,
  5% resistors

R16, R17, R18 150 ohm,
  ¼-watt, 5% resistors

R19 680 ohm, ¼-watt,
  5% resistors

R20, R21, R22 270 ohm,
  ¼-watt, 5% resistors

R23, R24 330 ohm,
  ¼-watt, 5% resistors

R25 10 megohm,
  ¼-watt, 5% resistors

R26 150 k ohm, ¼-watt,
  5% resistors

R27, R30 2.2 megohm,
  ¼-watt, 5% resistors

R28 33 k ohm, ¼-watt,
  5% resistors

R29 3.3 k ohm, ¼-watt,
  5% resistor

P1 100 k ohm
  potentiometer
  (trimpot)

C1 2.2 µF, 35-volt
  electrolytic capacitor

C2 1nF, 100-volt
  green capacitor

C3 22 µF, 35-volt
  electrolytic capacitor

C4 100 µF, 35-volt
  electrolytic capacitor

C5, C7 4.7 µF, 35-volt
  electrolytic capacitor

C6 10 nF, 100-volt
  green capacitor

C8 470 µF, 35-volt
  electrolytic capacitor

Q1, Q2, Q3, Q4, Q5, Q6
  BC547 NPN transistor—
  2N3904 or NTE123AP

Q7, Q8, Q9, Q10, Q11
  BC547 NPN transistor—
  2N3904 or NTE123AP

Q12 BC557 PNP
  transistor—2N3906 or
  NTE159

U1, U2 CD 4011 Quad
  NAND gate

L1, L2, L3, L4, L5
  15—Red 3 mm LEDs

L6, L7, L8, L9, L10
  15—Red 3 mm LEDs

L11, L12, L13, L14, L15
  15—Red 3 mm LEDs

D1, D2 1N4002
  silicon diodes

D3 1N914 or
    1N4148 diode

B1 9-volt transistor
    radio battery

Misc PC board, battery
      clip, IC socket,
      hardware, wire etc.

Everybody likes re-discovering something they did years ago. Here's a game we all played at school. One player thinks of a word and writes down the number of letters in that word in the form of boxes or dashes. The object of the game is for the opponent to suggest letters of the alphabet, and if they are correct, are placed on the dashes in the correct order so that the word gradually appears. To make the game more interesting, a side issue is introduced which effectively counts the number of incorrect guesses. Each time an incorrect letter is suggested, a systematic framework is created with straight lines in the form of a gallows. A stick man, representing a person being hung, see Figure 28-1.

The game is concluded when the correct word is created or the stick man is completed, whichever comes first. This is an electronic version of that game. The stick man and gallows are made with 15 LEDs and each time a Touch Plate is touched, one more section of the cartoon is illuminated. The last LEDs to be lit are 14 and 15, which represent the feet of the man. When these LEDs are at full brightness, the 8th LED begins to flash, indicating the man is "Hanged." The game can be played in two ways. The "normal" way involves the secret word and using the hangman to count the incorrect letters. The other suggestion is to take it in turns illuminating the LEDs until the flashing LED is set into oscillation. The player creating the first sign of continued flashing is the winner. In either game, you will have lots of fun, especially in a darkened room where the full effect of the LEDs will be produced.

The Hangman Game consists of seven main building blocks as shown in Figure 28-2. These are shown in the block diagram and are identified as follows: the Hangman Game is a hybrid circuit—meaning it is composed of two different species. We have combined transistors with ICs to achieve an update of an old game. The complexity of the circuit comes from the repetition of the transistor stages. Due to the number of biasing resistors required it is strongly suggested that

Figure 28-1 *Hangman Game*

you use a PC board. Not only has the layout of the board been carefully designed to make it look symmetrical when completed but it also allows the project to go together so much easier.

When the power is applied, the only building block to come into operation is the 2 kHz multivibrator, block 2. It is made up of gates c and d of U2 and feeds the push-pull buffer consisting of Q11 and Q12 to charge the 100 µF electrolytic. The oscillator runs at a fairly high frequency and this reduces the size of the coupling capacitor. This building block is called a voltage doubler and the voltage appearing at the output terminal is very close to double the 9-volt supply minus the voltage drops across the two diodes. Under no-load

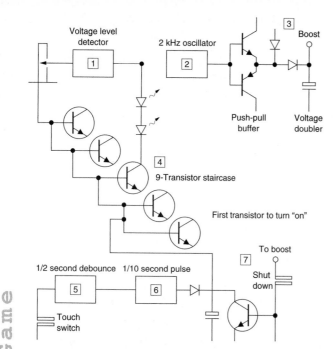

Figure 28-2 *Hangman block diagram*

conditions this voltage appears at the output as 14 volts. We call this "boost" and we have labeled it 12 volts BOOST because it reduces to 12 volts under full-load conditions.

The mechanics of the voltage doubling circuit are easy to follow, by referring to the main schematic shown in Figure 28-3. The multivibrator c and d produces a square wave which is fed to the bases of the two complementary transistors. When one transistor is turned hard on, the other is full off. For the first cycle, the output gate c is "low" and the BC557 is turned "on." The negative end of the 22 µF is taken to the negative rail and charges quickly via the top 1N4002 diode to 7.5 volts. At the same time the 100 µF electrolytic is charging to 7.6 volts via the two diodes. When the multivibrator swings "high," the top BC547 transistor turns "on" and the BC557 turns "off." The negative end of the 22 µF is now brought to the positive rail and its stored 7.5 volts will be added to that of the 100 µF electrolytic to bring the total voltage up to 15.2 volts minus .7 volts drop across the lower diode. In fact the voltage drop across the diodes has a double effect on reducing the voltage since they are used for each part of the voltage doubling action. They account for nearly 3 volts drop.

We must also include the Collector-Emitter voltage drop of each transistor as this reduces the maximum

voltage available on the 22 µF "boosting" electrolytic. Thus the resulting voltage out of the doubler is considerably less than you would expect. All these diode and transistor voltage drops are constant for any voltage doubler and would obviously be less noticeable when using higher voltages. This arrangement is capable of delivering 15 to 20 milliamps and since it does not have a very good regulation, the voltage under load drops to about 11 or 12 volts. This is just enough to illuminate LEDs 14 and 15 in the staircase circuit. LEDs 14 and 15 are positioned at the feet of the man being hung and are controlled by transistor Q10. The reason for providing a voltage doubler circuit is two-fold. It introduces a new building block into our "library" and adds interest to the project while providing an economical way of producing the necessary higher voltage rather than using a 12 volt battery.

How to Play

When a player places his finger on the Touch Switch, the first monostable multivibrator consisting of gates a and b in IC1 will be triggered into its unstable state and produce a spike through the 10 n capacitor to operate the second monostable. The output pin 10 is normally "high." It will go "low" for $1/10$ second then return HIGH. It will remain in this state until your finger is removed and re-applied. The time delay of the first monostable has been carefully chosen to be longer than the second so that the circuit is fully de-bounced and can only be triggered every half-second. The second monostable has a time delay completely independent of the first and produces a short pulse which charges the 470 µF electrolytic via the 3k3 resistor and diode. The purpose of the diode is to prevent discharging of the electrolytic once the monostable has fallen back into its stable state and produced a "low" on pin 10.

The charging of the electrolytic is exponential and each pulse from pin 10 produces sufficient energy to raise the voltage of the 470 µF approximately .75 volts to turn on one transistor at a time in the staircase. At the bottom of the staircase, one pulse will be sufficient to illuminate one of these steps but as they increase toward the top, more than one pulse will be required. Transistor Q1 is connected as an Emitter-follower, the load being the base resistors in the

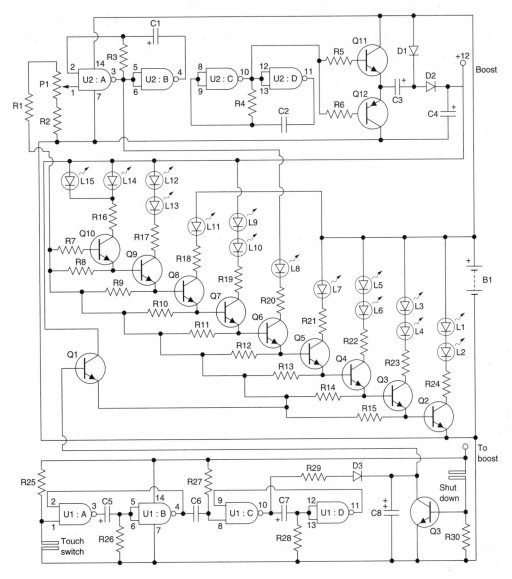

Figure 28-3 *Hangman Game*

staircase. The voltage across this load will be directly proportional to the voltage on the electrolytic (i.e. the Base voltage) minus the .75 volt Base-Emitter voltage drop. The main purpose of the transistor is to separate the 470 µF electrolytic from the load of the staircase. If we were to remove this transistor the circuit would function as before except that the load of the nine staircase transistors would tend to discharge the electrolytic rather quickly. So, in effect, the Emitter-follower transistor is providing an impedance matching arrangement to reduce the drain on the electrolytic, to a value about one-hundredth of a directly coupled version. The Base resistors have been chosen according to the voltage they will be required to drop. Many factors

influence the actual value selected for each Base resistor as the impedance of the circuit changes with rising voltage and this alters the conditions considerably.

The first transistor in the staircase is Q2. It will turn on when its Base voltage is .75 volt higher than the Emitter voltage. The second transistor, instead of being connected to the ground, is connected to the Base of the first transistor. It too will turn on when its Base voltage is .75 volt higher than its Emitter. This means the incoming voltage will need to be .75 volt + .75 volt or 1.5 volts before it will be fully turned on. This reasoning continues up the staircase so that the top transistor will require 6.75 volts to be turned on.

Extending back, the 470 μF electrolytic requires 6.75 + .75 volt or 7.5 volts and the output of the IC is .6 volt higher again to account for the diode drop. The IC needs an even higher output voltage to be able to supply a charging current through the 3.3 k current limit resistor. To be on the safe side, the supply rail should be 7.5 volts + .6 volt + 2 volts = 10.1 volts. The need for a high supply voltage is even more important when illuminating the top LEDs 14 and 15. So much of the supply has been lost in the Base-Emitter junctions that 11.5 volts is the absolute minimum voltage if we expect to get adequate brightness from the top LEDs. This will allow us just 4 volts for the dropper resistor and LEDs.

Note the power to the 15 LEDs comes from three different sources. LEDs 1 to 7 and 11 are supplied directly from the 9-volt supply. LED 8 derives its supply from a slow cycling oscillator. This is made up of gates a and b of U2. It forms a gated oscillator with pin 3 normally "high." The oscillator is triggered via pin 1. When it detects approximately half supply voltage through the high impedance resistor network comprising the 390 k, 100 k pot and 390 k resistor, it will flash LED 8 at about 2 cycles per second. LEDs 9 and 10 are series LEDs and need to be driven from a source slightly higher than 9 volts. To achieve this we must take them to the "boost" rail. LED 7, being a single LED, will just operate from the 9-volt rail. The remaining 4 LEDs need 12 volts to operate. LEDs 14 and 15 are paralleled together so that they can attain full brightness from the 12-volt rail.

Resetting the game is accomplished by discharging the 470 μF electrolytic via Q13. This transistor is normally biased in the off condition with the 2.2 μF Base-Emitter resistor. When you touch the Shut Down wires with your finger, a small forward bias is applied to the transistor and it turns "on" to bleed the electrolytic. The light-emitting diodes will gradually turn off as the voltage on the electrolytic falls. When all the LEDs are off, the quiescent current for the game is only about 100 microamps. This is so small that no on/off switch is required and even small "AAA" cells will last their normal shelf life.

An overlay makes construction so easy. It will take the best part of an hour to assemble the Hangman, even with all the parts ready at your fingertips. See Figure 28.2 if you are not sure where any of the components are placed. Use only DURACELL cells to power this game. Ordinary dry cells will not last very long as their voltage soon falls to 8 volts or less for a 9-volt battery At this voltage, the doubling circuit will be incapable of supplying sufficient voltage to light LEDs L14 and L15.

Construction

Before we begin building the Hangman Game, you will want to locate a clean, well-lit worktable or workbench on which to build your project. You will also want to gather some light tools, such as a pair of end-cutters, a pair of small needle-nose pliers, a magnifying glass, a pair of tweezers, a small flat-blade and Phillips screwdriver. Also locate a 27 to 33-watt pencil soldering iron, a roll of 60/40 rosin core solder and a small jar of "Tip Tinner," a soldering iron tip dresser compound, from your local Radio Shack store. The heated soldering iron tip is periodically inserted into the jar and rotated in the compound in order to clean and dress the soldering tip. Obtain an anti-static wrist strap, as this will prevent large static voltages from destroying the integrated circuits when handling them. Just moving about at your workbench or table, or getting up from your chair can create large static voltages. The wrist strap is grounded to an AC outlet, and is worn around the wrist while handling the integrated circuits. Heat up your soldering iron and we will begin.

Examine the printed circuit board for any holes which may be filled-in with solder. This happens during manufacture, when the copper tracks are being tinned. Clean these out with a soldering iron and needle or a length of copper wire. Whether you use a ready-made printed circuit board or a home-made board, some form of printed construction is essential for a circuit of this complexity.

Next you will need to refer to the chart in Table 28-1, which depicts the resistor color codes chart and how to use it. Resistors will have color band located on the body of the resistors which help you to identify the value of the resistor. The first color band is located at one end of the resistor body. The first color code denotes the resistor's first digit, while the second color code represents the second digit of the code. The third color band denotes the multiplier number of the resistor, and the fourth band indicates the tolerance value of the

Table 28-1

Resistor Color Code Chart

Color Band	1st Digit	2nd Digit	Multiplier	Tolerance
Black	0	0	1	
Brown	1	1	10	1%
Red	2	2	100	2%
Orange	3	3	1000 (K)	3%
Yellow	4	4	10000	4%
Green	5	5	100000	
Blue	6	6	1000000 (M)	
Violet	7	7	10000000	
Gray	8	8	100000000	
White	9	9	1000000000	
Gold			0.1	5%
Silver			0.01	10%
No color				20%

resistor. A silver band depicts a 10% tolerance value, while a gold band denotes a 5% tolerance value. No fourth band says the resistor has a 20% tolerance value. So, look through the resistors and check for a resistor with an orange band followed by a white band and yellow third multiplier band. This resistor with (3)(9) (10,000) will represent the 390,000 ohm resistor at R1. Locate the correct placement for R1 on the PC board and insert it into the printed circuit board. Solder the resistor to the circuit board, then follow-up by using your end-cutter to remove the excess resistor leads from the board. Repeat this process for the remaining resistors.

Next refer to the chart in Table 28-2, which illustrates the capacitor code chart used to help identify small capacitors. Larger capacitors, such as electrolytic types, will have their actual value printed on the body of the component. Oftentimes small capacitors do not have enough room to print the actual value on the part, so a code was devised to mark the capacitor. So if you see a capacitor marked with a (104), you can refer to the table and see that the capacitor will have a value of 0.1 µF. Note that electrolytic capacitors will have polarity marking on them. You will see either a black or white color band. Either on the color band or next to the color

band you will find a plus (+) or minus (−) symbol denoting the polarity of the component. You must carefully refer to the schematic and pictorial diagram when installing the capacitors to make sure you observe the correct orientation during installation. Failure to install the capacitors correctly can result in damage to the part or the circuit. Identify and install the remaining capacitors on to the circuit board. Solder the capacitors to the printed circuit board. Remember to cut the excess component leads with your end-cutter or diagonals.

Locate the 100 k potentiometer which will be placed at P1 on the circuit board. The trimmer potentiometer will have three leads. All three leads will be in-line, all in a row or two outer leads with a center offset lead. The center potentiometer lead is the adjustable wiper arm which controls the resistance output. Place the potentiometer at its correct location and insert the leads into the PC board. Now solder the potentiometer in place.

The circuit has three silicon diodes, two of the diodes are connected to the output of Q11 and Q12 and the remaining diode is connected to the Collector of Q13. Diodes, as you will remember, have polarity. The diode's symbol is an arrow or triangle pointing to a line. The arrow side of the diode is the anode and the symbol

Table 28-2

Capacitance Codebreaker Information

This table is designed to provide the value of alphanumeric coded ceramic, mylar and mica capacitors in general. They come in many sizes, shapes, values and ratings; many different manufacturers worldwide produce them and not all play by the same rules. Most capacitors actually have the numeric values stamped on them; however, some are color coded and some have alphanumeric codes. The capacitor's first and second significant number IDs are the first and second values, followed by the multiplier number code, followed by the percentage tolerance letter code. Usually the first two digits of the code represent the significant part of the value, while the third digit, called the multiplier, corresponds to the number of zeros to be added to the first two digits.

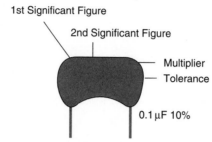

Value	Type	Code	Value	Type	Code
1.5 pF	Ceramic		1000 pF/0.001 µF	Ceramic/Mylar	102
3.3 pF	Ceramic		1500 pF/0.0015 µF	Ceramic/Mylar	152
10 pF	Ceramic		2000 pF/0.002 µF	Ceramic/Mylar	202
15 pF	Ceramic		2200 pF/0.0022 µF	Ceramic/Mylar	222
20 pF	Ceramic		4700 pF/0.0047 µF	Ceramic/Mylar	472
30 pF	Ceramic		5000 pF/0.005 µF	Ceramic/Mylar	502
33 pF	Ceramic		5600 pF/0.0056 µF	Ceramic/Mylar	562
47 pF	Ceramic		6800 pF/0.0068 µF	Ceramic/Mylar	682
56 pF	Ceramic		0.01	Ceramic/Mylar	103
68 pF	Ceramic		0.015	Mylar	
75 pF	Ceramic		0.02	Mylar	203
82 pF	Ceramic		0.022	Mylar	223
91 pF	Ceramic		0.033	Mylar	333
100 pF	Ceramic	101	0.047	Mylar	473
120 pF	Ceramic	121	0.05	Mylar	503
130 pF	Ceramic	131	0.056	Mylar	563
150 pF	Ceramic	151	0.068	Mylar	683
180 pF	Ceramic	181	0.1	Mylar	104
220 pF	Ceramic	221	0.2	Mylar	204
330 pF	Ceramic	331	0.22	Mylar	224
470 pF	Ceramic	471	0.33	Mylar	334
560 pF	Ceramic	561	0.47	Mylar	474
680 pF	Ceramic	681	0.56	Mylar	564
750 pF	Ceramic	751	1	Mylar	105
820 pF	Ceramic	821	2	Mylar	205

(Continued)

Table 28-2 (*Continued*)

PicoFarad (pF)	NanoFarad (nF)	MicroFarad (mF, μF or mfd)	Capacitance Code
1000	1 or 1n	0.001	102
1500	1.5 or 1n5	0.0015	152
2200	2.2 or 2n2	0.0022	222
3300	3.3 or 3n3	0.0033	332
4700	4.7 or 4n7	0.0047	472
6800	6.8 or 6n8	0.0068	682
10000	10 or 10n	0.01	103
15000	15 or 15n	0.015	153
22000	22 or 22n	0.022	223
33000	33 or 33n	0.033	333
47000	47 or 47n	0.047	473
68000	68 or 68n	0.068	683
100000	100 or 100n	0.1	104
150000	150 or 150n	0.15	154
220000	220 or 220n	0.22	224
330000	330 or 330n	0.33	334
470000	470 or 470n	0.47	474

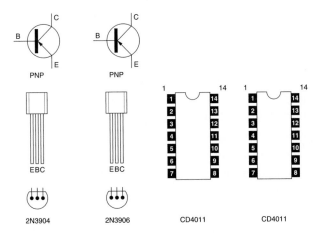

Figure 28-4 *Hangman semiconductor pin-out diagram*

on the PC board and then solder them in place. Don't forget to cut the excess component leads from the board.

The Hangman circuit has many transistors, most of which are NPN types with a single PNP type transistor at Q2. Transistors have three leads, a Base lead, a Collector lead, and an Emitter lead, see the semiconductor pin-out diagram in Figure 28-4, which shows the pin-outs view of the devices. Symbolically the flat line within the circle is the Base lead, while the lead with the arrow is the Emitter and the remaining lead is the Collector. If the arrow points away from the transistor then the transistor is an NPN device, and if the arrow points to the transistor then the device is a PNP type transistor.

The Hangman Game has two integrated circuits. It is advisable to install integrated circuit sockets before installing the actual integrated circuits. On rare occasions ICs can fail, and if they do it's very difficult

for the straight line is the cathode. Look the diodes over carefully and you will notice that there will be either a white or black band at one end of the diode body. The band denotes the diode's cathode end. Install the diodes

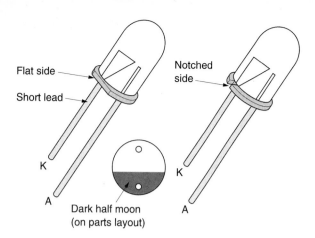

Figure 28-5 *LED identification*

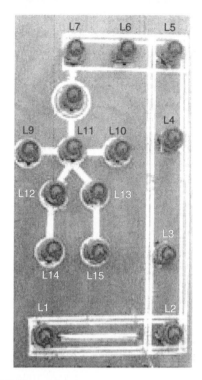

Figure 28-6 *LED template*

to remove an IC from the circuit board without damaging the PC board. So sockets are inexpensive insurance. The IC sockets will generally have a notch or cutout at one end of the socket. The pin to the left of the notch is pin 1. On U1, pin 1 connects to the wiper arm of the potentiometer at P1, while pin of U2 is connected to the touch switch. When you are ready to insert the IC into the socket, grab your anti-static wrist strap and place it on the hand that will be handling the chip. Remember to ground the anti-static wrist strap to a grounded outlet before using it. Each integrated circuit will either have a small indent circle or a cutout or notch at one end of the IC. This notch or cutout is the "locator" to help install the IC. Pin 1 will be just to the left of the cutout or notch. Carefully insert the IC into its socket.

Now fit the jumper wires including those making up the touch wires. You can use tinned copper wire or the ends of resistors. Do not use plain copper wire as its appearance will deteriorate and become tarnished after a period of time.

Next, you will want to install the 15 light-emitting diodes. These are very temperature sensitive and will be destroyed if allowed to get too hot during soldering. Hold them in position with your fingers during soldering to prevent them from getting too hot. The cathode side of the diodes all face one direction and are marked on the board with a dot. This is the short lead and if you look into the body of the LED you will see it is the largest portion inside the LED (see Figure 28-5). Notice that the LEDs are arranged in a specific pattern for the Hangman Game as shown in Figure 28-6. Mount

the LEDs at their respective locations, leaving about ¼ inch from the bottom of the LEDs to the circuit board.

Finally, grab the battery clip leads and solder them to the circuit board; the black or minus (−) lead will connect to the junction of R30, Q3, C8 and the "touch switch." The red or plus (+) battery clip lead connects to cathode end of the LEDs L1 through L7.

After you have completed the Hangman circuit, take a short deserved rest and when we continue we will inspect the circuit board for any possible "cold" solder joints or "short" circuits. Pick up the circuit board, with the foil side of the board facing upwards toward you. Now look over the solder joints on the PC board to make sure that all the solder joints appear clean, smooth and shiny. If any of the solder joints look dull, or "blobby", then you will have to unsolder the joint, remove the solder and then re-solder the joint. Finally, you will want to inspect the circuit board to make sure that there are no "shorts" on the board. Often when component leads are cut, they will "stick" to the underside of the circuit board due to the sticky residue left by the rosin core solder. Make certain that no "stray" leads are left on the board, they can "short" out the circuit traces when power is first applied. When your inspection is complete, you are ready to test your new Hangman Game. Attach a fresh 9-volt battery to the battery clip leads. Press the "touch switch" and your circuit should begin to "come-to-life."

If, for some unknown reason, your unit fails to operate, it can be diagnosed with the aid of a simple piece of test equipment. This is a LED and 330 ohm resistor connected to two jumper leads. A multi-meter will also be handy but not essential. Connect the cathode of the test LED to ground and use the resistor lead as a probe. Switch on the game and test these points on U2 for a "high": pins 3, 5, 6, 10, 11, 12, 13 and 14. Pins 10, 11, 12 and 13 will be supplying the "boost" circuit. Notice the LED will light up very brightly when connected to the positive of the $100\,\mu F$ electrolytic. Also touch the common Emitter terminal of the transistors for a "high." When you have "boost," U1 will be getting its voltage (boost voltage) and will be ready for test. Before checking the second IC, test each of the LEDs by taking the flying lead of the resistor to each of the resistors connected to the Collectors. The test LED will light up as well as a single or pair of LEDs on the game. If only one LED of a pair lights up as well as the test LED, you will know that one LED has shorted during assembly due to soldering. If no LEDs light up, one of the LEDs is open circuit, and if only the test LED lights up, both LEDs are shorted. If one section of LEDs is not being turned on, the transistor could be getting incorrect biasing or suffering from a Base-Emitter short.

The $470\,\mu F$ electrolytic can be charged externally via a 10 k resistor connected to the positive of the battery. This will sequentially turn on the LEDs. Once the LEDs are alight, they should stay illuminated for five minutes or so without the voltage on the electrolytic draining away. Any gradual decline will indicate a leakage path. Remove the shut-down transistor and the staircase Base resistors starting from Q10, to locate the fault. Q1 should also be checked.

The second $\frac{1}{10}$ second touch switch delay output is detected on pin 4. It is normally "high" and goes low for $\frac{1}{10}$ second. Pin 10 is the pulse output and will flash briefly when gates c and d are pulsed via the 10 n capacitor. You will be able to detect this pulse on the test LED. If LEDs 14 and 15 do not illuminate, the battery will be slightly "low." The circuit is fairly critical on voltage. The supply cannot deviate more than 1 volt either side of 9 volts. Use this same reasoning to investigate any other of the blocks you feel are not operating satisfactorily. Once you have run over the whole circuit, connect up the battery firmly and mount the board on a project box. The board will accept two small screws to finish the job. All that is left now is find an opponent and meet your challenge.

Chapter 29

Cliff Hanger Game

Parts List

Parts Bin

R1 220 k ohm, 5%, ¼-watt
resistor

R2 15 k ohm, 5%, ¼-watt
resistor

R3 680 k ohm, 5%, ¼-watt
resistor

R4 820 ohm, 5%, ¼-watt
resistor

R5, R11 510 ohm, 5%,
¼-watt resistor

R6, R13 10 k ohm, 5%,
¼-watt resistor

R7, R9, R10 4.7 k ohm,
5%, ¼-watt resistor

R8, R14 1 k ohm, 5%,
¼-watt resistor

R12 220 ohm, 5%,
¼-watt resistor

P1 47 k ohm trimmer
potentiometer

C1, C4 10 µF, 35-volt
electrolytic capacitor

C2, C3, C5, C6 0.01 µF,
35-volt ceramic
capacitor (103)

D1 1N4004 silicon diode

D2, D3 1N4148 silicon
diode

L1 green LED

L2, L3, L4, L5 red LEDs

L6, L7, L8 red LEDs

L9 yellow LED

Q1, Q2 MPSA20 transistor—
2N3904 or NTE123AP

U1, U2 LM555 timer IC

U3 CD4017 CMOS counter IC

S1 Pushbutton (normally
open)

Sx SPST ON/OFF switch
(optional)

B1 9-volt transistor
radio battery

Misc PC board, IC socket,
battery clip, wire, etc.

Have you ever met the challenge of mountain climbing? If not, this is your chance to do it electronically. The "Cliff Hanger" highlighted in Figure 29-1 will let you experience the thrill of mountain climbing and your skills to reach the top safely! Easy? Not at all! But it can be done. In this game you will have to press the pushbutton at the precise moment to advance to the next level on your way up to the top of the mountain. If the

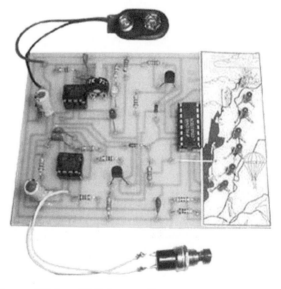

Figure 29-1 *Cliff Hanger Game*

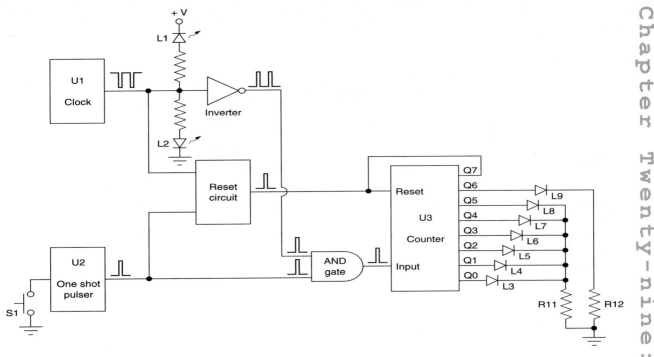

Figure 29-2 *Cliff Hanger Game block diagram*

control button is pressed at the wrong time, you will fall to the bottom. You will not break any bones but you will have to start all over again. This game also features a skill-adjust control for the would-be advanced mountain climbers! The Cliff Hanger Game operates from a single 9-volt battery.

The Cliff Hanger Game is illustrated in the block diagram at Figure 29-2 and in the schematic in Figure 29-3. The clock integrated circuit at U1 generates pulses of a high duty cycle (88%) that causes the red LED to be on most of the time, while the green LED (L3) will be "on" for only a short time. We have

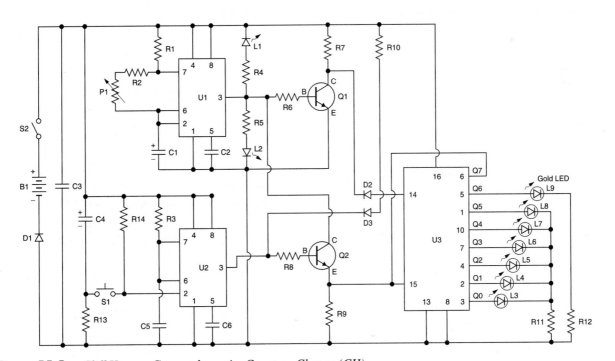

Figure 29-3 *Cliff Hanger Game schematic. Courtesy Chaney (CH)*

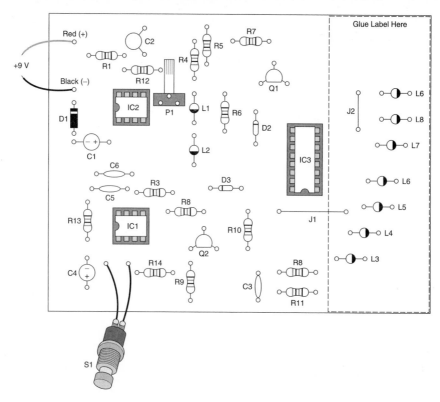

Figure 29-4 *Cliff Hanger pictorial diagram*

used an inverter, made by Q1, R6 and R7, to invert the clock pulses before sending them to the input of the AND gate. In the circuit we also have a one-short pulser (U2) which generates only one short pulse every time pushbutton S21 is pressed. Notice that the output of the pulser is connected to the other input of the AND gate. If pushbutton S1 is pressed when the green LED is "on," the AND gate will receive two simultaneous positive pulses at its inputs and will generate a pulse on its output that is applied to the input of U3, the CD 4017 decade counter with decoded outputs. IC U3 will shift a positive from one of its outputs to the next one (from Q0 to Q1 to Q2 to Q3, etc.) every time it receives a pulse on its input. This positive on the output of U3 will turn "on" the LED connected to it. Notice that LED L3, the one at the bottom of the mountain, is connected to output Q0 of U3, and that LED L9, the one at the top, is connected to output Q6 of U3. Therefore LEDs L3 to L9 will turn "on" successively as the pushbutton S1 is pressed at the "right" time. If pushbutton S1 is pressed at the wrong time, when the green LED is off and the red LED is "on," the reset circuit made by Q2, R8 and R9 will send a positive pulse to the reset of U3. This will reset U3 and put a positive on output Q0 that

will turn on LED L3, the one at the bottom. Notice that Q7 of U3 is connected to the rest input to reset this IC, if this output becomes positive. The AND gate in this circuit is formed by diodes D2 and D3 and resistor R10.

The Cliff Hanger Game was constructed on a 3¼ inch × 4 inch printed circuit board. The parts layout is shown on the pictorial diagram at Figure 29-4. Notice that the LED display is at the right side of the circuit board. Before we begin building the Cliff Hanger Game you will need to secure a clean, well-lighted worktable, so that you can place all the diagrams, the printed circuit and all the parts in front of you. Once you have secured a good spot, you will want to grab a 30 to 37-watt pencil tip soldering iron, some 60/40 lead/tin solder, a wet sponge and some "tip dresser." First heat up your soldering iron and clean off the tip once it's hot and then wipe the heated iron tip across the "tip dressing." Now you are about ready to begin. Open the parts bag. With a large project, you may want to place the components in small trays or egg cartons to hold all the small parts from flying off the worktable. Locate the resistors from the parts and now refer to Table 29-1, which lists the resistor color code information. Resistors will generally have at least three or four color bands

Table 29-1

Resistor Color Code Chart

Color Band	1st Digit	2nd Digit	Multiplier	Tolerance
Black	0	0	1	
Brown	1	1	10	1%
Red	2	2	100	2%
Orange	3	3	1000 (K)	3%
Yellow	4	4	10000	4%
Green	5	5	100000	
Blue	6	6	1000000 (M)	
Violet	7	7	10000000	
Gray	8	8	100000000	
White	9	9	1000000000	
Gold			0.1	5%
Silver			0.01	10%
No color				20%

which help identify the resistor values. The color band will start from one edge of the resistor body, this is the first color code which represents the first digit, the second color band depicts the second digit, while the third band is the resistor multiplier value. The fourth color band represents the resistor tolerance value. If there is no fourth band then the resistor has a 20% tolerance, if the fourth band is silver, then the resistor has a 10% tolerance value, and if the fourth band is gold then the resistor has a 5% tolerance value. Try to locate a resistor with the first band having a red color, the second band with a red color and a third band with a yellow color. This first resistor will be R1, with a value of 220 k ohms. Place R1 on the circuit board in its proper location and repeat the procedure for each of the resistors until all have been installed. Then solder all the resistors in their proper locations and clip the excess leads.

Note that potentiometers will have three in-line leads or two leads at either edge of the potentiometer body with an offset center lead which is the wiper or adjustable lead. Once you have identified where the potentiometer goes on the circuit board, you can go ahead and install the potentiometer P1, then remove the excess component leads.

Now let's move to installing the six capacitors; note that there are two electrolytic capacitors and four ceramic capacitors. When installing the ceramic capacitors refer to Table 29-2, which lists capacitor (code) values vs. actual capacitor values. Note that the four ceramic capacitors in this project have a code marking of (104) which corresponds to a value of 0.01 μF. The ceramic disk capacitors have no polarity markings, so they can be mounted without regard to direction. The electrolytic capacitors will have their values printed on the capacitor body. You will also note that there will be either a white or black band on electrolytic capacitors. This band will denote the minus (−) or plus (+) capacitor lead. Refer to the schematic to determine the polarity of the capacitor in the circuit, before inserting the electrolytic capacitors into the circuit board. The polarity of the capacitor must match its notation on the schematic and circuit board. Capacitors installed backwards are often the reason circuits do not work upon power-up and can sometimes cause damage to the circuit, so pay close attention to the correct orientation of the capacitors when installing them. Install all of the capacitors and solder them to the circuit board, remembering to cut the excess capacitor leads flush to the edge of the circuit board.

Table 29-2

Capacitance Codebreaker Information

This table is designed to provide the value of alphanumeric coded ceramic, mylar and mica capacitors in general. They come in many sizes, shapes, values and ratings; many different manufacturers worldwide produce them and not all play by the same rules. Most capacitors actually have the numeric values stamped on them; however, some are color coded and some have alphanumeric codes. The capacitor's first and second significant number IDs are the first and second values, followed by the multiplier number code, followed by the percentage tolerance letter code. Usually the first two digits of the code represent the significant part of the value, while the third digit, called the multiplier, corresponds to the number of zeros to be added to the first two digits.

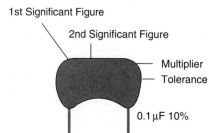

1st Significant Figure
2nd Significant Figure
Multiplier
Tolerance
0.1 µF 10%

Value	Type	Code	Value	Type	Code
1.5 pF	Ceramic		1000 pF/0.001 µF	Ceramic/Mylar	102
3.3 pF	Ceramic		1500 pF/0.0015 µF	Ceramic/Mylar	152
10 pF	Ceramic		2000 pF/0.002 µF	Ceramic/Mylar	202
15 pF	Ceramic		2200 pF/0.0022 µF	Ceramic/Mylar	222
20 pF	Ceramic		4700 pF/0.0047 µF	Ceramic/Mylar	472
30 pF	Ceramic		5000 pF/0.005 µF	Ceramic/Mylar	502
33 pF	Ceramic		5600 pF/0.0056 µF	Ceramic/Mylar	562
47 pF	Ceramic		6800 pF/0.0068 µF	Ceramic/Mylar	682
56 pF	Ceramic		0.01	Ceramic/Mylar	103
68 pF	Ceramic		0.015	Mylar	
75 pF	Ceramic		0.02	Mylar	203
82 pF	Ceramic		0.022	Mylar	223
91 pF	Ceramic		0.033	Mylar	333
100 pF	Ceramic	101	0.047	Mylar	473
120 pF	Ceramic	121	0.05	Mylar	503
130 pF	Ceramic	131	0.056	Mylar	563
150 pF	Ceramic	151	0.068	Mylar	683
180 pF	Ceramic	181	0.1	Mylar	104
220 pF	Ceramic	221	0.2	Mylar	204
330 pF	Ceramic	331	0.22	Mylar	224
470 pF	Ceramic	471	0.33	Mylar	334
560 pF	Ceramic	561	0.47	Mylar	474
680 pF	Ceramic	681	0.56	Mylar	564
750 pF	Ceramic	751	1	Mylar	105
820 pF	Ceramic	821	2	Mylar	205

(Continued)

Table 29-2 (*Continued*)

PicoFarad (pF)	NanoFarad (nF)	MicroFarad (mF, μF or mfd)	Capacitance Code
1000	1 or 1n	0.001	102
1500	1.5 or 1n5	0.0015	152
2200	2.2 or 2n2	0.0022	222
3300	3.3 or 3n3	0.0033	332
4700	4.7 or 4n7	0.0047	472
6800	6.8 or 6n8	0.0068	682
10000	10 or 10n	0.01	103
15000	15 or 15n	0.015	153
22000	22 or 22n	0.022	223
33000	33 or 33n	0.033	333
47000	47 or 47n	0.047	473
68000	68 or 68n	0.068	683
100000	100 or 100n	0.1	104
150000	150 or 150n	0.15	154
220000	220 or 220n	0.22	224
330000	330 or 330n	0.33	334
470000	470 or 470n	0.47	474

The Cliff Hanger Game has three single silicon diodes, two 1N4148 diodes and a power diode at D1. Silicon diodes will have a polarity marking on them which must be observed when installing the diodes. The diode body will generally have a black or white band at one edge of the diode. This black band denotes the diode's cathode or minus (–) lead. Refer to the schematic when installing the diodes and make sure that you can identify where the cathode lead is on the schematic; for example, diode D1 has its cathode toward the minus (–) battery clip lead, diode D2 has its cathode pointing toward Q1's Collector lead, and D3 has its cathode pointing to U2 pin 3. Locate the diodes and place them on the circuit board and solder them in place. Next cut the excess diode leads from the board.

There are two transistors in the Cliff Hanger Game project, both are PNP types. Refer to the diagram shown in Figure 29-5 for insight on how to identify the transistor pin-outs. Transistors generally have three leads: a Base lead, a Collector lead and an Emitter lead. The diagram illustrates the two most common transistor

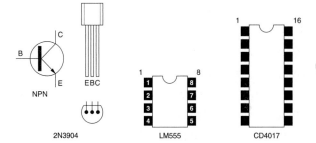

Figure 29-5 *Cliff Hanger Game semiconductors*

package types, the TO-92 and the TO-5 case. The TO-92 package has a flat side, and all three leads in a row with the Emitter at one end and the Collector lead at the opposite end of the package. The TO-5 transistor package is usually metal, with the Base lead offset in the center with a Collector lead opposite the Emitter lead. The Emitter lead corresponds to the metal tab on the bottom of the top-hat. Once you have identified the transistor, refer to the schematic and observe the pin-outs for the transistor on the diagram. The Base lead enters the flat line in the symbol perpendicularly,

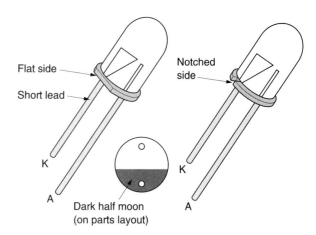

Figure 29-7 *LED identification*

Figure 29-6 *Cliff Hanger LED templates*

the Emitter is shown with the arrow pointing downward. The Collector is the other diagonal line opposite the Emitter lead. When inserting the transistors into the printed circuit holes make sure that you line up the transistor leads with the proper holes on the board before inserting them. When the transistors have been installed on the board, you can solder them in place, then cut the excess leads.

Next, let's install the LEDs, then cut out the mountain label from the sheet supplied (see Figure 29-6). Line up the dots marked on the paper with the holes in the circuit board. Glue the label to the board in the location shown in the parts layout diagram. Using a pin, carefully poke holes in the label where the holes of the PC board are located. Through these holes you will insert the leads of the LEDs.

Note that LEDs L1 and L2 are not mounted under the mountain label but are in the middle of the board. The LED at L1 is a green LED and the LED at L9 is yellow, while the rest of the LEDs from L3 to L8 are red LEDs. Refer to the LED mounting diagram depicted in Figure 29-7; this diagram illustrates the LED package, and note that one edge of the LED has a flat edge. The shorter lead nearest the flat edge is the cathode or

minus (–) LED lead. The cathode lead also corresponds to the dark half moon on both the LED diagram and the schematic. Solder all of the LEDs to the circuit board, then remove the excess leads from the board. Don't forget to install jumpers J1 and J2.

Locate the three integrated circuits. Carefully handle the integrated circuits using good anti-static techniques. If you have a grounded anti-static wrist band use this to handle the integrated circuits. Every IC package will have some form of location marking to help position the IC on the circuit board. You will find either a small indented circle at one end of the IC package, or a small cutout or notch at one end. Pin 1 of the IC will be just slightly to the left of these markings. Make sure you can identify the location of pin 1 on each of the sockets before attempting to insert the integrated circuits. Study the schematic and mark each socket, so you can identify pin 1. Next, install the integrated circuits into their respective sockets. Remember that there are three integrated circuits in this project. When inserting the ICs into their sockets be sure to note that pin 1 of U1 is connected to ground, and pin 1 of U2 is connected to ground, while pin 1 of U3 is connected to LED L8. Observe that pins 4 and 8 of U1 are connected to the (+) 9-volt bus, while pins 4 and 8 are connected to (+) 9-volts bus as well. Notice that pin 16 of U3 is connected to the (+) 9-volt bus.

Next, prepare two 6 inch lengths of #22 ga stranded-insulated wires, and solder the free ends of the two wires to the pushbutton switch at S1. The remaining wire ends will be soldered to the circuit as shown in the layout and schematic diagrams. The switch is connected between pin 2 of U2 and the junction of R13 and the minus (–) side of C4.

If you elected to install the optional SPST on/off switch as S2, you will wire it in series with the red battery clip which is connected to pins 4 and 8 on U1. The black battery clip lead is soldered to the cathode of diode D1.

Your Cliff Hanger Game is now completed, but before we test it out, you should take a short break. After the break we will inspect the circuit board for any "cold" solder joints and any "short" circuits. Pick up the circuit board and place the circuit board with the copper foil side up facing toward you. Inspect the solder joints on your PC board, you will want to make sure that all the solder joints look clean, smooth and shiny. If any of the solder joints look dull or "blobby" then you will need to remove solder from that particular joint with a solder-sucker or brain material and then re-solder the joint so that it looks clean, smooth and shiny. "Cold" solder joints will cause the circuit to fail prematurely, so it is best to attend to these problems quickly. Next, you will want to inspect the PC board for any possible "short" circuits which can be caused from any "stray" cut component lead which may have attached themselves to the underside of the board when the component leads were initially cut. "Short" circuit can damage the circuit when power is first applied to the circuit. So you will want to find any "short" circuits before connecting up the battery.

Now, for the "moment of truth." Find your 9-volt battery and snap on the battery clip. If you installed the optional power switch, turn it to the "on" position. Now rotate the shaft of P1 all the way toward the side facing the LEDs. This is the position of the lowest skill level. Now press pushbutton S1 twice to RESET the game. LED L3, at the bottom of the mountain, will turn "on," and the green and red LEDs, L1 and L2 will begin blinking.

If all went well, you can start playing "climbing the mountain," to get to its "top" and get the "gold" (i.e. the yellow LED), by pressing pushbutton S1 precisely at the same moment the green LED (L1) is on. If you do, the next LED up the mountain will light up. If you press pushbutton S1 at the wrong time, you will remain at the bottom if you are already there, or you will "fall down" if you are at a higher level. The goal is to get to the "top," which is LED L9, and get the gold. But it is not easy! You can change the level of difficulty of this game by adjusting the potentiometer at P1. To make the game easier turn P1 toward the LED side, and to make the game more difficult turn the control at P1 in the other direction. Go ahead and challenge your friends to a game of Cliff Hanger and have some fun!

In the event that your Cliff Hanger Game does not work, you will have to go back and re-check a few things. First, remove the battery! Carefully read the color codes on each resistor, and make sure that you have placed the correct resistor in the correct PC board hole. Next, take a look at the two transistors and make sure that you have installed them correctly. Carefully compare the transistors with the pin-out diagram and make sure that each of the transistor leads goes to the correct place on the PC board. Now make sure that the polarity sensitive capacitors are installed with respect to the polarity marking band at one end of the capacitor body, and finally make sure that you have installed the LEDs correctly. If in doubt, check the LEDs mounting diagram to be able to identify the flat edge of the LED and its negative or ground lead.

The installation of the transistors and integrated circuits are critical to the proper operation of the Cliff Hanger Game, so pay particular attention when installing them. Be sure to refer to the schematic and pictorial diagram and the parts pin-out diagrams. If you are having difficulty identifying these components or their pin-outs ask a knowledgeable electronics friend to help you. Once you have re-checked for any possible errors, you can go ahead and re-attach the battery, making sure that the battery clip leads are soldered to the correct point on the circuit board. Good luck and gaming!

Chapter 30

Alien Attack Game

Parts List

R1, R9, R10, R12 1k ohm, 5%, ¼-watt resistor

R14, R16, R17 1k ohm, 5%, ¼-watt resistor

R2 820 ohm, 5%, ¼-watt resistor

R3, R11, R13, R15 15k ohm, 5%, ¼-watt resistor

R4 330 k ohm, 5%, ¼-watt resistor

R5 22 k ohm, 5%, ¼-watt resistor

R6, R8 4.7 k ohm, 5%, ¼-watt resistor

R7 10 k ohm, 5%, ¼-watt resistor

R18, R19, R20 100 ohm, 5%, ¼-watt resistor

P1 5 k trimmer potentiometer

C1, C3 47 µF, 35-volt electrolytic capacitor

C2 1 µF, 35-volt electrolytic capacitor

C4 0.01 µF, 35-volt disk capacitor (103)

C5, C6, C7, C8 0.1 µF, 35-volt disk capacitor (104)

D1, D2, D3, D4, D5 1N4148 silicon diode

D6, D7, D8, D9, D10, D11 1N4148 silicon diode

Q1, Q2, Q3 2N3904 NPN transistor or NTE123AP

Q4 2N5060 SCR or NTE 5400

U1 LM556 Timer IC

U2, U3 CD4017 CMOS IC

L1, L11, L15, L19 yellow LEDs

L2 green LED

L3, L4, L5, L6, L7, L8, L9, L10 red LEDs

L12, L13, L14, L16, L17, L18 red LEDs

S1 on/off switch power

S2, S3, S4 momentary pushbutton (normally open)

Misc PC board, IC sockets, wire, battery clip, enclosure, etc.

The Aliens are coming in their warships and are dropping their bombs on you! The only way to defend yourself is by neutralizing their bombs before they reach the ground. The bombs fall at different speeds and in different directions. The Alien Attack Game depicted in Figure 30-1, will provide hours of fun and challenges for both young and old. The Alien Attack Game operates from a 9-volt battery and is safe for children.

The heart of the Alien Attack Game depicted in schematic diagram Figure 30-2, is U2, a CMOS CD4017 BCD counter. Integrated circuit U2 places a positive voltage, sequentially on each of the outputs from 00 to 09, as clock pulses arrive at pin 14. This

Figure 30-1 *Alien Attack Game*

positive voltage turns "on" LEDs L2 to L19 sequentially, from top to bottom, giving the illusion of a falling bomb. Integrated U3, another CD4017, selects which of the three rows of LEDs (L8 to L11, L12 to L15 or L16 to L19) is going to be activated. This is accomplished by putting a positive voltage sequentially, on its outputs, every time LED L2 turns "on," i.e. a bomb drops. Notice that the input of U3 at pin 14 is connected directly to the anode of L2. The Base of transistors Q1 to Q3 are connected to the outputs of U3 through one resistor and the diodes D2 to D11. Transistors Q1 to Q3, when activated by the outputs of U3, apply a negative voltage to the cathodes of the LEDs of the three rows of LEDs. In this manner, when a row is selected by U3, the LEDs in this row turn "on" when a positive voltage is applied to pins 5, 6, 9 or 11 of U2. The action switches S2 to S4 RESET, the

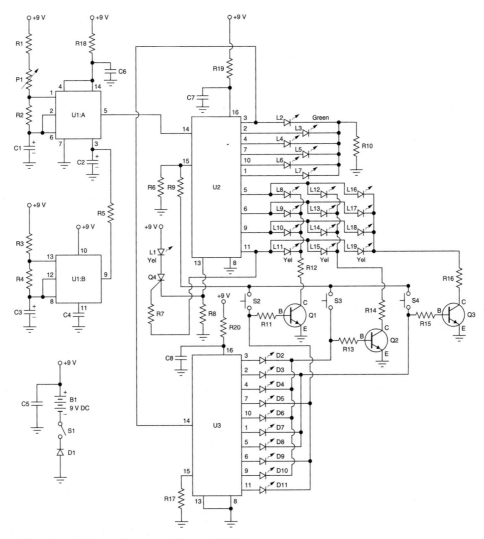

Figure 30-2 *Alien Attack Game. Courtesy Chaney (CH)*

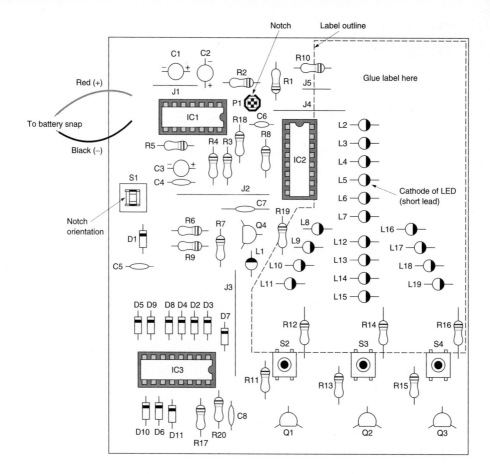

Figure 30-3 *Alien Attack Game pictorial diagram*

counting sequence of U2, by applying a positive voltage to the RESET pin 15 of U2, only if they are pressed when the corresponding row of LEDs is selected. This RESET action of the counter creates the illusion of "intercepting and destroying" the alien bombs. Finally, the input of U2 receives clock pulses from a voltage controlled oscillator (VCO) made from two halves of U1. The frequency of the pulses produced by the VCO depends on the voltage applied to pin 3 of U1-A by the low frequency oscillator created with U1-B. The change in frequency of the pulses produced by the VCO will cause an increase or decrease in the rate and speed of the "dropping bombs."

The Alien Attack Game is a more complex project skill 3 category game. A printed circuit board is a must for this project, and since there are many semiconductors and LEDs, a pictorial diagram is shown in Figure 30-3. If you are ready, let's begin building the Alien Attack Game.

Prepare a large, clean work area, so that you can spread out all the diagrams and parts for the project. Locate a 27 to 30-watt solder iron or pencil for this project, some 22 ga 60/40 rosin core solder, a wet sponge and some "Tip Tinner" used to clean the solder iron tip. An anti-static wrist band for handling integrated circuits is also a good idea. There is a velcro band which wraps around the wrist you will be using to handle the integrated circuits. The free wire end of the wrist band connects to a power ground at a 110-volt outlet. Before you start, make sure you have all the project parts in front of you and seat yourself down. Moving around or across carpets while you are assembling a circuit can cause a static charge which can destroy integrated circuits. Next, locate all the resistors for the project and place them in front of you. Refer to Table 30-1, which lists the resistor color codes and their values. You will note that each resistor will have either three or four color bands which start at one end of the

Table 30-1
Resistor Color Code Chart

Color Band	1st Digit	2nd Digit	Multiplier	Tolerance
Black	0	0	1	
Brown	1	1	10	1%
Red	2	2	100	2%
Orange	3	3	1000 (K)	3%
Yellow	4	4	10000	4%
Green	5	5	100000	
Blue	6	6	1000000 (M)	
Violet	7	7	10000000	
Gray	8	8	100000000	
White	9	9	1000000000	
Gold			0.1	5%
Silver			0.01	10%
No color				20%

resistor. The first color band is the first digit of the resistor code as seen in the table. The second color band is the second digit of the code. The third color band is the resistor's multiplier value. If the resistor has no fourth color band, then the resistor will have a 20% tolerance value or range. If the fourth color band is silver then the resistor will have a 10% tolerance value, and if the fourth color band is gold then the resistor will have a 5% tolerance value. So look through the resistors in front of you for a group of resistors which have brown as the first band, with black as the second band, and red as the third band. This color code will denote that the resistor has a 1000 ohm value. You will note that you should have seven of these resistors. Go ahead and insert these resistors into their proper positions on the circuit board. Once the seven 1000 ohm resistors have been inserted into the circuit board, you can solder them in place. Locate the rest of the resistors and place them on the circuit board and solder them in place. After soldering the resistors in place, take your end-cutters and cut the excess resistor leads, flush to the circuit board.

Find the trimmer potentiometer in the parts pile. Generally trimmer potentiometers will have three leads,

all in a row or two leads at the edges with an offset center lead which corresponds to the wiper or movable lead. The potentiometer is mounted near the top of the circuit board next to U1. Solder the potentiometer to the circuit board, and then remove the excess leads with your end-cutter.

Now let's move on to installing the capacitors into the circuit. Refer to Table 30-2, this table illustrates the capacitor code which may be marked on some of the disk or mylar capacitors. Sometimes capacitor values are marked on the body of the components and other times a code may be placed on the capacitor body instead of the value. You will note that capacitor C4 is a disk capacitor and it will have a code (103) which denotes a value of 0.01 µF. Capacitors C5, C6, C7 and C8 will all have (104) marked on them; this corresponds to a value of 0.1 µF. You will also see some larger electrolytic capacitors in the parts pile. These larger tubular capacitors should have their actual values marked on them. The electrolytic capacitors will also have a black or white color band. In the color band you will notice that there will be a plus (+) or minus (−) marking. These capacitors are polarity sensitive and must be inserted into the circuit correctly with respect

Table 30-2
Capacitance Codebreaker Information

This table is designed to provide the value of alphanumeric coded ceramic, mylar and mica capacitors in general. They come in many sizes, shapes, values and ratings; many different manufacturers worldwide produce them and not all play by the same rules. Most capacitors actually have the numeric values stamped on them; however, some are color coded and some have alphanumeric codes. The capacitor's first and second significant number IDs are the first and second values, followed by the multiplier number code, followed by the percentage tolerance letter code. Usually the first two digits of the code represent the significant part of the value, while the third digit, called the multiplier, corresponds to the number of zeros to be added to the first two digits.

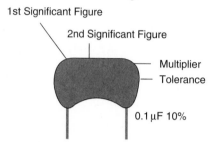

1st Significant Figure

2nd Significant Figure

Multiplier

Tolerance

0.1 µF 10%

Value	Type	Code	Value	Type	Code
1.5 pF	Ceramic		1000 pF/0.001 µF	Ceramic/Mylar	102
3.3 pF	Ceramic		1500 pF/0.0015 µF	Ceramic/Mylar	152
10 pF	Ceramic		2000 pF/0.002 µF	Ceramic/Mylar	202
15 pF	Ceramic		2200 pF/0.0022 µF	Ceramic/Mylar	222
20 pF	Ceramic		4700 pF/0.0047 µF	Ceramic/Mylar	472
30 pF	Ceramic		5000 pF/0.005 µF	Ceramic/Mylar	502
33 pF	Ceramic		5600 pF/0.0056 µF	Ceramic/Mylar	562
47 pF	Ceramic		6800 pF/0.0068 µF	Ceramic/Mylar	682
56 pF	Ceramic		0.01	Ceramic/Mylar	103
68 pF	Ceramic		0.015	Mylar	
75 pF	Ceramic		0.02	Mylar	203
82 pF	Ceramic		0.022	Mylar	223
91 pF	Ceramic		0.033	Mylar	333
100 pF	Ceramic	101	0.047	Mylar	473
120 pF	Ceramic	121	0.05	Mylar	503
130 pF	Ceramic	131	0.056	Mylar	563
150 pF	Ceramic	151	0.068	Mylar	683
180 pF	Ceramic	181	0.1	Mylar	104
220 pF	Ceramic	221	0.2	Mylar	204
330 pF	Ceramic	331	0.22	Mylar	224
470 pF	Ceramic	471	0.33	Mylar	334
560 pF	Ceramic	561	0.47	Mylar	474
680 pF	Ceramic	681	0.56	Mylar	564
750 pF	Ceramic	751	1	Mylar	105
820 pF	Ceramic	821	2	Mylar	205

Table 30-2 (*Continued*)

PicoFarad (pF)	NanoFarad (nF)	MicroFarad (mF, µF or mfd)	Capacitance Code
1000	1 or 1n	0.001	102
1500	1.5 or 1n5	0.0015	152
2200	2.2 or 2n2	0.0022	222
3300	3.3 or 3n3	0.0033	332
4700	4.7 or 4n7	0.0047	472
6800	6.8 or 6n8	0.0068	682
10000	10 or 10n	0.01	103
15000	15 or 15n	0.015	153
22000	22 or 22n	0.022	223
33000	33 or 33n	0.033	333
47000	47 or 47n	0.047	473
68000	68 or 68n	0.068	683
100000	100 or 100n	0.1	104
150000	150 or 150n	0.15	154
220000	220 or 220n	0.22	224
330000	330 or 330n	0.33	334
470000	470 or 470n	0.47	474

to these markings, in order for the circuit to work properly. Inserting electrolytic capacitors backwards can cause circuit damage. Pay particular attention to the band markings and insert the capacitors, then solder them in place. Remember to cut the excess capacitor leads from the circuit board.

Next, let's move on to installing the silicon diodes. Silicon diodes are a polarity sensitive component. You will notice a black or white band on the body of the diode. The band denotes the diode's cathode marking. The cathode marking corresponds to the flat line on the diode symbol on the schematic, while the anode of the diode is shown as the triangle. Locate all the diodes and place them on the circuit board. Note that all the diodes except for D10, D6 and D11 will have their cathodes facing the top of the board. Once the diodes have all been inserted, you can go ahead and solder them in, then remove the excess lead lengths.

Now try to locate the transistors for the project—there will be three of them. Refer to the semiconductor pin-out diagram in Figure 30-4, which illustrates the

pin-out and symbols. Transistors usually have three leads: a Base lead which corresponds to the flat line on the symbol, the Collector shown coming to the Base at an angle, and the Emitter lead opposite the Collector lead. Transistors Q1 to Q3 will probably look like the diagram and have a TO-92 type case. You will note that the Emitter is at one end with the Base lead in the center and Collector at the opposite end of the device. Note the Emitter will have an arrow either pointing toward or away from the Base lead. If the arrow points to the Base lead then the transistor is a PNP type, and if the arrow points away from the Base the transistor is an NPN device. Note that all of the transistors are mounted at the lower right side of the circuit board. The Alien Attack Game also has a silicon controlled rectifier or SCR at Q4. From the diagram you will see that the SCR looks more like a diode with a line coming off from the cathode; this is the Gate lead. An SCR is shown as a diode with an anode and cathode and the Gate. The TO-92 case of the SCR will have the cathode at one end with the Gate lead in the center and the anode at the

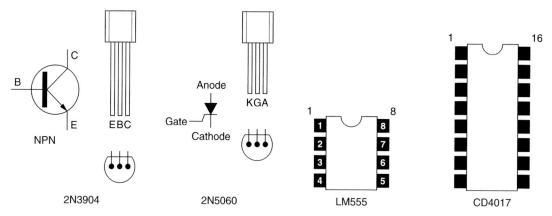

Figure 30-4   *Alien Attack Game semiconductors*

opposite end of the device. If you have any difficulties identifying the transistor or SCR pin-outs, consult with a knowledgeable electronics enthusiast for help. Once the transistors and the SCR have been placed on the circuit board, you can solder them in place. Cut the excess SCR and transistor leads when finished soldering the components on the board.

Locate the three integrated circuit sockets, and note that U1 is a 14 pin device, while U2 and U3 are 16 pin devices. Using integrated circuit sockets is a good idea. In the event of circuit failure at some date, integrated circuit sockets will permit you to easily replace the parts without having to unsolder the IC from the circuit board. Most people have great difficulty in trying to remove integrated circuits from circuit boards, usually resulting in damaged circuit boards, so it is wise to use sockets to avoid problems later. Most IC sockets have small cutouts or notches at one end of the socket; just to the left of the notch you will find pin 1. Line-up pin 1 on the sockets with where pin 1 should be placed for each of the integrated circuits. Solder each of the IC sockets to the circuit board.

Let's move on to installing the LEDs into the circuit board. Installing the LEDs will take some patience, since there are many of them and they must be installed correctly for the circuit to work properly. Cut out the template label provided for the LEDs, shown in Figures 30-5 and 30-6, and note you can use either of them. Glue the label to the circuit board where the LEDs are to be inserted, just above the pushbutton switches. Using a pin, push out the holes, so the LEDs can be mounted through them. Now, refer to diagram Figure 30-7 which shows the LED pin-outs. You will

Figure 30-5   *Alien Attack Game LED template # 1*

notice that each LED has a flat edge on it. The flat edge corresponds to the shorter lead which is the cathode lead. The cathode or shorter lead is also represented by the dark half moon on the diagram.

So here comes the tricky part, refer to the pictorial diagram, which illustrates the LED layout. Note that the yellow LEDs are mounted at L1, L11, L15 and L19. The green LED is amounted at L2, while the remaining LEDs are all red. When installing the LEDs, make

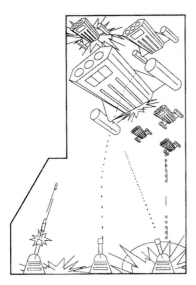

Figure 30-6 *Alien Attack Game LED template # 2*

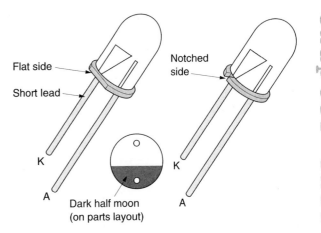

Figure 30-7 *LED identification*

sure that you observe the cathode lead placement which corresponds to the dark half moon. All the cathodes or shorter leads will face the same direction. Solder all the LEDs to the circuit board, then remove the excess leads by cutting them flush to the edge of the circuit board.

Next, go ahead and solder the jumpers on the circuit board. Locate all the switches for the project and place them in front of you. There will be a single on/off slide switch and three momentary pushbutton switches. The on/off switch S1 will be mounted at the left side of the circuit board, and the pushbutton switches will be mounted below the LEDs, toward the bottom right side of the circuit board. Position the switches in their respective places and solder them in place.

Now locate and identify the integrated circuits. Integrated circuits will have some form of markings on them, which help identify their pin locations. Integrated circuits may have a small indented circle or a small cutout or notch at one end of the IC package. The location of pin 1 of the IC will be just to the left of the cutout or notch or indented circle. Make sure you identify which of the IC socket pins is pin 1 before inserting the IC into its respective socket. You will want to make sure pin 1 of the IC matches what will be pin 1 on the IC socket, so refer to schematic as necessary. Using the grounded wrist strap insert the ICs into their respective sockets.

Remember that pin 1 of U1 connects to the one end of the potentiometer P1 as seen on the schematic.

Integrated circuit U2 has pin connected to LED, L7, while U3 has pin connected to D7.

Now locate the 9-volt battery clip and solder to the circuit board; it is placed at the top left side of the board, above S1, while the red or positive (+) clip lead will be connected to C5. The minus (–) or black clip lead is soldered to the ground bus of the circuit.

Now that you have completed the Alien Attack Game circuit, take a short rest and when you return we will inspect the circuit board for any possible "cold" solder joints and "short" circuits. Turn the circuit board over, so that the copper foil side of the board is facing up toward you. Examine the foil side of the board for possible "short" circuits. The solder joints should all look clean, bright and shiny. If any of the solder joints look dull or "blobby" then remove the solder from that particular joint and re-solder the joint so that it looks clean, bright and shiny.

Next examine the foil side of the circuit board for any "short" circuits caused by any remaining "cut" component leads which may have been left after the leads were cut off. Sometimes stray component leads can "short" across copper foil pads and "short" out causing damage to the circuit once powered-up. If the circuit board looks clean with no extraneous wires across circuit pads, then you are ready to apply power and test the circuit for the first time.

Install the battery snap, observing the proper connections. After the battery is connected, rotate the shaft of potentiometer P1 to its center position. Next, place three fingers of the same hand on top of the pushbutton switches. Operate the on/off switch S1, to the "on" position. As you do this the Aliens will start to drop "bombs" on you. You can intercept and destroy the "bombs" by pressing the right switch or S4 after the "bomb" chooses one of the three paths. If the "bomb" reaches one of the three yellow LEDs at the bottom of each path, you have failed to destroy the "bomb" before touching ground, the "bomb" has hit you, LED 1 turns "on" and the game stops. To play again, operate switch S1. As you play the game, you will notice that the pace at which the aliens drop the "bombs" suddenly changes and you will have to intercept them either faster or slower depending upon the transitions. You can also adjust the overall speed of the game with trimmer potentiometer P1. To slow the game, rotate the shaft of P1 clockwise; to speed up the game, rotate the shaft of P1 counterclockwise.

Your Alien Attack Game is now ready for play. Challenge you friends or siblings to some championship games and see who can get the highest score, who can kill the most aliens before they "get" you!

In the event that the Alien Attack Game does not work upon power-up, then you will have to remove the 9-volt battery and look for possible errors. First you will want to inspect the circuit board to make sure that you have placed the correct resistor in its proper place, check the color on each resistor and make sure they are in the correct location. If the resistors are all in the correct place, you can move on to inspecting the

capacitor placement, to make sure that you have installed the correct capacitor in the correct location. You will also need to inspect to see if the polarity markings on the electrolytic capacitors have been observed. Once this is accomplished you can again move on looking to see if the diodes have all been inserted correctly with the diode bands pointing in the proper direction. Remember, diodes and electrolytic capacitors have polarity marking and they must be observed.

OK, let's move on to checking for proper installation of the transistors and SCR. Make sure that you can identify which leads are the Base, Collector and Emitter, re-check these leads and where they are inserted on the board. The Emitters on all of the transistors are connected to ground. The Base of each transistor is connected to a switch point on the pushbutton switches. Each of the Collectors will be connected to a resistor. The SCR will have its anode connected to plus 9 volts, while the cathode is connected to resistor R8. The SCRs gate is connected to resistor R7.

Finally, you will have to make sure that all of the LEDs are mounted in their respective locations. Remember, the cathode leads of the LEDs are the shorter leads nearest the flat edge of the LED body. The cathode corresponds to the dark half moon on the layout diagram. Once you have re-checked everything, you again connect the 9-volt battery and switch S1 to the "on" position to test the circuit once again. Good luck!

Chapter 31

Starburst Light Display

Parts List

Parts Bin

R1, R2, R3, R4   1 k ohm, 5%, ¼-watt resistor

R5, R6, R7, R8, R9, R10   220 ohm, 5%, ¼-watt resistor

R11, R13, R15   30 k ohm, 5%, ¼-watt resistor

R17, R19, R21   30 k ohm, 5%, ¼-watt resistor

R12, R14, R16   100 k ohm, 5%, ¼-watt resistor

R18, R20, R22, R24   100 k ohm, 5%, ¼-watt resistor

R25, R27, R28   10 k ohm, 5%, ¼-watt resistor

R26, R29   5 k ohm, 5%, ¼-watt resistor

P1   100 k ohm trimmer potentiometer

C1, C2, C3, C4, C5   4.7 µF, 35-volt electrolytic capacitor

C6, C7, C8, C9, C10   4.7 µF, 35-volt electrolytic capacitor

Q1, Q2, Q3, Q4, Q5   C458 NPN transistor—NTE85 replacement

Q6, Q7, Q8, Q9, Q10   C458 NPN transistor—NTE85 replacement

U1   CD4017 CMOS Decade counter IC

L1 to L25 red LEDs

S1   SPST power on/off switch

B1   9-volt transistor radio battery

Misc   PC board, IC socket, wire, enclosure, hardware, battery clip, etc.

The Starburst Display shown in Figure 31-1 is a real attention grabber; you can use it to highlight a display, a special object, or product. You could build the Starburst circuit and hang it on your bedroom wall! The Starburst Display circuit utilizes 25 LEDs, the center LEDs and the display spread out much like a fireworks display, producing an eye-grabbing display. You'll have lots of fun building and using this cool display circuit.

The heart of the Starburst Display circuit illustrated in Figure 31-2 is the oscillator circuit formed with transistors Q8 and Q9. Capacitors C9 and C10, along with resistors in the oscillator circuit, determine the operating frequency. You can adjust the speed of the

Figure 31-1   *Starburst Display Project*

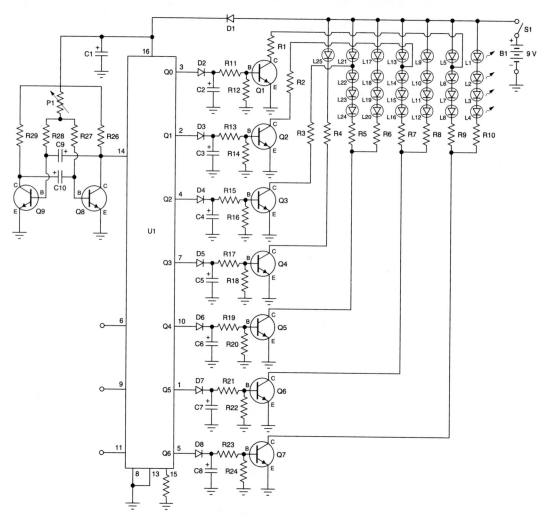

Figure 31-2 *Starburst Display circuit. Courtesy Future Kit (FK)*

Starburst Display repetition by using potentiometer P1. The output from the oscillator is fed directly to the decade counter at U1. The seven outputs from the decade counter are fed to transistors Q1 through Q7, which are used to drive the banks of LEDs. Capacitors C2 through C8 are utilized to delay the outputs from each of the decade counter outputs, which smooth the appearance of the display. The first output of the CD4017 is the Q0 output at pin 3, followed by the next output Q1 at pin 2 and so on. The display sequence of the LEDs is depicted in the chart shown in Table 31-1. The sequence operation of the Starburst is unique and operates by first turning "on" LED L5 with transistor Q1, then transistor Q2 turns "on" LED L13. Next transistor Q3 turns "on" LED L21, followed by transistor Q4 which turns "on" LED L25, in the center of the display. Next transistor Q5 turns "on" all of the LEDs in the center or inner ring as shown in the

sequence chart. Next transistor Q6 is used to turn "on" the LEDs in the middle ring and finally transistor Q7 turns "on" the LEDs in the outer ring. The display shuts down briefly and then begins the sequence all over again. The repetition rate or speed of the display is controlled by the potentiometer P1.

The pictorial or layout diagram shown in Figure 31-3 illustrates the component and LED placement on the circuit board. The Starburst LED Display board was constructed on a 3¾ inches × 3¾ inches circuit board, and the electronic driver circuit was built on a 3¾ inch × 2 inch circuit board. The LED display board houses only the 25 LEDs and six 220 ohm resistors, while the electronics driver board houses the electronic circuit used to drive the Starburst Display.

Before we begin constructing the Starburst Display circuit let's take a few moments to locate a large table or workbench so you can spread out all of the project

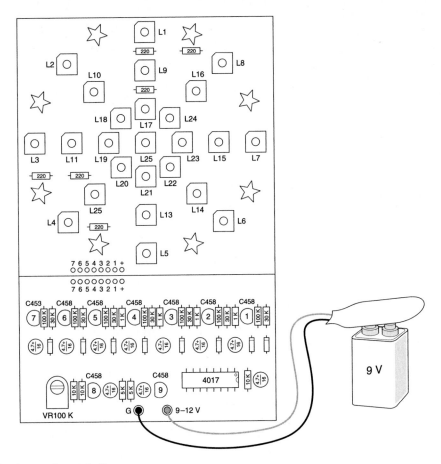

Figure 31-3 *Starburst pictorial diagram*

parts, diagrams, charts and tools. You will need to locate a low power pencil tip soldering iron; a 27 to 33-watt iron will work fine. Locate a 22 ga roll of 60/40 rosin core solder, a wet sponge and a small jar of "Tip Tinner," a soldering tip cleaner/dresser. You will need a small pair of end-cutters, a small pair of needle-nose pliers, a set of small screwdrivers, a magnifying glass and a pair of tweezers. Place all the project components in front of you, along with the schematic, pictorial

diagram, tables and charts. Locate an anti-static wrist strap, which is used to protect integrated circuits when handling them. Moving around while handling integrated circuits can cause static charges to build up and destroy integrated circuits, so it is advisable to wear an anti-static wrist band while handling the integrated circuits.

With all the components in front of you, heat up your soldering iron and we will begin building the Starburst circuit. Refer now to Table 31-2, which illustrates the

Table 31-1

Starburst Display Sequence Chart

Sequence	LEDs Turned "On"
1	L5
2	L13
3	L21
4	L25
5	L17, L18, L19, L20, L21, L22, L23, L24
6	L9, L10, L11, L12, L13, L14, L15, L16
7	L1, L2, L3, L4, L5, L6, L7, L8

Table 31-2

Resistor Color Code Chart

Color Band	1st Digit	2nd Digit	Multiplier	Tolerance
Black	0	0	1	
Brown	1	1	10	1%
Red	2	2	100	2%
Orange	3	3	1000 (K)	3%
Yellow	4	4	10000	4%
Green	5	5	100000	
Blue	6	6	1000000 (M)	
Violet	7	7	10000000	
Gray	8	8	100000000	
White	9	9	1000000000	
Gold			0.1	5%
Silver			0.01	10%
No color				20%

resistor color code chart. Resistors will have three or four color bands on the body of the component. The colors will begin at one end of the resistor body. The first color band will be the first digit of the resistor's code, a number from 0 to 9. The second color band represents the second digit of the resistor value, and the third color band represents the resistor's multiplier value. A fourth color band depicts the resistor's tolerance. If the fourth color band is silver then the tolerance will be 10%, and if the fourth color band is gold then the resistor has a 5% tolerance value. No fourth color band indicates a 20% tolerance value. Look through the project resistors. Resistor R1 will have a (brown) band followed by a (black) band and then a (red) band followed by a (gold) band. Resistor R1 will then have a 1000 ohm value (1) (0) (000). Identify the remaining resistors and place them on the circuit board, in their proper locations. Finally solder the resistors to the circuit board, and then remove the excess component leads by cutting the leads flush to the circuit board with a pair of end-cutters.

Capacitors will be of two major types—electrolytic types and disk or mylar capacitors. Electrolytic capacitors will often be larger capacitors which will have a polarity marking on them. You will see either a white or black band on an electrolytic capacitor. Near or

in the band, you will see either a plus (+) or minus (−) designation. You will want to compare this marking against the schematic diagram when placing the capacitor on the circuit board. Installing an electrolytic capacitor backwards in the circuit could damage the circuit and or the component, so it is important to install it correctly. Circuits will often have smaller capacitors; these capacitors may have their value printed on the component body, but often the component is too small to have its actual value printed on it. In this situation the capacitor will have a three-digit code printed on it. Next, refer to the chart in Table 31-3, which illustrates the capacitor codes. For example, a three-digit code (104) denotes a 0.1 µF or 100 nF value capacitor. This project uses only electrolytic capacitors, so pay particular attention to the polarity marking when installing them. Install all of the capacitors, and then solder them in place on the PC board. Remember to clip the excess solder leads after soldering in the capacitors.

The Starburst Display project uses a number of silicon diodes, but remember that diodes also represent a polarity sensitive device. Diodes have an anode and a cathode lead. Looking at the schematic, note that the anode is the symbol of an arrow or triangle which points toward the straight line, which is the cathode symbol. Diodes will generally have a white or black

Table 31-3
Capacitance Codebreaker Information

This table is designed to provide the value of alphanumeric coded ceramic, mylar and mica capacitors in general. They come in many sizes, shapes, values and ratings; many different manufacturers worldwide produce them and not all play by the same rules. Most capacitors actually have the numeric values stamped on them; however, some are color coded and some have alphanumeric codes. The capacitor's first and second significant number IDs are the first and second values, followed by the multiplier number code, followed by the percentage tolerance letter code. Usually the first two digits of the code represent the significant part of the value, while the third digit, called the multiplier, corresponds to the number of zeros to be added to the first two digits.

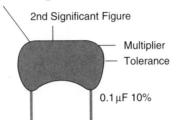

1st Significant Figure
2nd Significant Figure
Multiplier
Tolerance
0.1 µF 10%

Value	Type	Code	Value	Type	Code
1.5 pF	Ceramic		1000 pF/0.001 µF	Ceramic/Mylar	102
3.3 pF	Ceramic		1500 pF/0.0015 µF	Ceramic/Mylar	152
10 pF	Ceramic		2000 pF/0.002 µF	Ceramic/Mylar	202
15 pF	Ceramic		2200 pF/0.0022 µF	Ceramic/Mylar	222
20 pF	Ceramic		4700 pF/0.0047 µF	Ceramic/Mylar	472
30 pF	Ceramic		5000 pF/0.005 µF	Ceramic/Mylar	502
33 pF	Ceramic		5600 pF/0.0056 µF	Ceramic/Mylar	562
47 pF	Ceramic		6800 pF/0.0068 µF	Ceramic/Mylar	682
56 pF	Ceramic		0.01	Ceramic/Mylar	103
68 pF	Ceramic		0.015	Mylar	
75 pF	Ceramic		0.02	Mylar	203
82 pF	Ceramic		0.022	Mylar	223
91 pF	Ceramic		0.033	Mylar	333
100 pF	Ceramic	101	0.047	Mylar	473
120 pF	Ceramic	121	0.05	Mylar	503
130 pF	Ceramic	131	0.056	Mylar	563
150 pF	Ceramic	151	0.068	Mylar	683
180 pF	Ceramic	181	0.1	Mylar	104
220 pF	Ceramic	221	0.2	Mylar	204
330 pF	Ceramic	331	0.22	Mylar	224
470 pF	Ceramic	471	0.33	Mylar	334
560 pF	Ceramic	561	0.47	Mylar	474
680 pF	Ceramic	681	0.56	Mylar	564
750 pF	Ceramic	751	1	Mylar	105
820 pF	Ceramic	821	2	Mylar	205

(Continued)

Table 31-3 (*Continued*)

PicoFarad (pF)	General Capacitance NanoFarad (nF)	Code Information MicroFarad (mF, μF or mfd)	Capacitance Code
1000	1 or 1n	0.001	102
1500	1.5 or 1n5	0.0015	152
2200	2.2 or 2n2	0.0022	222
3300	3.3 or 3n3	0.0033	332
4700	4.7 or 4n7	0.0047	472
6800	6.8 or 6n8	0.0068	682
10000	10 or 10n	0.01	103
15000	15 or 15n	0.015	153
22000	22 or 22n	0.022	223
33000	33 or 33n	0.033	333
47000	47 or 47n	0.047	473
68000	68 or 68n	0.068	683
100000	100 or 100n	0.1	104
150000	150 or 150n	0.15	154
220000	220 or 220n	0.22	224
330000	330 or 330n	0.33	334
470000	470 or 470n	0.47	474

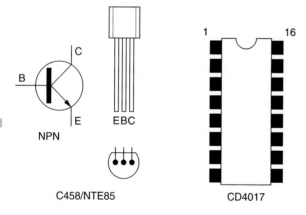

Figure 31-4   *Starburst semiconductor pin-out diagram*

band at one end of the diode body which indicates the diode's cathode polarity. So when installing the diode pay attention to direction of the diode both in the schematic and the layout or pictorial diagram. Place the diodes in their respective locations on the circuit board and solder them in place. Finally cut the excess leads flush to the circuit board.

Next we will install the nine NPN transistors. Refer to the transistor and IC pin-out diagram shown in Figure 31-4, which illustrates the location of the transistor leads. Remember that transistors have three leads, a Base lead, a Collector lead, and an Emitter lead. Referring to the schematic, you will see that the Base lead is the straight line on the symbol. Both the Collector and the Emitter join at the Base lead. The Emitter lead will have the arrow on it. If the arrow points away from the transistor then the device is a NPN transistor. If the arrow points toward the transistor, then the transistor is an PNP device. When installing the transistors note that you have to refer to the pin-out diagram as well as the schematic and the pictorial diagram. The pin-out diagram will show the position of the leads with respect to the package; usually the TO-92 plastic transistor case will have the Emitter at one end, the Base lead in the center and the Collector lead at the opposite end. Most transistor pin-out diagrams are shown from the bottom view. Note that when installing

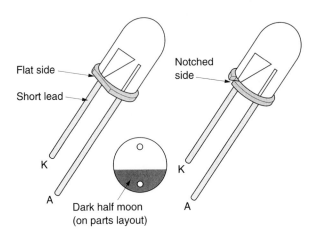

Flat side
Short lead
Notched side
K
A
K
A
Dark half moon
(on parts layout)

Figure 31-5 *LED identification*

the transistors the Base lead will be connected to one end of the 30 k ohm resistor at the outputs of U1, while the Emitter leads are connected to ground. The Collector leads from transistors Q1 through Q7 are connected to the LEDs, transistors Q8 and Q9 for the system oscillator. The Base leads are connected to capacitors C9 and C10, the Emitter leads are connected to ground, and the Collector leads are connected to resistors R26 and R29. If you are having difficulties identifying the transistor leads ask a knowledgeable electronics enthusiast for help. Place the transistors in their respective locations and solder them in place. Finally cut the excess component leads flush to the circuit board. Installing the transistors correctly is crucial to the proper operation of the circuit.

The next component we will install will be the integrated circuit, the CD4017 decade counter IC. The Starburst Display project uses a single integrated circuit. It is recommended that an integrated circuit socket be installed prior to installing the IC. The IC socket is good insurance, in the event of a circuit failure at a later date, down-the-road. It would then be a simple matter to just pull out the defective IC and replace it with a new one. Most mortals cannot unsolder an IC from the PC board without damaging the printed circuit board. Note that most integrated circuit sockets and integrated circuits will have some sort of identifying marking on them to help orient them on the circuit board. IC sockets will have a notch or cutout at one end of the socket. Pin of the socket will be to the left of the cutout or notch. Solder the IC socket onto the circuit board.

At one end of the IC package, you should see either a small indented circle, a cutout or a notch. Immediately to

the left of the indented circle or notch, you will find pin 1. When installing the IC into the socket, you will have to match pin 1 of the IC socket with pin 1 on the IC. Usually there will also be a notch on the socket, which will help locate the pions of the IC. When installing the IC or IC socket make sure that you align pin 1 of both the IC and socket so that pin 1 connects to diode D7 and that pin 16 is connected to the +9-volt power bus. Using your anti-static wrist band, which should be ground to a power outlet, you can insert the IC into its socket. Have a knowledgeable friend help you if you are confused.

Finally, we are going to install the LEDs, note that there are 25 LEDs in this project. You will have to pay particular attention to mounting them correctly, if you want the circuit to work properly. Refer to the diagram depicted in Figure 31-5, which illustrates the pin-outs of the LEDs. LEDs, like diodes have an anode and a cathode lead, as shown. The anode lead is an arrow pointing to the cathode lead, just the same as a diode representation. The cathode lead is generally the shorter of the two leads, and the cathode lead is usually noted, next to the flat edge of the plastic LED package. The LEDs must be oriented correctly for the display to work correctly, so take your time installing them. Refer to the pictorial diagram shown in Figure 31-3, which illustrates the placement of the LEDs on the PC board. Note that LED L5 is mounted at the bottom of the display (with LED L13 adjustable L5,) with L21 above L13. LED L25 is mounted at the center of the Starburst display. When the LEDs light up L5 lights first with L13 next followed by L21 and then L25. Next LEDs L17, L18, L19, L22, L23 and L24 light up in a circle. The Starburst then lights up the next circle formed by LEDs L9, L10, L11, L12, L13, L14, L15 and L16. Finally the Starburst advances to the final stage of the outer ring formed by LEDs L1, L2, L3, L4, L5 L6, L7 and L8. Solder the LEDs to the circuit board, then remove the excess component leads.

Locate the 9-volt battery clip and solder the black or negative (−) lead to the ground bus on the circuit board. If you use the optional power toggle switch at S1, you should wire the switch in series with the red or positive (+) lead of the 9-volt battery clip, as shown.

If you are building the kit version of the Starburst display project then you will notice that there are seven jumpers that are used between the display board and the electronics driver board, as some builders may want to separate the display board from the electronics driver board.

Why not take a short, well-deserved break, and when we return we will inspect the circuit board for any possible "short" circuits or "cold" solder joints. OK, let's begin by placing the circuit board in front of you with the foil side of the board facing upwards toward you. First, we will look for possible "cold" solder joints. All of the solder joints on the circuit board, should look clean, smooth and shiny. If any of the solder joints look dull, or "blobby," then unsolder the solder from the PC solder pad and re-solder the joint with new solder, so that the joint looks clean, smooth and shiny. Next we will inspect the circuit board for any "short" circuits which might have formed from "stray" component leads left over from when you cut the excess component leads. Often the sticky rosin core leaves residue which can "trap" component leads, or wire to the circuit board, which can cause "short" circuit paths between copper lands or component pads. Once you have inspected the board and you are satisfied that there are no "short" circuits or "cold" solder joints, you can attach the battery clip to a fresh 9-volt battery. Toggle the switch S1 to the "on" position and you should begin to see the Starburst pattern display on the circuit board. The LEDs from the bottom of the circuit board will light up first, followed by the inner circular rings as the display goes outward to the outer edges. Once the lights reach the outer edge of the display board, the circuit will repeat itself and the display will start all over again. Note that the potentiometer control P1, will adjust the speed of the display. You adjust the control clockwise to increase the speed rate of the display.

In the event that your Starburst Display circuit does not operate correctly, you will have to remove the 9-volt battery and re-inspect the circuit board for any errors which may have occurred during construction. Sometimes mistakes do occur and they can often be discovered and corrected quickly and hopefully without damage to the circuit.

If one of the LEDs does not light-up, you may have installed one of them backwards. This is a common error and it is easy to fix. Simply unsolder the LED from the board, remove or "suck" the solder from the joint, then rotate the LED 180 degrees and then re-solder the LED back into the PC board. Re-connect the battery and test the circuit to see if it now works. If the circuit appears "dead" with no response then your troubleshooting skills will be tested.

First, you will want to inspect the circuit board to make sure that all of the resistors have the correct value and that they are in the correct location. This is a common problem and easily solved. Take your time and if in doubt, use an ohmmeter to make sure the correct value is displayed for a resistor in question. Next, you will have to move to inspect that the capacitors have been installed correctly. Remember that capacitors have polarity and it must be observed for correct operation. Since this circuit uses only electrolytic capacitors, you must take note of the polarity band when installing them. Refer to the schematic diagram and layout diagram when checking your capacitor placement. Now, check the diode placements on the circuit board. Remember that the silicon diodes have polarity, which must be observed for the circuit to work correctly. The white or black band on the diode indicates the cathode lead. The cathode leads of the diodes point toward the electrolytic capacitors, while the diode's anode are connected directly to the outputs of the IC.

Re-check all of the transistor placements on the circuit board. Note that the Base leads of transistors Q1 through Q7 connect to the 30 k ohm resistors which are connected to the diode/capacitor junctions. The Emitter leads are connected to the common ground and the Collector leads connect directly to the LED display. Re-check the transistor pin-out diagram and compare it to the schematic and pictorial diagram and make sure that you correctly understand the pin-out diagram and that the reference is usually from the bottom of the transistor.

Finally, you will have to inspect the installation of the LEDs on the circuit board. Observe the LED pin-out diagram and the layout diagram for the LEDs to make sure you understand the pin-outs of the device. Remember that the anode is the arrow on the LED symbol and that the straight line on the symbol is the cathode. In the schematic the cathodes all face downwards toward the IC outputs, while the anodes generally connect to the 9-volt power bus. If you discovered that you installed an LED backwards, remove the solder from the connection, reverse the LED and re-solder the LED back onto the circuit board.

After inspecting the circuit board carefully, reconnect the 9-volt battery and you should see the Starburst display spring to life. Hopefully your Starburst display circuit now works fine and you are ready to dazzle your friends and family. The Starburst circuit is a great attention getter for any project of display. Have fun and enjoy!

Chapter 32

Electronic Voice Changer

Parts List

Parts Bin

R1, R12, R13 10 k ohm, 5%, ¼-watt resistor

R2 150 k ohm, 5%, ¼-watt resistor

R3, R5 220 k ohm, 5%, ¼-watt resistor

R4 470 ohm, 5%, ¼-watt resistor

R6, R9, R10 22 k ohm, 5%, ¼-watt resistor

R7, R8 47 k ohm, 5%, ¼-watt resistor

R11 100 k ohm, 5%, ¼-watt resistor

R14 4.7 k ohm, 5%, ¼-watt resistor

R15 47 ohm, 5%, ¼-watt resistor

P1 100 k ohm potentiometer (trimpot)

P2 50 k ohm potentiometer (trimpot)

P3 1 megohm potentiometer (trimpot)

C1, C4, C8, C10 100 nF, 35-volt Greencap capacitor

C2 47 µF, 35-volt electrolytic capacitor

C3 10 µF, 35-volt electrolytic capacitor

C5, C6, C9 220 nF, 35-volt Greencap capacitor

C7 10 nF, 35-volt Greencap capacitor

C11 4.7 nF, 35-volt Greencap capacitor

D1 1N4004 silicon diode

Q1 PN100 NPN transistor—2N3904 or NTE123AP

U1 LM833 Quad Dual op-amp IC

U2 CD4066B Quad Analog Switch IC

U3 LM555 IC Timer

Mic electret microphone

J1 ⅛ mini phone jack

S1 SPST power switch

B1 9-volt battery

Misc PC board, IC sockets, battery clip, wire, hardware, etc.

Ever wonder how they produce those strange metallic "robot" sounds in science fiction movies or TV shows? You can make your own voice sound very strange using the Electronic Voice Changer circuit illustrated in Figure 32-1. This project is great fun for kids of all ages, and also great party fun!

The voice changer circuit, shown in schematic in Figure 32-2, takes your voice from the electret they are alternately reversed in polarity using a switching signal with steady but adjustable frequency. This has the effect of "shifting" the frequencies in your voice up or down, depending upon the frequency of the switching signal. So your voice is effectively converted into that of a weird, metallic-sounding robot.

The weak audio signals from the microphone are amplified by U1:a, one half of the LM833 op-amp. The op-amp second amplifies the input signal 469 times by using the 220 k and 470 ohm resistors. The amplified output from U1:a is coupled through potentiometer P1. The output of P1 is coupled through a 100 nF capacitor at C6. Potentiometer P1 allows you to control the audio level to the next stage of the circuit.

The balanced modulator is controlled by U2:a and U2:b which are electronic switches, two sections of a CD4066B analog switch IC. The CMOS switches act like a normal switch but have no moving parts, instead they provide a connection between their input and output pins, when their control pin 13 is "high." When the voltage control pin 13 is "low" the switch is "off." So U2:a is turned "on" and "off" by changing the control voltage pin 13, while U2:b is controlled by the DC voltage on pin 5.

Why do we need two of these IC switches? Because with a balanced modulator, we need two versions of the input signal. The two versions of the audio have to be opposite polarity, so one goes positive while the other goes negative.

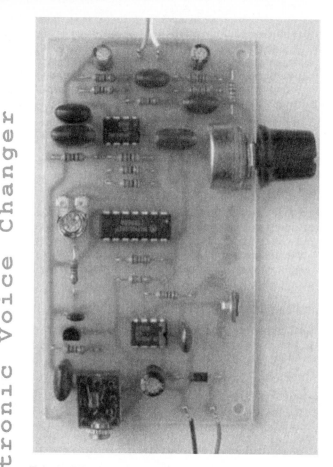

Figure 32-1   *Voice Changer*

microphone and amplifies it and then feeds the voice signal to something called a balanced modulator, where

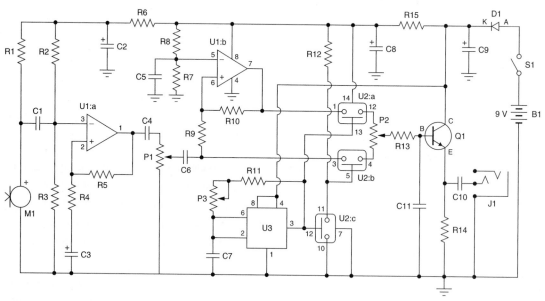

Figure 32-2   *Electronic Voice Changer. Courtesy Jaycar (JC)*

In this modulator, we feed the original audio signal from U1:a and P1 directly to pin 3 of switch U2:b, as you can see. At the same time we also feed it through U1:b, the second op-amp, in the LM833 IC. Integrated circuit U1:b has a lot of negative feedback because of the 22 k ohm resistors connected between its output pin 7 and negative input pin 6, and in series with the same signal input. This means that it provides no voltage amplification at all, but reverses the audio signal polarity or "inverts" them.

As a result of this polarity inversion in U1:b, the audio signals fed to pin 1 of U2:a are the same signals fed to pin 3 of switch U2:b, except that they're inverted. Which is what is needed to make a balanced modulator work properly.

The modulation switching frequency is generated by U3, which is a low cost IC timer. This is connected as a free running astable oscillator, which oscillates at a frequency controlled by trimpot P3. The 100 k ohm series resistor and the 10 nF capacitor are connected from pins 2 and 6 to the negative line (0V). With the component values shown, the oscillator frequency can be varied between 28 kHz and 280 Hz by varying trimpot P3.

Now just as the balanced modulator switches need two different versions of the audio input signals, they also need to be controlled by two versions of the switching signals as well. One is opposite in polarity to the other. So we feed the switching signal from output pin 3 of U3 directly to U2:a's control pin 13, but at the same time we feed it to the control pin 12 of switch U2:d.

This third switch is connected as an inverter, because pin 10 is connected to ground (0V) line, while pin 11 is connected to the +9 volt line via a 10 k "pull-up" resistor. So when the switch is "off," pin 11 will be at the +9 volt level or "high." But when the switch is turned "off" pin 11 will be connected directly to 0 volt or "low." And because this switch is being turned "on" and "off" by the switching up and down—with opposite polarity. When pin 12 is "high" pin 11 will be "low," and vice-versa.

By connecting pin 11 to control pin 5 of switch U2:b, we therefore end up turning U2:b on whenever switch U2:a is turned "off," and turning U2:b "off" when U2:a is turned "on." So the switches are driven in correct opposite polarity.

To complete the balanced modulator, we combine the switch output signals from pins 2 and 4 of U2 via trimpot P2. This is used for "balancing" the two outputs, so we can cancel out any of the unwanted original audio signals of switching signals which may appear in the switched outputs. If this isn't done, they can be audible and make the modulator's output sound "muddy." Small amounts of unwanted signals tend to appear in the switch outputs because inverted audio signals fed to U2:a and U1:b may not be exactly the same voltage as that fed to U2:b and also the switching signals fed to the two switches may not have an exact 1 to 1 on/off "mark/space" ratio.

Potentiometer P2 allows us to cancel out the two signals by varying the amount of resistance in series with the two switches, to compensate for these minor circuit imbalances.

As you can see, the output signal from P2 is fed to the base of transistor Q3, via a simple low pass filter formed by the 10 k series resistor and 4.7 nF capacitor. These filter out any remaining "switching spikes," which might be heard as a buzzing sound in the output.

Transistor Q3 is connected as "emitter follower" which provides current amplification to make the shifted audio signals stronger and better able to pass through a cable to your amplifier or recorder. The 100 nF capacitor couples the AC output signal through to the output, but blocks the DC voltage which is present at the emitter of Q3. Note that the output of this circuit must be followed by an audio amplifier, so you could elect to use the 8 watt audio amplifier project in another chapter.

You are probably ready to build the Electronic Voice Changer circuit, but before we go there, let's prepare your work area, gather some tools, diagrams and parts. Locate a large table or workbench area to assemble the project. Make sure you have ample lighting and ventilation. You will want to procure some small tools, such as a mini screwdriver, a pair of dykes or wire-cutters and maybe a pair of a needle-nose pliers. Also locate a small pencil type soldering iron. You will want a 27 to 33-watt pointed tip for the soldering iron. Find some #24 to 26 ga 60/40 rosin core solder. Next, locate a small jar of "Tip Tinner," soldering iron tip cleaner/dresser and an anti-static wrist band from your local Radio Shack store. The "Tip Tinner" is used to dress and clean the soldering iron tip, and the anti-static wrist band is needed for handling the integrated circuits. Static can be built up from moving around the

Table 32-1

Resistor Color Code Chart

Color Band	1st Digit	2nd Digit	Multiplier	Tolerance
Black	0	0	1	
Brown	1	1	10	1%
Red	2	2	100	2%
Orange	3	3	1000 (K)	3%
Yellow	4	4	10000	4%
Green	5	5	100000	
Blue	6	6	1000000 (M)	
Violet	7	7	10000000	
Gray	8	8	100000000	
White	9	9	1000000000	
Gold			0.1	5%
Silver			0.01	10%
No color				20%

workbench area, or walking across a rug or shifting in and out of a chair. Most anti-static wrist bands wrap around the wrist you use to handle the integrated circuits. Newer anti-static wrist bands do not have be grounded to an AC outlet.

Locate all the parts for the project and place them in front of you along with the printed circuit for the project. Get the schematic diagram, the layout or pictorial diagram and the resistor and capacitor color code charts. With the parts in front of you, locate the resistors. Refer to the chart in Table 32-1 which illustrates the resistor color chart and how it works. Resistors all have color bands on the body of the resistor. The first color band will be at one end of the resistor body. The first color band corresponds to the first digit in the code chart. The second color band represents the second digit of the color code and the third color band represents the resistor's multiplier code. A fourth silver band denotes a 10% tolerance resistor and a gold fourth band notes a 5% tolerance value, while the absence of a fourth band indicates a tolerance of 20%. So let's try to find resistor R1 in the parts pile. Look for a resistor with its first band colored brown, its second color band black, and its third color band orange. This resistor will be a 10 k ohm resistor for resistor R1. Repeat this procedure and try to identify all

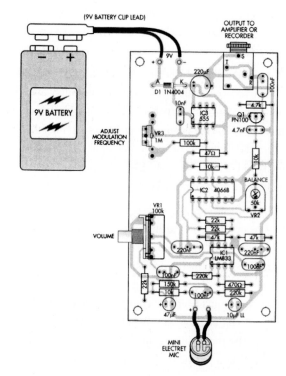

Figure 32-3 *Voice Changer pictorial diagram*

of the other resistors. Once you have identified all the resistors, you can find out where they are located on the circuit board. Refer to the schematic diagram in Figure 32-2 and the pictorial diagram in Figure 32-3.

These diagrams will help you locate and identify the parts before installing them on the PC board. Once all the resistors have been identified and their places on the circuit board determined, then you can go ahead and mount the resistors on the board. Finally solder the resistors to the PC board and cut the excess component leads with your wire-cutters. Cut the component leads flush to the circuit board.

Locate all three of the potentiometers for the circuit, P1 is a 100 k ohm pot entiometer, while P2 is a 50 k ohm potentiometer and P3 is a 1 megohm pot entiometer. Potentiometers usually have three leads, sometimes all three leads are in-line or all in straight row, sometimes there are three leads with two outer leads and an offset center lead. The center lead on a potentiometer is the adjustable wiper arm lead. Install the potentiometers onto the circuit board, then follow-up by removing the excess component leads.

Now we can move on to locating and identifying the capacitors in the project: refer to the chart in Table 32-2, this table will help you identify the capacitors. Often capacitors will have a value marked on the body of the capacitor, but smaller capacitors may not have their value on them but a code. The capacitor codes are generally three digit codes as can be seen on the chart. A code marked (104) will denote a capacitor with a $0.1\,\mu F$ value or 100 nF. So use the code chart and identify the capacitors for the project as needed. The Voice Changer circuit also uses electrolytic capacitors. These capacitors are polarity sensitive and usually have their values marked on the capacitor. Electrolytic capacitors will have a white or black band on the body of the component. Within the colored band will be a plus (+) or minus (−) marking, and this denotes the polarity of the capacitor. When installing the electrolytic capacitors refer to the schematic and pictorial diagram to see how the capacitor is mounted with respect to the polarity markings. Failing to observe the correct polarity marking and installing the capacitor backwards can often damage the circuit and/or the component as well. Once you have identified and located the capacitors, you can install them on the circuit board, and then solder them in their correct location. Remember to cut the excess component leads flush to the circuit board.

The Voice Changer circuit has one silicon diode at D1. Diodes are also polarity sensitive, polarity must be observed when mounting the diode. Note the symbol for a diode is a triangle or arrow pointing to a line. The triangle is the anode of the diode, while the line is represented as the cathode of the diode. The diode's cathode end will have a black or white band. Refer to the schematic and layout diagram when mounting the diode to make sure you have mounted the diode correctly with respect to the polarity markings.

The Voice Changer circuit utilizes a single transistor placed at Q1. Transistors are generally three lead devices with a Base, a Collector and an Emitter lead. Refer to the diagram shown in Figure 32-4, which depicts the transistor and integrated circuit pin-outs. Note the bottom view with the Emitter lead at one end, the Base lead in the center and the Collector lead at the opposite end from the Emitter. Now refer to the schematic diagram and note that the Base lead of the transistor is connected to the 10 k ohm resistor which is connected to trimpot P2. The Collector lead of Q1 is connected to the junction of C9 and D1, while the Emitter lead is connected to junction of C10 and R14 at the audio output jack J1.

Before installing the integrated circuits, consider installing IC sockets. Integrated circuit sockets will greatly aid you in the event of a circuit failure at some later point in time if it should ever occur. Sockets are inexpensive and good insurance, and will allow you to easily remove the IC rather than trying to unsolder the chip from the circuit board which will often result in a damaged circuit board. Referring to the schematic and layout diagram, identify which pin on the socket should be designated pin 1. Most integrated circuits will have some form of identification markings on the IC to help orient it on the circuit board. Solder the IC sockets to the PC board now. Notice that you will find either a small indented circle, a notch or a small cutout at one end of the IC package. Just to the immediate left of the notch or circle, you will find pin 1. Now, line up pin 1 on the IC with pin 1 on the circuit board and insert the IC into the socket, using the anti-static wristband. Repeat the process for all of the integrated circuits. If you are having trouble or are confused about the IC pin-outs don't be afraid to ask someone for help if you are not sure of yourself.

Now, you can install the electret microphone. You could choose to mount the mic on the PC board or run some short stiff wire between the mic and the circuit board if desired. When mounting the mic, be sure to

Table 32-2
Capacitance Codebreaker Information

This table is designed to provide the value of alphanumeric coded ceramic, mylar and mica capacitors in general. They come in many sizes, shapes, values and ratings; many different manufacturers worldwide produce them and not all play by the same rules. Most capacitors actually have the numeric values stamped on them; however, some are color coded and some have alphanumeric codes. The capacitor's first and second significant number IDs are the first and second values, followed by the multiplier number code, followed by the percentage tolerance letter code. Usually the first two digits of the code represent the significant part of the value, while the third digit, called the multiplier, corresponds to the number of zeros to be added to the first two digits.

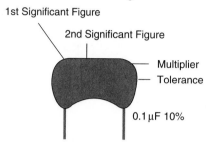

Value	Type	Code	Value	Type	Code
1.5 pF	Ceramic		1000 pF/0.001 µF	Ceramic/Mylar	102
3.3 pF	Ceramic		1500 pF/0.0015 µF	Ceramic/Mylar	152
10 pF	Ceramic		2000 pF/0.002 µF	Ceramic/Mylar	202
15 pF	Ceramic		2200 pF/0.0022 µF	Ceramic/Mylar	222
20 pF	Ceramic		4700 pF/0.0047 µF	Ceramic/Mylar	472
30 pF	Ceramic		5000 pF/0.005 µF	Ceramic/Mylar	502
33 pF	Ceramic		5600 pF/0.0056 µF	Ceramic/Mylar	562
47 pF	Ceramic		6800 pF/0.0068 µF	Ceramic/Mylar	682
56 pF	Ceramic		0.01	Ceramic/Mylar	103
68 pF	Ceramic		0.015	Mylar	
75 pF	Ceramic		0.02	Mylar	203
82 pF	Ceramic		0.022	Mylar	223
91 pF	Ceramic		0.033	Mylar	333
100 pF	Ceramic	101	0.047	Mylar	473
120 pF	Ceramic	121	0.05	Mylar	503
130 pF	Ceramic	131	0.056	Mylar	563
150 pF	Ceramic	151	0.068	Mylar	683
180 pF	Ceramic	181	0.1	Mylar	104
220 pF	Ceramic	221	0.2	Mylar	204
330 pF	Ceramic	331	0.22	Mylar	224
470 pF	Ceramic	471	0.33	Mylar	334
560 pF	Ceramic	561	0.47	Mylar	474
680 pF	Ceramic	681	0.56	Mylar	564
750 pF	Ceramic	751	1	Mylar	105
820 pF	Ceramic	821	2	Mylar	205

Table 32-2 (*Continued*)

PicoFarad (pF)	NanoFarad (nF)	MicroFarad (mF, µF or mfd)	Capacitance Code
1000	1 or 1n	0.001	102
1500	1.5 or 1n5	0.0015	152
2200	2.2 or 2n2	0.0022	222
3300	3.3 or 3n3	0.0033	332
4700	4.7 or 4n7	0.0047	472
6800	6.8 or 6n8	0.0068	682
10000	10 or 10n	0.01	103
15000	15 or 15n	0.015	153
22000	22 or 22n	0.022	223
33000	33 or 33n	0.033	333
47000	47 or 47n	0.047	473
68000	68 or 68n	0.068	683
100000	100 or 100n	0.1	104
150000	150 or 150n	0.15	154
220000	220 or 220n	0.22	224
330000	330 or 330n	0.33	334
470000	470 or 470n	0.47	474

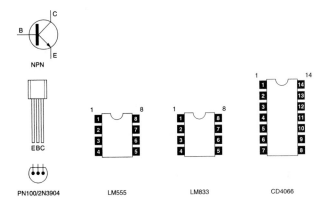

Figure 32-4 *Electronic Voice Changer semiconductors*

observe the plus (+) marking on the microphone housing, before hooking it up.

Locate the output jack at J1 on the schematic and install it on the circuit board. Note that the ground lead is connected to the common or ground bus on the circuit. Use a meter to determine which lead is the center or (hot) lead before installing in on the PC board.

Finally, you can connect the 9-volt battery clip. The black or minus (–) battery clip lead is connected to the ground bus of the circuit board, while the red or plus (+) battery clip lead is connected in series with a SPST toggle switch. The free end of the toggle switch is then connected to the anode of the diode at D1.

The Electronic Voice Changer circuit is now complete and ready to test; but before we go about testing the circuit, take a well-deserved break and when we return we will look over the PC board for "cold" solder joints and possibly "short" circuits.

Returning to the circuit board, turn over the circuit board so that the foil side of the PC board is upwards and facing you. First, let's take a look at all the solder joints on the circuit board. We are looking for possible "cold" solder joints. The solder joints should all look clean, smooth and shiny. If any of the solder joints look dull, "blobby" or dirty, then remove the solder from the joint and re-solder the joint until the joint looks clean, smooth and shiny. Finally we are going to inspect the circuit board for possible "short" circuits. When cutting excess component leads, it is not uncommon for a "cut" lead to stick to the underside of the board often from the sticky rosin from the solder. Look for any "bridges" of wire between the solder pads on the circuit board. Once you are satisfied that the board passed all your inspection tests, you can move on to testing the circuit electronically.

Connect up a 9-volt battery to the battery clip, turn on the switch at S1. If there is no smoke, then you can adjust P1 and P2 to their center position, and turn P3 to its most counterclockwise position. Connect up the Electronic Voice Changer circuit to an audio amplifier using a length of coax cable.

Turn up the volume on your external amplifier, and you should begin to hear a faint buzzing sound. Adjust P2 carefully until this sound becomes as faint as possible. You should find that this happens at only one setting of P2, because this trimpot adjusts the modulator's "balance" and the buzzing sound becomes faint only when the circuit is balanced.

Next begin speaking into the microphone, and gently turn up the modulator's gain control at P1. As you do, you should begin to hear a frequency shift version of your voice, from the speaker connected to your external amplifier. The sounds coming from the Voice Changer should sound pretty weird, but exactly how weird it sounds can be varied by adjusting P3. So try different settings and see how your voice sounds at each setting.

Unfortunately sometimes electronic circuits don't always work the first time we assemble them. If your Electronic Voice Changer does not work, don't despair, but remove the battery and we will take one more look at inspecting the circuit board for possible errors. Let's have a look at the circuit board more closely and see what we can find. First, take a look at the resistors and verify that you have the right resistor at the right location. Make sure that you can identify each resistor and its value. Have a friend help you; often another pair of eyes can catch a mistake easier. After inspecting the resistors, we will move on to examine the capacitor placements. You may have placed the wrong capacitor in the wrong location, or you may have installed the electrolytic capacitor backwards. Examine the color polarity band to make sure that the band faces in the proper direction. Next, have a look at the diode, and make sure that you have installed the diode in the right direction. As you remember, diodes have polarity which must be observed.

Now look a the transistors and make sure that you can identify each of the leads and that they are placed into the correct PC holes on the circuit board. It is a common error to reverse the Emitter and Collector leads. Compare the actual part with the schematic and pin-out diagrams to make sure of your placement. Finally you will need to make sure that the integrated circuits have been installed correctly. You need to make sure that pin 1 of the IC is truly connected to pin 1 of the IC socket, or the pin that you designate as pin 1. Study the schematic carefully so that you understand which location on the circuit board corresponds to pin 1 of the IC. Have a knowledgeable electronic enthusiast help you if necessary.

Once you have examined the circuit board carefully and thoroughly, you can re-connect the amplifier and 9-volt battery and re-test the circuit once again to make sure that it works correctly. In the rare event that the circuit still does not work, you may have to replace the integrated circuits and/or transistor.

Good luck and have fun with your new Electronic Voice Changer, it can be used for unlimited fun at parties. If you are into Karaoke the Voice Changer will add a new twist. If you have a mini radio station, you may want to try the Voice Changer for new effects. Connecting the Voice Changer to a public address system is sure to surprise everyone. Have fun!

Chapter 33

Racquetball Game

Parts List

Parts Bin

R1 220 k ohm, 5%,
   ¼-watt resistor

R2 22 k ohm, 5%, ¼-watt
   resistor

R3, R5, R7, R11, R13
   4.7 k ohm, 5%, ¼-watt
   resistor

R4 50 trimmer
   potentiometer

R6 47 k ohm, 5%,
   ¼-watt resistor

R8, R14 1 k ohm, 5%,
   ¼-watt resistor

R9, R10 10 k ohm,
   5%, ¼-watt resistor

R12 15 k ohm, 5%,
   ¼-watt resistor

C1 47 µF, 35-volt
   electrolytic capacitor

C2, C4, C5 1 µF,
   35-volt electrolytic
   capacitor

C3 2.2 µF, 35-volt
   electrolytic capacitor

C6 10 µF, 35-volt
   electrolytic capacitor

C7, C8 0.01 µF, 35-volt
   disk capacitor (103)

D1, D2, D3, D4, D5
   1N4148 silicon diode

D6 1N4004 silicon diode

U1 LM556 IC (National
   Semiconductor)

U2 CD4029 CMOS IC

U3 CD4028 CMOS IC

U4 CD4001 CMOS IC

L2, L3, L4, L6, L7 red
   LEDs

L8, L9, L10, L11 red
   LEDs

L1 yellow LED

L5 green LED

S1 pushbutton switch
   (normally open)

S2 on/off switch

B1 9-volt transistor
   radio battery

Misc PC board, wire, IC
   sockets, battery clip,
   wire, enclosure, etc.

The Racquetball Game allows you to "play" against the computer opponent. The racquet ball is represented by LEDs, and you have to "hit" the ball by pressing a button. If you do, the ball changes direction, travels up the court, bounces off the wall, then returns for you to "hit" once again. But that's not all! Suddenly the speed of the ball changes and you have to play faster or slower to keep up with the game. The project includes an adjustable speed control to allow beginners to play expertly. You have to keep "on your toes" though, because if you miss the ball, the game stops and you will have to start over again. The circuit operates from a single 9-volt battery and is safe for kids, see Figure 33-1.

At the heart of the Racquetball Game, shown in Figure 33-2, is the CD4029 IC. The CD4029 IC is an up/down counter. When this IC is counting up the "ball" (lighting sequence of LEDs) is moving up, from L1 to L11. When the CD4029 is counting down, the ball is moving down, from L11 to L1. The RS-flip-flop, made

Figure 33-1 *Racquetball Game*

up of NOR gates G1 and G2, changes the direction of the counting process by applying a high (counting up) or a low (counts down) to pin 10 of CD4029 IC. The outputs of the CD4029 are connected to the "1 of 10" decoder IC, the CD4028, which puts a positive voltage on one of its ten outputs depending on the binary number it receives at its inputs (i.e. pins 10, 11, 12 and 13). When the counting process goes up, the CD4028 will apply a positive voltage to its outputs O0 (pin3) to O9 (pin 5) sequentially, lighting LEDs L1 to L11 and creating the sensation that the "ball" is going up.

A similar opposite process occurs when the counting process is going down. The counting down process can be reversed by activating the one-shot pulser, by pressing pushbutton S1, when LEDs L2, L3 or L4 are on. When this happens, the AND gate, whose inputs are connected to L2, L3, and L4, and to the one-shot pulser, will apply a positive voltage to pin 6 of G2, which will cause the output of the flip-flop (pin 3) to become negative, reversing the counting process. The input of CD4029 receives clock pulses from a voltage controlled oscillator (VCO) made with one-half of the LM556 IC. The frequency of the pulses produced by the VCO depends on the voltage applied to pin 3 by the low-frequency oscillator made with NOR gates G3 and G4. This change in the frequency of the pulses produced by the VCO and applied to the input of the up/down counter will cause the "ball" to move faster or slower.

The Racquetball Game is a skill 2 type game, and is more complex and will require more skill and will take longer to construct than some other projects in this book. The Racquetball Game circuit was built on a 4½ inch × 3 inch printed circuit board. There are four

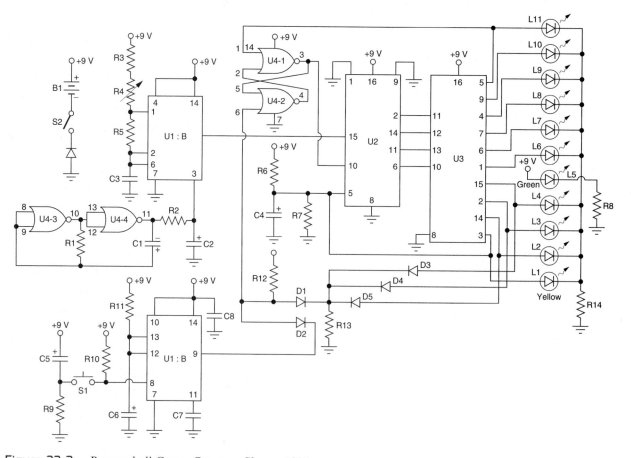

Figure 33-2 *Racquetball Game. Courtesy Chaney (CH)*

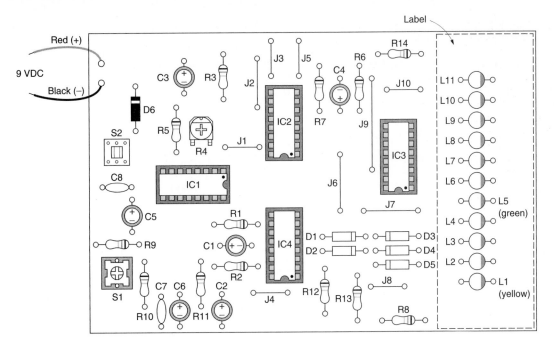

Figure 33-3 *Racquetball Game pictorial diagram*

integrated circuits, six diodes and 12 LEDs; the increased part numbers hints at the increased complexity. Let's begin constructing the exciting Racquetball Game. First find a clean, well-lit and comfortable work area where you can spread out all the parts on a table along with the diagrams. Locate a 27 to 33-watt pencil tip soldering iron with either a small flat edge tip or a small pointed tip. You will want also to locate a roll of 22 ga 60/40 rosin core solder, a small jar of "Tip Tinner", a soldering iron tip dresser/cleaner, and an anti-static wrist strap, all of which are available at your local Radio Shack store. The anti-static wrist strap will help protect the integrated circuits from damage due to high voltage static charge buildup when handling the ICs. Static charges can readily buildup from just moving around the workbench or table or from getting up from your chair, so it is best to take this precaution. You will also want to procure a pair of needle-nose pliers, a pair of diagonals or end-cutters, a magnifying glass and a pair of tweezers. Try to locate a couple of small screwdrivers, a small flat-blade and a small Phillips head driver. Heat up your soldering iron, dress the soldering iron tip by rolling the tip in the jar of "Tip Tinner" and then clean the tip with a wet sponge.

Locate the pictorial or parts layout diagram, shown in Figure 33-3, and place it in front of you along with the

schematic. Refer to the soldering discussion, heat up your soldering iron and you'll be ready to begin.

Locate four integrated circuit sockets and install them on the Racquetball Game PC board. Integrated circuit sockets are a great idea and will save immeasurable time and effort in the event of a circuit failure which might occur at some point in time. If you find that an integrated circuit becomes defective, you will only have to "pull" out the old one and quickly replace it with a new one, rather than trying to unsolder an IC and trying to replace it. Most people are not skilled enough to replace an integrated circuit without damaging the circuit board.

Locate all the resistors for the project and put them in front of you. Now refer to Table 33-1 which illustrates the color code values for resistors. Each resistor has three or four color bands. The first color band will be the first digit and will be located at one end of the resistor package. The second color band represents the second digit. The third color band is the resistor multiplier, and if there is a fourth color band then it represents the resistor's tolerance value. So let's take a look at resistor R1: its first color band is red, its second color band is also red, while the third multiplier band is yellow. If your resistor has no fourth color band then the resistor has a tolerance of 20%. If the resistor has a silver color band, then the tolerance is 10%.

Table 33-1

Resistor Color Code Chart

Color Band	1st Digit	2nd Digit	Multiplier	Tolerance
Black	0	0	1	
Brown	1	1	10	1%
Red	2	2	100	2%
Orange	3	3	1000 (K)	3%
Yellow	4	4	10000	4%
Green	5	5	100000	
Blue	6	6	1000000 (M)	
Violet	7	7	10000000	
Gray	8	8	100000000	
White	9	9	1000000000	
Gold			0.1	5%
Silver			0.01	10%
No color				20%

Finally, if the resistor has a gold band then the tolerance of the resistor is 5%. Identify all of the remaining resistors in the project using the color code chart as a reference and install all of the resistors onto the circuit board. Now, solder the resistors to the board. Next, use your end-cutters to cut the excess resistor leads flush to the edge of the circuit board.

This project has a number of capacitors, many of which are electrolytic types. Electrolytic capacitors all have polarity markings on the body of the component. The polarity markings are usually noted by a white or black band at either end of the package. Sometimes the polarity band will have a (−) imprinted in the marker band and other times the band may have a (+) marking. Be sure that you can identify if the marking is a plus (+) or a minus (−) when installing the capacitor into the circuit board. If the polarity band is represented with a minus (−) band then make sure that the end of the capacitor "points" to the ground connection on the circuit board when installing the device. The circuit can be damaged if an electrolytic capacitor is installed backwards, so take time and install them carefully. Capacitors C7 and C8 are non-electrolytic types and will either be marked with their respective value or with (103) as the value which denotes the value of 0.01 μF (see the chart in Table 33-2).

Next, let's move on to installing the diodes. Diodes are also polarity sensitive devices and must be installed correctly for the circuit to work properly. Schematically the diode's anode is the arrow or triangle pointing to the flat line which is the cathode. Usually diodes have black bands at the cathode end of the diode. The cathode is usually the (−) side of the diode and corresponds to the ground or more minus (−) side of the circuit. Note that most of the diodes are 1N4148 silicon diode types but diode D6 is a different type. Install the diodes, then solder them in place in their respective locations. Remember to cut the excess diode leads from the board.

Locate the variable 50 k ohm trimmer potentiometer at R4, it will have three in-line leads or two leads with a center offset lead, which is the adjustable wiper arm lead. The trimmer potentiometer is located at the top right side of the circuit board. Install the trimmer potentiometer and then solder it into place.

Before mounting the LEDs, you will want to locate the game field "template," shown in Figure 33-4. Punch out all the LED mounting holes on the "template." The "template" is then laid over the PC board. Find the "ball player" and mount L1, the yellow LED at the first hole position, next to the "ball player." Continue mounting the LEDs up the field. There will be three more red

Table 33-2

Capacitance Codebreaker Information

This table is designed to provide the value of alphanumeric coded ceramic, mylar and mica capacitors in general. They come in many sizes, shapes, values and ratings; many different manufacturers worldwide produce them and not all play by the same rules. Most capacitors actually have the numeric values stamped on them; however, some are color coded and some have alphanumeric codes. The capacitor's first and second significant number IDs are the first and second values, followed by the multiplier number code, followed by the percentage tolerance letter code. Usually the first two digits of the code represent the significant part of the value, while the third digit, called the multiplier, corresponds to the number of zeros to be added to the first two digits.

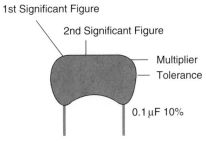

Value	Type	Code	Value	Type	Code
1.5 pF	Ceramic		1000 pF/0.001 µF	Ceramic/Mylar	102
3.3 pF	Ceramic		1500 pF/0.0015 µF	Ceramic/Mylar	152
10 pF	Ceramic		2000 pF/0.002 µF	Ceramic/Mylar	202
15 pF	Ceramic		2200 pF/0.0022 µF	Ceramic/Mylar	222
20 pF	Ceramic		4700 pF/0.0047 µF	Ceramic/Mylar	472
30 pF	Ceramic		5000 pF/0.005 µF	Ceramic/Mylar	502
33 pF	Ceramic		5600 pF/0.0056 µF	Ceramic/Mylar	562
47 pF	Ceramic		6800 pF/0.0068 µF	Ceramic/Mylar	682
56 pF	Ceramic		0.01	Ceramic/Mylar	103
68 pF	Ceramic		0.015	Mylar	
75 pF	Ceramic		0.02	Mylar	203
82 pF	Ceramic		0.022	Mylar	223
91 pF	Ceramic		0.033	Mylar	333
100 pF	Ceramic	101	0.047	Mylar	473
120 pF	Ceramic	121	0.05	Mylar	503
130 pF	Ceramic	131	0.056	Mylar	563
150 pF	Ceramic	151	0.068	Mylar	683
180 pF	Ceramic	181	0.1	Mylar	104
220 pF	Ceramic	221	0.2	Mylar	204
330 pF	Ceramic	331	0.22	Mylar	224
470 pF	Ceramic	471	0.33	Mylar	334
560 pF	Ceramic	561	0.47	Mylar	474
680 pF	Ceramic	681	0.56	Mylar	564
750 pF	Ceramic	751	1	Mylar	105
820 pF	Ceramic	821	2	Mylar	205

(Continued)

Table 33-2 (*Continued*)

PicoFarad (pF)	NanoFarad (nF)	MicroFarad (mF, µF or mfd)	Capacitance Code
1000	1 or 1n	0.001	102
1500	1.5 or 1n5	0.0015	152
2200	2.2 or 2n2	0.0022	222
3300	3.3 or 3n3	0.0033	332
4700	4.7 or 4n7	0.0047	472
6800	6.8 or 6n8	0.0068	682
10000	10 or 10n	0.01	103
15000	15 or 15n	0.015	153
22000	22 or 22n	0.022	223
33000	33 or 33n	0.033	333
47000	47 or 47n	0.047	473
68000	68 or 68n	0.068	683
100000	100 or 100n	0.1	104
150000	150 or 150n	0.15	154
220000	220 or 220n	0.22	224
330000	330 or 330n	0.33	334
470000	470 or 470n	0.47	474

LEDs at L2, L3 and L4, followed by the green LED at L5. Next mount the remaining red LEDs from L6 to L11.

Refer to the LED layout diagram shown in Figure 33-5, which depicts the LED pin-outs. Note that the LED should have one flat edge, this flat edge is next to the negative (–) or minus lead. This minus (–) lead will usually also be the shortest lead and is reflected in the diagram as the dark half moon. Be sure to look at the pictorial diagram once again to see the layout of the LEDs and their placement. Observe that the first ascending LED is yellow, it is followed by three red LEDs. LED L5 is a green LED followed by six red LEDs.

Next locate the two switches in the Racquetball Game. Switch S2 is the power "on/off" switch and S1 is the momentary "START" pushbutton switch, located in the lower left side of the circuit board. Find the 9-volt battery clip and solder it into place on the top left edge of the circuit board. The red or plus (+) battery clip lead is soldered to the top-most circuit pad on the top of the PC board, with the minus (–) battery lead just underneath the plus lead.

Look for the four integrated circuit packages, and refer to the semiconductor pin-out diagram, shown in Figure 33-6. Integrated circuits U2 and U4 are 16 pin packages, and U1 and U3 are 14 pin packages. Carefully handle the IC packages, with your anti-static wrist strap, to avoid any electrostatic potential transfer to the integrated circuits. When installing the integrated circuits be extremely careful to avoid moving or walking on a carpet. Sit or stand in one place while installing these components, since moving generates high voltage electrostatic potential which may damage the IC. All integrated circuits have some type of locating marks and you will need to observe these before inserting the integrated circuits into the circuit board. Some integrated circuits will have a small indented circle at one end of the IC package, other IC packages may have a notch or cutout at one edge of the IC package. Pin of the IC will be to the left of either of these markings. Make sure that you can identify pin 1 of the IC and that you can also identify where pin 1 is located on the IC socket before inserting the IC. It is very important that all of the integrated circuits are correctly installed to prevent damage to the circuit when

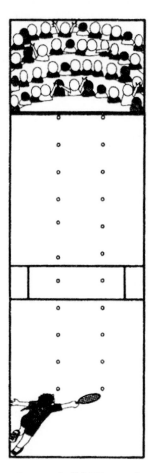

Figure 33-4 *Racquetball LED template*

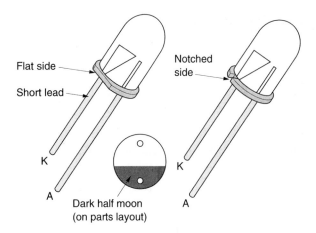

Figure 33-5 *LED identification*

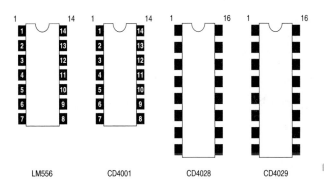

LM556   CD4001   CD4028   CD4029

Figure 33-6 *Racquetball semiconductor pin-out diagram.*

power is first applied. If you are having difficulties identifying the pin-outs of the integrated circuits, be sure to ask a knowledgeable electronics enthusiast for help.

Take a few moments and rest before inspecting the PC board for "cold" solder joints and possible "short" circuits. Turn the PC board over, so that the foil side is facing up toward you. Now look at all the solder joints and observe them carefully. All the solder joints should look clean, smooth and shiny. If any of the solder joints do not look good, you will have to remove the solder and re-solder the joint so that its looks clean, smooth and shiny with no big blobs of solder at the circuit pads. "Cold" solder joints will usually deteriorate in time and will cause the circuit to fail prematurely, so it's best to discover them early and to fix them quickly. Once you have inspected the board for "cold" solder joints, you can move on to inspect the board for possible "short" circuits. Take another look at the foil side of the circuit board; this time you are looking for "stray" component

leads which may have attached themselves when the component leads were cut. Often these remaining "cut" leads can attach themselves to the underside of the board and just stick there, usually due to the residue from the rosin core solder. Applying power with these "shorts" can damage the circuit when power is first applied.

Now is the "moment of truth"! Rotate potentiometer R4 to the counterclockwise limit. Attach your 9-volt battery to the battery clip, turn the circuit to the "on" position using switch S2 and the circuit will be powered-up and ready to test. As you do this some of the LEDs will sequence and then all of the red LEDs will be "off" and only the yellow and green LEDs will remain "on."

If all goes well you are now ready to play the game; however, sometimes circuits don't always work the first time power is applied. In this event you will need to turn "off" the circuit and remove the battery and re-inspect the circuit for any errors.

If one or more of the LEDs does not light-up, then you will have to reverse the leads. Unsolder the LED pins and remove the LED from the board. Then re-insert it carefully rotating the LED 180 degrees and solder it back into the circuit. If that was your only problem, then you were lucky, but if your circuit didn't light-up at all then you will have to delve deeper into troubleshooting your circuit. First you will need to carefully inspect the resistor placement to make sure that you have the correct resistor in the correct location on the circuit board. Compare color codes with each of the resistors to make sure that they are in the right location. Next, let's move on to inspecting the diodes to make sure that they are in the correct location and that they are correctly oriented on the circuit board. Remember that all diodes have a polarity band which denotes the cathode or minus (–) side of the device. Make sure that the diode band is oriented so that the cathode points to the most negative side of the circuit. Next, be sure that the integrated circuits have been inserted into their respective sockets and that they are oriented correctly in their socket. Finally, observe the mounting of all of the LEDs. Make sure that the flat edges are pointed in the same direction, all toward the common or negative (–) bus of the circuit on the right side of the schematic.

Once your inspection has been completed, and hopefully the problem corrected, you can connect up the battery again and begin to test the circuit as described earlier. This time the circuit will no doubt operate correctly and "all will be well."

So let's get ready to "play ball." In the Racquetball Game, the red LEDs will light up in sequence giving you the effect of a ball moving back and forth. To play the game, you will have to "hit the ball," by pressing pushbutton S1, every time the ball is coming toward the yellow LED and it is between the green and the yellow LED. If you press pushbutton S1 at any other time, it will not have any effect on the "ball." Also if you do not press pushbutton S1 and the ball reaches the yellow LED, you missed the ball, and the game stops. The green LED indicates the "line of hitting" and it always remains "on." After the ball has passed the green LED, it is ready to be "hit." To start playing first turn the game "off" if it was "on," by pressing and releasing the "on/off" pushbutton at S2. Get ready to play the game by placing your finger on top of S1. Now press and release S2 to turn "on" the game; as you do this the "ball" will start moving. Press pushbutton S1 when the "ball" is coming toward the yellow LED and has passed the green LED. If you do this, you "hit" the "ball" and it will change direction and go back, "bounce" off the wall and return. If you continue hitting the "ball" at the right time, you will continue playing. If you miss the "ball," the game will stop. To start playing again, turn S2 "off" and then back on again. The tricky part of this game is that the speed of the "ball" changes so you have to be very careful not to miss the "ball." Also the level of difficulty of this game can be adjusted by the use of the trimmer potentiometer at R4. With the R4 all the way counterclockwise the game will be slow and easy, but as you turn R4 clockwise the overall speed of the game increases and becomes more difficult. This is a fun game and can be enjoyed by gamers of all ages; it will provide many hours of fun for its players. Challenge your friends to a game of Racquetball!

Chapter 34

Theremin Project

Parts List

Parts Bin

R1, R13, R14, R28, R29
100 k ohm, 1%, ¼-watt
resistor

R2, R5, R6, R7, R8 1 k
ohm, 1%, ¼-watt
resistor

R10, R18, R27 1 k ohm,
1%, ¼-watt resistor

R3 1.2 k ohm, 1%,
¼-watt resistor

R4 8.2 k ohm, 1%,
¼-watt resistor

R9, R12 3.3 k ohm,
1%, ¼-watt resistor

R11 6.8 k ohm, 1%,
¼-watt resistor

R15, R21 10 k ohm,
1%, ¼-watt resistor

R16 150 ohm, 1%,
¼-watt resistor

R17 10 ohm, 1%,
¼-watt resistor

R19 1 megohm, 1%,
¼-watt resistor

R20 2.2 k ohm, 1%,
¼-watt resistor

R22, R24, R26
100 ohm, 1%,
¼-watt resistor

R23 680 ohm, 1%,
¼-watt resistor

R25 270 k ohm, 1%,
¼-watt resistor

P1 10 k log
potentiometer

P2 2 k ohm
potentiometer-trimmer

C1, C19, C22 68 pF,
35-volt, ceramic
capacitor

C2, C4, C5, C6 0.1 µF,
35-volt, MKT polyester
capacitor

C13, C16, C20 0.1 µF,
35-volt, MKT
polyester capacitor

C3, C18, C21 220 pF,
35-volt, ceramic
capacitor

C7, C8, C12 0.047 µF,
MKT polyester
capacitor

C9, C10, C14 10 µF,
35-volt electrolytic
capacitor

C23, C26, C27 10 µF,
35-volt electrolytic
capacitor

C11, C24 470 µF, 35-volt
electrolytic capacitor

C15 2.2 µF, 35-volt
electrolytic capacitor

C17 560 pF, 35-volt
ceramic capacitor

C25 100 µF, 35-volt
electrolytic capacitor

D1 1N914 silicon diode

D2, D3 1N4004 silicon
diode

Q1, Q2, Q3 2N5484 JFET
    transistor

Q4 BC548 NPN
    transistor—2N3904 or
    NTE123AP

U1 MC1496 balance
    modulator IC

U2 LM358 dual op-amp IC

U3 LM386N-1 1-watt
    audio amplifier IC

U4 LM7805 IC regulator

T1, T2, T3 Transformer—
    2nd IF coils—(white)
    455kHz

T4 Transformer—3rd IF
    coil—(black) 455kHz

SPR 8 ohm speaker

S1 SPST power switch

J1 3-circuit coaxial
    power jack

J2 RCA jack

Misc PC circuit board,
    IC sockets, wire,
    hardware

Figure 34-1 *Theremin Project*

The Theremin, a forerunner of modern music synthesizer music, can produce sounds with the slightest of hand gestures, You can create eerie science fiction movie sounds or make melodious music simply by hand movements. The Theremin was invented by Leon Theremin, of Leningrad, Russia, in 1924, and it represented a revolutionary change in thinking about how music could be produced, challenging traditional stringed, brass and percussion musical instruments. The Theremin is composed of electronic oscillators, which allow control over both pitch and amplitude by moving the hands over sensor plates. Its design eventually led to the development of the Moog Synthesizer and electronically synthesized music in general. But the invention was not only instrumental in the development of electric music, it also had an impact on a free-form style of playing music. The free gesture hand control afforded by the Theremin pre-empted the modern Sensor Chair synthesizer controller where the whole body is a part of the musical generation process.

Before this, Jimi Hendrix was creating new sounds by generating feedback between his guitar and the amplified sound and then moving his body to modulate the amplitude. It freed him from the restriction of generating music solely by plucking the guitar strings.

The Theremin was commercially manufactured by the Radio-victor Corporation of America (RCA) around 1929. It comprised a large box to which were attached an antenna and wire loop. The antenna provided the control for the pitch while the loop enabled the volume to be adjusted. Moving the right hand toward the antenna would reduce the pitch while moving the hand away from the antenna would increase the pitch.

In some ways this is similar to playing a trombone whereby the slide is moved back and forth to vary the pitch. The left hand would reduce the volume as it was brought near the sensor loop. As you would expect, the original Theremin circuit used tubes.

The pitch control antenna stood vertically, while the volume loop sat horizontally. This meant that there was little interaction between the two controls especially since the pitch control hand would be held side on to the antenna and the volume control hand would be horizontal. Thus the interaction between the two controls would be minimized.

In modern times there has been quite a bit of renewed interest in the Theremin. The Theremin project in Figure 34-1 has exactly the same operational characteristics as the original RCA Theremin, but does not use tubes. Instead it uses just three low cost ICs and a handful of other components. Not only that, our new Theremin is considerably smaller than the original design although you could build a large one if that's what you

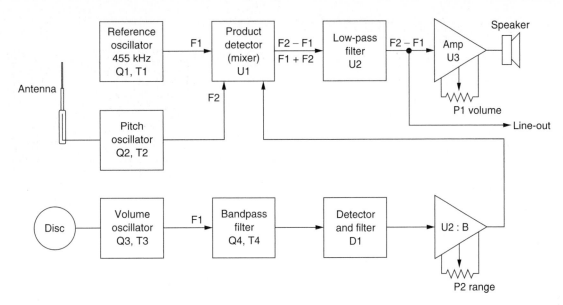

Figure 34-2   *Theremin block diagram*

fancy. Our Theremin project was built into a medium sized plastic box with the antenna and sensor-loop mounted on opposite sides of the box. It includes a small loudspeaker for practice sessions and a line output for connection to a sound system. The only manual controls are an on/off switch and volume control. The unit is powered by a 9 V or 12 V DC "wall wart" power supply.

The block diagram of Figure 34-2 shows the basic arrangement of the Theremin circuit. It comprises three oscillators which all operate at about 455 kHz. The reference and pitch oscillators are mixed together to generate a beat signal which becomes the audible tone while the volume oscillator is used to change the level of the tone output.

The reference oscillator operates at a fixed frequency and is mixed with the pitch oscillator in the product detector (U1). The pitch oscillator changes in frequency depending upon the amount of capacitance to earth presented by your hand when it is near the antenna. As noted, the frequency will fall when the hand is brought near to the antenna and rises when it is further away. The product detector essentially multiplies the reference oscillator (f1) with the pitch oscillator (f2) to produce sum (f1 + f2) and difference (f2 − f1) frequencies. The sum (f1 + f2) signal is at around 900 kHz which is easily filtered out with a low pass filter and we are left with the difference signal of f2 − f1 which comprises audio frequencies from 1.4 kHz down to below 10 Hz. So if the pitch oscillator frequency is 456 kHz and the

reference oscillator is at 455 kHz, we will obtain a 1 kHz audio output from the low pass filter.

The audio output from the low pass filter is applied to a power amplifier which can drive a loudspeaker. The overall volume from the amplifier is set by the volume control P1. The volume oscillator is controlled by the sensor loop which is also affected by hand capacitance. As you bring your hand closer to the loop, the frequency of the volume oscillator decreases. This is fed to a bandpass filter which has a center frequency (fc) which is higher than the volume oscillator frequency. So if the volume oscillator is operating at frequency fl the level will be low. As the frequency increases, the level will increase as it approaches the center frequency of the filter.

This signal level is detected using a diode and filtered to produce a DC voltage. The following amplifier increases the DC voltage and the level shifter sets the voltage so that it can control the product detector output level over a suitable range via its transconductance input.

The schematic diagram for the Theremin is shown in Figure 34-3, uses three JFETs (Junction Field Effect Transistors), four pre-wound IF (Intermediate Frequency) coils, three ICs, one detector diode, a three-terminal regulator and associated resistors and capacitors. As you can see, all three oscillators are identical with the exception of the 100 ohm drain resistor for Q3. Each oscillator comprises a junction FET (JFET) Q1 and a standard IF transformer, as used in low-cost AM radio receivers. The transformer

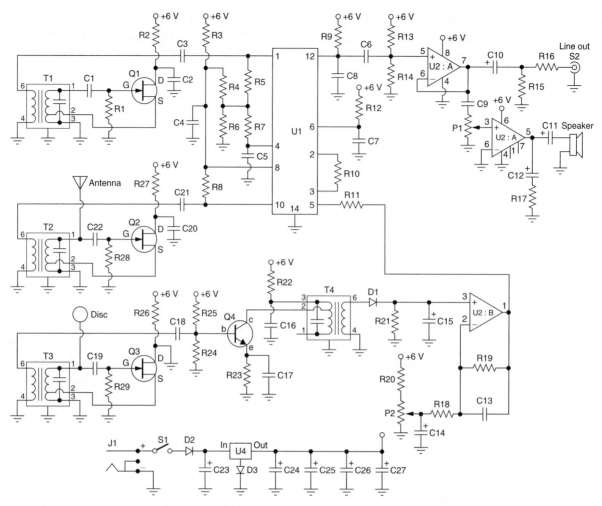

Figure 34-3   *Theremin. Courtesy Jaycar Electronics (JC)*

comprises a tapped winding which has a parallel-connected capacitor to form a tuned circuit. The secondary winding couples the oscillator signal to the following circuitry.

Each JFET drives a portion of the primary winding lie, between the tap connection pin 2 and ground, while the signal across the full winding is applied back to the gate via a 68 pF capacitor. This is the positive feedback which ensures oscillation.

To make them controllable by hand capacitance, the pitch and volume oscillators have the antenna disconnected to the top of the tuned coil where they will have the most effect. The reference oscillator and pitch oscillator outputs are applied to an MC1496 balanced mixer, at UI. Resistors between the +6 V supply and ground set the bias voltages for the inputs of the balanced mixer, while the 1 k resistor between pins 2 and 3 sets the gain of the circuit.

Integrated circuit U1 provides a balanced output where outputs are filtered with a 3.3 k pull-up resistor and 0.047 pF capacitor to produce roll-off above about 1 kHz. This heavily attenuates frequencies at 455 kHz. The output from pin 12 is AC-coupled to op-amp U2a which simply buffers the signal before it is applied to volume control P1. U2a's output signal also goes to the line output terminal. The signal from U2a is AC coupled to both P1 and the line output to prevent DC voltage flowing through the pot and the line output. Integrated circuit U3 is an LM386, a 1W amplifier which drives the loudspeaker via a 470 pF electrolytic capacitor. The 0.047 pF capacitor and series 10 k resistor form a Zobel network to prevent spurious oscillation from the amplifier.

The output from the volume oscillator at the secondary winding of T3 is AC-coupled to the base of transistor Q4. This is connected as a common emitter amplifier with the collector load being a parallel-tuned circuit comprising an IF coil with internal capacitor. T4 and the associated capacitor are tuned to a frequency just above the maximum available from the volume oscillator. The emitter resistor is bypassed with a 560 pF capacitor which provides roll-off below about 400 kHz. The output level from transformer T4 will vary in proportion to the frequency from the volume oscillator. This is because the filter provides a sharp roll-off below its tuning frequency and small changes in frequency which are below the center frequency will cause large changes in the filter response. The action of this circuit is a simple frequency modulation (FM) detector. The high frequency signal from T4 is rectified by diode D1 and filtered to provide a DC signal which is amplified by op-amp U2b by up to 1000, depending on the setting of P2. U2b's output is then fed to pin 5 of U1 to vary the level of the audio signal.

When the reference and pitch oscillators are locked together at 441 kHz, the result is that no output tone is produced. Normally, the reference oscillator remains fixed while the pitch oscillator is varied by hand capacitance.

Power for the circuit comes from a DC power pack which is regulated by U4, a 5 V regulator. The output is "jacked up" by a nominal 0.7 V by diode D2 to give a nominal +6 V which will actually be around +5. The input and output terminals of U4 are decoupled with electrolytic capacitors to aid in supply filtering and to prevent instability in the regulator

Construction

The Theremin was constructed on a 133 × 88 mm PC board, and was mounted in a plastic box which measures 158 × 95 × 53 mm. Although the assembly description resolves around the plastic case with its small speaker, there is no reason why you couldn't build it into a much larger wooden case in keeping with the original musical instrument. A larger amplifier and loudspeaker would also be a considerable benefit in the overall sound quality.

Before we begin building the Theremin project, let's take a few minutes to locate a large table or workbench on which to construct the project. Look for an area that is well-lit and has good ventilation to allow for solder fumes to dissipate. You will want to obtain a 27 to 33-watt solder iron with either small flat-blade or a small pointed tip. Find a spool of #22 ga 60/40 rosin core solder, a wet sponge, a magnifying glass and a pair of tweezers. A small flat-blade and small Phillips screwdrivers would round out your tool requirements. You should also try to obtain a small jar of "Tip Tinner," a soldering iron dresser/cleaner, and an anti-static wrist strap from your local Radio Shack store. The anti-static wrist strap is worn while you are handling the integrated circuits to prevent damage due to high voltage static charges caused by moving around the workbench or getting up and down from your chair. The anti-static wrist strap is grounded through a 110-volt outlet plug. Locate the Theremin schematic, and pictorial diagram, and the various charts and tables. Place the project parts in front of you, plug-in your soldering iron and we will begin construction.

You can begin construction by checking the PC board for any defects such as shorts between tracks, breaks in the copper tracks and incorrectly drilled holes. You will need slightly larger than the standard 1 mm sized hole for the coil earth pins on the sides of the shielding cans, and while holes for the PC stakes should be sized to suit their diameter, they should be a tight fit. Check that the PC board clips neatly into the plastic case's integral side pillars. It may need to be filed down to make a snug fit. The Theremin wiring details are shown in the pictorial diagram Figure 34-4. Insert the two links and then the resistors. Use Table 34-1 as a guide to selecting and placing each resistor. Each resistor will have at least three or four color bands. Note that the color bands will start at one end of the resistor body. The first column in the resistor table represents the resistor's first color band, while the second column in the chart represents the second color band on the resistor. The third column in the chart represents the resistor's multiplier value, while the fourth column shows the resistor's tolerance value. Look through the parts pile and see if you can locate a resistor with a black color band (1) a brown band (0) and a yellow band (10000) multiplier. Resistor R1, with a black-brown-yellow band with a gold fourth band would represent a 100,000 resistor with a 5% tolerance value. You can use a digital multi-meter to measure and verify the resistor value if desired. Go ahead and install resistor R1 and solder it in place. Trim the excess lead length with your end-cutter.

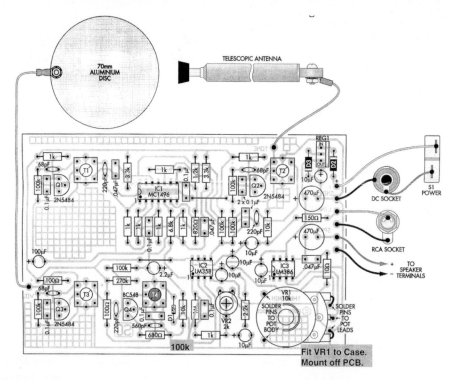

Figure 34-4 *Theremin Project*

Next, you can mount and install the project's capacitors. Ceramic and MKT type capacitors usually are coded with a three-digit code. You can check this against the values shown in Table 34-2. A 0.1 µF capacitor will be marked (104). Electrolytic capacitors are polarized and must be oriented with respect to the correct polarity in order for the circuit to work properly. An electrolytic capacitor will have a plus (+) or minus (–) marking on one side of the capacitor body. You may see a line or band with a minus or plus marking on it. Go ahead and mount the capacitors onto the circuit board, then solder them in place on the PC board. Remember to cut the excess component leads from the PC board.

The Theremin project contains four coils, which are all pre-wound types each with an integral tuning capacitor and are mounted in a small square can. Be sure to place the ones with the white slugs (the threaded ferrite core) in the TI–T3 positions and the coil with the black slug in the T4 position.

Next locate and identify the silicon diodes. Diodes are two-lead devices which schematically have an arrow pointing to a straight line. The arrow represents the anode, while the straight line is the cathode. Looking at a silicon diode, you will notice a black or white colored band at one end of the body of the diode. The band represents the cathode lead of the diode. Locate and install all of the silicon diodes onto the circuit board. Solder the diodes to the PC board, then remove the excess component leads.

Now refer to the semiconductor pin-out diagram shown in Figure 34-5. Now, go ahead and identify the transistors. Most transistors will have three leads, a Base, a Collector lead and an Emitter lead. The Base lead is usually in the center of the transistor symbol, at the flat line. The Collector and Emitter point to the flat line. The Emitter will have an arrow either pointing to or away from the transistor. If the arrow points away from the transistor, then the device is an NPN, if the arrow on the Emitter points to the transistor then the device is a PNP transistor. The FET are a bit different and they will have a Drain lead, a Source lead and a Gate lead. The diagram depicts the transistor and FET from a bottom view. Make sure you can positively identify the pin-outs before installing the transistors and FETs. Go ahead and install the JFETs (2N5484), the BC548 or equivalent transistor at Q4, and the three-terminal regulator.

Locate the four integrated circuits from the parts pile. Integrated circuit sockets are low cost insurance against a possible circuit failure at a later date. Since most electronic enthusiasts cannot unsolder an IC from a PC

Table 34-1

Resistor Color Code Chart

Color Band	1st Digit	2nd Digit	Multiplier	Tolerance
Black	0	0	1	
Brown	1	1	10	1%
Red	2	2	100	2%
Orange	3	3	1000 (K)	3%
Yellow	4	4	10000	4%
Green	5	5	100000	
Blue	6	6	1000000 (M)	
Violet	7	7	10000000	
Gray	8	8	100000000	
White	9	9	1000000000	
Gold			0.1	5%
Silver			0.01	10%
No color				20%

successfully without destroying the circuit board, integrated circuit sockets are highly recommended. An IC socket will have a notch or cutout at one end of the socket. Note that pin 1 of the socket will be just left of the notch. Go ahead and install the IC sockets, then solder them in place. When installing the integrated circuits into their respective sockets, be sure to wear your anti-static wrist band to avoid damaging the sensitive integrated circuits. Just moving up and down in a fabric chair or walking around your worktable can generate high voltages, which can destroy integrated circuits.

When placing the IC into its respective socket, be sure to line-up pin 1 of the IC with pin 1 of the IC socket. Make sure that U2 and U3 are placed in the correct positions, taking particular care with their correct orientation. Place all of the integrated circuits into their respective sockets. If you are having any difficulties identifying or installing any of the transistors and integrated circuits, ask a knowledgeable electronics friend for some help.

Finally, mount potentiometer P1 and trim-pot P2. Potentiometer P1 was mounted off the PC board, on the

project case, while trimmer potentiometer P2 was mounted on the PC board.

After constructing the Theremin, you will need to check your circuit board for possible "cold" solder joints and for possible "short" circuits. Pick up the circuit board with the foil side facing upwards toward you. Have a look at all the solder joints on the PC board. Make sure that all the solder joints look clean, smooth and shiny. If you find that any of the solder joints look dull or "blobby," then you should remove the solder from the joint with a "wick" or solder-sucker and then re-solder the joint, so that it looks clean, bright and shiny. "Cold" solder joints can cause the circuit to fail prematurely, so it is a good idea to find them and repair them at once.

Next, we will move on to inspecting the PC board for "short" circuits. "Short" circuits can often be caused by "stray" excess component leads which may have gotten stuck to the PC board due to the sticky nature of rosin core solder. "Short" circuits can also be caused from solder blobs, or any other metal which may stick to the underside of the PC board. You will want to eliminate "shorts" before applying power to the circuit in order to prevent damage to the circuit.

Table 34-2

Capacitance Codebreaker Information

This table is designed to provide the value of alphanumeric coded ceramic, mylar and mica capacitors in general. They come in many sizes, shapes, values and ratings; many different manufacturers worldwide produce them and not all play by the same rules. Most capacitors actually have the numeric values stamped on them; however, some are color coded and some have alphanumeric codes. The capacitor's first and second significant number IDs are the first and second values, followed by the multiplier number code, followed by the percentage tolerance letter code. Usually the first two digits of the code represent the significant part of the value, while the third digit, called the multiplier, corresponds to the number of zeros to be added to the first two digits.

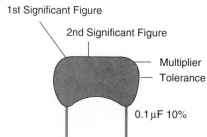

Value	Type	Code	Value	Type	Code
1.5 pF	Ceramic		1000 pF/0.001 µF	Ceramic/Mylar	102
3.3 pF	Ceramic		1500 pF/0.0015 µF	Ceramic/Mylar	152
10 pF	Ceramic		2000 pF/0.002 µF	Ceramic/Mylar	202
15 pF	Ceramic		2200 pF/0.0022 µF	Ceramic/Mylar	222
20 pF	Ceramic		4700 pF/0.0047 µF	Ceramic/Mylar	472
30 pF	Ceramic		5000 pF/0.005 µF	Ceramic/Mylar	502
33 pF	Ceramic		5600 pF/0.0056 µF	Ceramic/Mylar	562
47 pF	Ceramic		6800 pF/0.0068 µF	Ceramic/Mylar	682
56 pF	Ceramic		0.01	Ceramic/Mylar	103
68 pF	Ceramic		0.015	Mylar	
75 pF	Ceramic		0.02	Mylar	203
82 pF	Ceramic		0.022	Mylar	223
91 pF	Ceramic		0.033	Mylar	333
100 pF	Ceramic	101	0.047	Mylar	473
120 pF	Ceramic	121	0.05	Mylar	503
130 pF	Ceramic	131	0.056	Mylar	563
150 pF	Ceramic	151	0.068	Mylar	683
180 pF	Ceramic	181	0.1	Mylar	104
220 pF	Ceramic	221	0.2	Mylar	204
330 pF	Ceramic	331	0.22	Mylar	224
470 pF	Ceramic	471	0.33	Mylar	334
560 pF	Ceramic	561	0.47	Mylar	474
680 pF	Ceramic	681	0.56	Mylar	564
750 pF	Ceramic	751	1	Mylar	105
820 pF	Ceramic	821	2	Mylar	205

Table 34-2 (*Continued*)

PicoFarad (pF)	NanoFarad (nF)	MicroFarad (mF, µF or mfd)	Capacitance Code
1000	1 or 1n	0.001	102
1500	1.5 or 1n5	0.0015	152
2200	2.2 or 2n2	0.0022	222
3300	3.3 or 3n3	0.0033	332
4700	4.7 or 4n7	0.0047	472
6800	6.8 or 6n8	0.0068	682
10000	10 or 10n	0.01	103
15000	15 or 15n	0.015	153
22000	22 or 22n	0.022	223
33000	33 or 33n	0.033	333
47000	47 or 47n	0.047	473
68000	68 or 68n	0.068	683
100000	100 or 100n	0.1	104
150000	150 or 150n	0.15	154
220000	220 or 220n	0.22	224
330000	330 or 330n	0.33	334
470000	470 or 470n	0.47	474

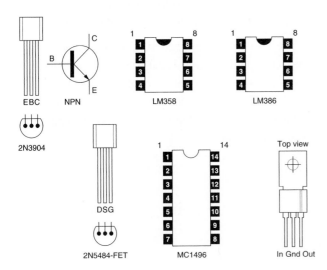

Figure 34-5 *Theremin semiconductor pin-out diagram*

Project Mounting

You are now ready to work on the plastic case. You will need to drill holes in the sides of the box for the DC panel socket, the RCA socket and for the antenna securing screw. The volume plate is made from light gauge aluminum 70 mm in diameter and is connected via a lead and solder lug to the PC board. We mounted the volume plate on the prototype so that it could slide into the case when not in use. This requires a narrow slot to be cut into the side of the case. We made the slot by drilling a series of small holes and then filing it to the correct size.

If you don't want to go to the trouble of making the slot you can permanently attach the volume plate to the lid of the case or mount it so that it can swivel over the lid for easy storage. Attach the front panel label to the lid and drill the holes for the switch and some holes for the loudspeaker. First we drilled a series of holes on the top lid for the speaker sounds to come out. We mounted the speaker by smearing super glue around its perimeter and then placing it inside the top lid. Follow the wiring details of Figure 34-3. You can use hookup wire to connect the power and audio output connectors, switch S1 and the loudspeaker to the PC board. The antenna is attached with an M3 × 15 screw plus an M3 nut which is secured to the case with another M3 nut. The eyelet lug is held beneath the nut and the wire connects to the PC board as shown. A connection is made to the volume plate via an eyelet using an M3 × 10 mm screw through a hole which is held using a nut. Drill a hole for potentiometer P1 on the top lid of the project case. The potentiometer at P1, was mounted off the PC on the top of the project case. It was discovered that the volume potentiometer worked best if it was mounted away from the PC board. Three insulated wires were used to connect the potentiometer to the circuit board.

Place the circuit board in the case once all the connections have been made. Finally, apply power and check to make sure that there is +6 V between the case of one of the transformer coils T1–T4 and pin 8 of U2 and pin 6 of U3. You will want to make sure that when you power-up the circuit that the foil side of the board does not have any metal underneath it, which can cause the board to "short" out. The voltage should be between +5.6 V and +5.8 V. Adjust P2 so that pin 1 of U2 goes to about +4.3 V and wind P1 slightly clockwise from its fully anticlockwise position. Use a plastic alignment tool to rotate the slug in transformer T2 slightly until a tone is heard in the loudspeaker. Then adjust it to obtain a good frequency range when your hand is brought near to the extended antenna. The note should be at its highest when your hand is away from the antenna and should fall to a very low frequency (just a growl) when your hand is very close to the antenna. If the effect is the reverse of this (higher frequency as your hand is brought close to the antenna) then adjust the slug in the opposite direction until the effect is correct. Note that you must do this adjustment away from the effects of

metallic objects or the Theremin will require re-tuning when removed from these grounding sources. In fact, the Theremin will give more consistent results if it is mounted on a raised stand which keeps the unit at least 60 mm from any surfaces. The stand should be made from a non-metallic material.

The volume operation is set by adjusting the slug in T4 until the voltage at the cathode of diode D1 is at +1.7 V. Then carefully adjust P2 so that the volume is at its maximum when your hand is away from the volume plate. Bringing your hand close to the plate too should reduce the volume. You may need to set P2 so that the volume just goes to its minimum level when rotated anticlockwise. You then slowly adjust it clockwise until the volume just snaps into full level. If the volume does not reduce with your hand approaching the plate and the level remains essentially constant or if the level rises, then T4 is adjusted with the slug too far clockwise. This means that the circuit is operating with the volume oscillator equal to or higher than the tuned frequency. Adjust the slug of T4 anticlockwise so that the volume plate operates correctly. Note that when the lid is fitted to the case, the tuning will change. We drilled a hole in the lid to allow P2 to be adjusted with the lid in place. Also, the adjustment of P2 will set the sensitivity of the Theremin volume plate to hand movement. The more precisely P2 is adjusted, the greater will be the sensitivity.

The Theremin is designed to drive the internal speaker to produce its mystical sounds, but note there is also a line-out available at the output of the LM358 at pin 7, which can be fed into a home stereo system or fed into an audio board mixer for a musical amplifier system if desired.

Chapter 35

Multi-Chip Programmer

Parts List

Parts Bin

R1, R3, R4 470 ohm, ¼-watt, 5% resistor

R2 4.7 k ohm, ¼-watt, 5% resistor

R5, R6 10 k ohm, ¼-watt, 5% resistor

C1 10 µF, 35-volt electrolytic capacitor

C2 22 µF, 35-volt electrolytic capacitor

D1, D2, D4, D6 1N4148 silicon diode

D3 8.2-volt Zener diode

D5 5.1-volt Zener diode

L1 LED red

L2 LED yellow

L3 LED green

Q1, Q2 BC547 NPN transistor—2N3904 or NTE123AP

Misc PC board, IC socket, wire, etc.

The Multi-Chip Programmer, depicted in Figure 35-1, will burn a wide range of PIC chips. The Multi-Chip Programmer project is provided in order to program the microprocessor chips used in the Simon Memory Game project, the Tic-Tic-Toe project and the Tetris Game project. The Multi-Chip Programmer was designed to be powered by your computer serial port; a cable connects your computer's serial port with the programmer via a 4-conductor phone cord.

Figure 35-1 *Multi-Chip Programmer*

Programming or burning these PIC microprocessor chips is straightforward, by first running the chip burning program, called IC-Prog, selecting the chip to be programmed and then load the particular HEX program listing for the game you desire, then run the "burn" program and download the program to the PIC chip.

How the Circuit Works

In the Multi-Chip Programmer circuit the supply voltage for the chip comes from the RS-232 feature of the serial port. Some of the lines making up the RS-232 are capable of rising to a positive voltage (about 8 to 12 V) and falling to a negative voltage (about – 8 V to – 12 V). There are also lines that fluctuate from 0 V to +5 V. If all computers had a line that fluctuated between + 12 V and – 12 V, the programmer circuit would be very simple. But unfortunately some computers fluctuate between +8 V and –8 V. To make a circuit that works on all ports was a challenge. The circuit we have used was designed by JDM

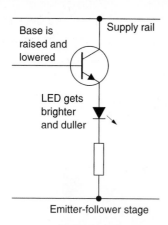

Figure 35-2 *Emitter follower*

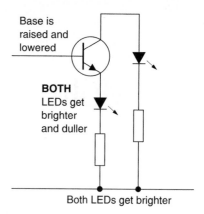

Figure 35-3 *Varying LED brightness*

(http://www.jdm.homepage.dk/) and full credit is given to him.

The chip requires a voltage of 13 V on the MCLR pin, which is between 12 V and 14 V, in order to tell the chip to go into program mode. The chip does not require any current on this line, just a voltage so the program mode can be invoked (begun). If one of the lines from the computer goes to +8 V, and another goes to – 8 V, they can be combined to get a total of about 16 V. This is more than enough to create the necessary 13 V. This is the basis of how the circuit works and the reason for the diodes and Zener diodes. But it's more complicated than that. The voltage-delivering lines are also the lines that provide the signals to and from the chip during programming and reading modes. So, the circuit becomes quite complex. The lines delivering the signals are also the lines that charge the electrolytic capacitors.

Normally, a transistor in emitter-follower mode is connected with the collector to the supply rail and the base is raised and lowered from 0 V to supply voltage. The voltage on the emitter is 0.7 V lower than the base, but it has a higher current capability than that delivered by the base. It's quite simple, the current comes from the collector! The normal emitter-follower circuit is shown in Figure 35-2.

But, suppose the transistor is connected with a LED on the emitter and collector as shown in Figure 35-3. This time, the current for the LED cannot come from the supply rail (via the collector) and thus the base must supply the current. It is easy to see that the lower LED is turned on via the current from the base. But the interesting feature is the LED in the collector circuit will also come on with the same brightness as the LED in the emitter circuit.

The base-collector junction is reverse biased and will perform exactly like the base-emitter junction. The base must supply the current for both LEDs This is how the first transistor in the circuit is operating. The base is supplying current to charge the 10 μF electrolytic and the 8.2-volt Zener is allowing the 10 μ electrolytic to charge to 8.2 V higher than the 22 μF electrolytic and supplying a voltage-reference for the MCLR pin. The 10 μF electrolytic does not deliver its energy to the MCLR pin, and no current flows between the emitter and collector leads of the transistor in this arrangement.

With this basic theory understood, you will be able to see how the Multi-Chip Programmer works. But before we get to the main programmer circuit, Figure 35-4 shows how the voltages on the chip are developed with reference to the GND line.

Figure 35-5 shows how the "chip" actually "sees" these voltages. You simply add +5 V to each of the

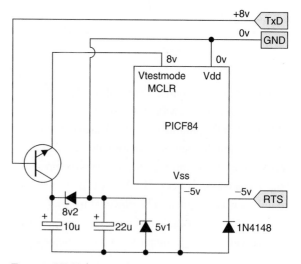

Figure 35-4 *Programmer voltages*

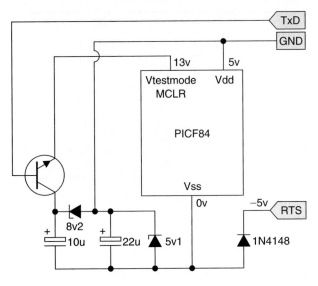

Figure 35-5 *Programmer voltage response*

voltages to make $V_{ss} = 0$ V. This makes $V_{dd} = 5$ V and the programming voltage = 13 V.

The main Multi-Chip Programmer schematic diagram is depicted in Figure 35-6. The transistor on the data line also operates in an unusual way. It functions in a bi-directional mode, since the data must be transmitted *into* the chip when burning and *from* the chip when reading.

When delivering data to the chip, the DTR line goes "High" and the transistor is in emitter-follower mode. The input of the chip will be high-impedance and the emitter voltage will be "High," being pulled up by the

10 k resistor and fed by the voltage from the DTR line. When the DTR line goes "Low," the collector of the transistor will go "Low," because the only voltage supplying the circuit comes from the base. Since the resistance on the base is 10 k, and the resistance between collector and ground is 2.2 k, the voltage division will produce about 1 V on the collector. Since the collector voltage goes "Low," the emitter voltage will also go LOW as the transistor is in exactly the same arrangement as shown in Figure 35-3, above. Thus, by taking the DTR line "High-Low," the data line of the chip will be taken "High-Low."

When the transistor is being read, the data appears on the data line. This time the DTR line is kept "High" and when the data line of the chip goes "Low," current is drawn through the collector lead. This current flows through the 2.2 k resistor and produces a voltage drop across it. This voltage drop is enough to bring the collector voltage down to about 1 V or less and the CTS line reads this as a "Low."

When the data line of the chip goes "High," current does not flow through the 2.2 k resistor and CTS reads the line as a "High." There are three more signal diodes in the circuit (the fourth diode has been explained as it charges the 22 μF when RTS is "Low"). The diode on the MCLR line takes MCLR LOW when TxD goes "Low," while the other diode on this line prevents the voltage on TxD from going below 0 V. The two diodes on the RTS line prevent the line going above 5 V or below 0 V.

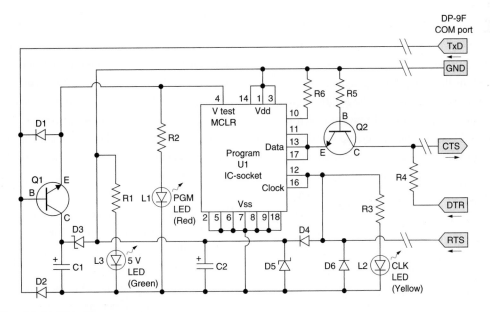

Figure 35-6 *Multi-Chip Programmer*

Click on the diagram below to see an animation of the chip being set up for programming and data being clocked in. This is only a simple representation as the chip looks for six initial clock cycles and, depending on the data it receives during these six cycles, the chip will go into one of nine different modes. For instance, it can go into a mode called Load Configuration where the next 16 clock cycles will load the Configuration Memory with the necessary data bits.

The MCLR line must then be taken "Low" and "High" again and the chip is ready to receive a different loading mode. One of these modes is Load Data for Program Memory and after six clock cycles the next set of 16 cycles will consist of a zero start-bit, 14 bits of data and a zero stop bit. As you can see, it takes a lot of cycles to get each byte of data into the chip, but this is always the case when information is being serial-fed.

Once you know how the circuit works, you will feel much more comfortable about working on it and/or modifying its operation.

At the moment we don't have any access to the software so you will not be in a position to modify the operation of the program. But since it works perfectly, I don't see any need for modification.

Construction

All the components fit on the single-sided board and the project is connected via a 4-pin US telephone plug to the serial port of a computer. It gets all its voltages from this port as well as the programming signals.

Before we begin construction of the Multi-Chip Programmer, let's take a moment to secure a well-lit worktable or workbench area in which to build the project. You will want to obtain a 33 to 40-watt pencil tip soldering iron, with a sharp pointed tip, a small spool of 60/40 rosin core solder and a wet sponge. You should also try to locate a small jar of "Tip Tinner," a soldering iron tip dresser/cleaner, as well as an anti-static wrist strap from your local Radio-Shack store. You will also need to collect a few hand tools for construction. Locate a small pair of needle-nose pliers, a pair of end-cutters, a magnifying glass, a pair of

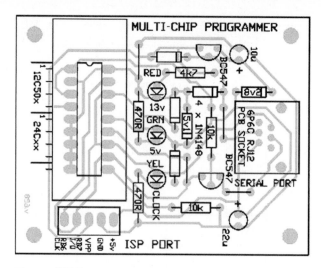

Figure 35-7 *Multi-Chip Programmer pictorial diagram*

tweezers and a small flat-blade and a Phillips screwdriver.

Locate the Multi-Chip Programmer pictorial PC layout diagram illustrated in Figure 35-7 and once you have collected all the necessary tools and project parts, we can begin building the Multi-Chip Programmer project. Place all the project parts in front of you and heat up your soldering iron. Now refer to Table 35-1, which illustrates the resistor color chart. All resistors have color codes on the body of the resistor, which start at one end of the resistor body. The first color band represents the first digit of the resistor code. The second color band represents the second digit of the resistor's code and the third color band denotes the multiplier value used to compute the value. A fourth color band depicts the resistor's tolerance value. A silver band denotes a 10% tolerance value, while a gold band denotes a 5% tolerance. Absence of a fourth color band show the resistor has a 20% tolerance. So look through the project resistors for R1, and try to see if you can locate a resistor with a 470 ohm value. The first band will be yellow (4), the second band will be violet (7) and the third band will be brown, a multiplier of (0). Once you locate this resistor you can place it on the PC board at its proper location. Next you can solder the resistor to the circuit board, then follow-up by cutting the excess resistor lead flush to the edge of the circuit board with your end-cutters. Now locate the remaining resistors using the resistor color code chart. Install the remaining resistors, then go ahead and solder them in place on the PC board.

Table 35-1

Resistor Color Code Chart

Color Band	1st Digit	2nd Digit	Multiplier	Tolerance
Black	0	0	1	
Brown	1	1	10	1%
Red	2	2	100	2%
Orange	3	3	1000 (K)	3%
Yellow	4	4	10000	4%
Green	5	5	100000	
Blue	6	6	1000000 (M)	
Violet	7	7	10000000	
Gray	8	8	100000000	
White	9	9	1000000000	
Gold			0.1	5%
Silver			0.01	10%
No color				20%

Next we will move on to installing the capacitors used in this project. The programmer project has only two capacitors, which are electrolytic types. Electrolytic capacitors will have either a black or white color band along one side of the capacitor body. Near this color band you will also see a plus (+) or minus (−) marking which denotes the polarity of that particular terminal. When installing electrolytic capacitors, make sure that you observe the polarity markings, for example the plus (+) terminal of C1 connects to Q1 and D3. Install C1 and then solder the capacitor in place, then cut the excess component leads flush to the edge of the PC board. Locate C2 and install it on the PC board, remembering to cut the excess component leads.

The programmer project also contains four silicon diodes, and two Zener diodes. Symbolically a diode has two terminals. One terminal is an arrow or triangle which points to the straight line, this is the anode lead of the diode. The straight line on the diode symbol is the cathode. Each diode will have either a black or white color band at one end of the diode body. This color band is the diode's cathode. Note that the cathode of D1 is connected to the Base of Q1, also note D2's cathode is also connected to Q1. Install all the diodes on the PC board and then solder them in place. Notice that there are

two Zener diodes, one at D3 and one at D5. Make sure you can identify the four signal diodes and the 5.1 V and 8.2 V Zener diodes. The symbols for Zener diodes are a bit different from regular silicon diodes and are specified by voltage cutoff or voltage knee ratings. Remember to cut the extra remaining lead lengths from the diode.

The circuit contains two NPN transistors at Q1, Q2. Transistors have three leads, a Base lead, a Collector lead, and an Emitter lead. The Base lead is the straight line symbol, while the Emitter lead has the small arrow pointing away from the transistor. The diagram in Figure 35-8 illustrates the pin-outs of the transistors; note that the reference view of the component is from the bottom view. Install the transistors on the PC board at their respective locations and then solder them in place. Solder them quickly to prevent heat running up the leads and damaging the semiconductor junction. Remove the excess transistor leads after soldering.

The Multi-Chip Programmer project contains four LEDs, some of which are different colors. See Figure 35-9 for LED pin-outs. The schematic shows where they are installed onto the board. Push all the LEDs until they touch the board and make sure they are upright. Solder the leads very quickly to prevent damage to the light-emitting crystals inside the body of the LED.

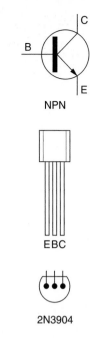

NPN

EBC

2N3904

Figure 35-8   *Multi-Chip Programmer transistors*

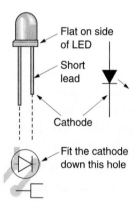

Flat on side of LED

Short lead

Cathode

Fit the cathode down this hole

Figure 35-9   *LED identification*

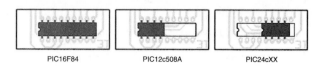

PIC16F84   PIC12c508A   PIC24cXX

Figure 35-10   *PC Chip IC placement*

of the notch or cutout, you will find pin 1. When installing "blank" micro-processor chip to be programmed into the socket, you will need to identify a small notch or cutout or indented circle at one end of the IC package. Again pin 1 will be just to the left of the notch or cutout. Be sure to line-up pin 1 of the IC with pin 1 of the IC socket when inserting the microprocessor IC into the socket. If the chip you are programming is not a "full" 18-pin chip be sure to line-up the pin 1 on the chip with pin 1 on the socket, see Figure 35-10. When handling the microprocessor ICs be sure to wear your grounded anti-static wrist strap to avoid damaging the IC. High voltage static charges can easily build up from walking around your workbench and from getting up and down in your chair.
A little insurance can go a long way in protecting your investment. Finally install the 6P6C RJ12: 4-pin phone jack on the multi-chip Programmer PC board. This telephone jack is used to connect the serial cable which connects the programmer to the PC for programming.

Once construction of the printed circuit has been completed, you can take a short break and when we return we will inspect the PC board for possible "cold" solder joints and "short" circuits. First, we will inspect the PC board for "cold" solder joints. Pick up the circuit board with the foil side of the board facing upwards toward you. Take a look at all of the solder joints, they should all appear to look clean, smooth and shiny. If you find that any of the solder joints look dull or "blobby" then simply unsolder the joint, remove the solder and then re-solder the joint so that it looks clean, smooth and shiny. Next, we will inspect the circuit board for possible "short" circuits. Often "cut" or "stray" component leads will attach themselves to the underside of the board, and get "stuck" due to the sticky nature of the rosin core solder. Inspect the board and remove any component leads or wires that may have attached to the underside of the board during construction.

Usually microprocessor chip programmers utilize a ZIF or zero insertion force socket. These types of sockets will present no physical resistance when trying to place the PIC chip into the socket. ZIF (zero insertion force) sockets are quite expensive and not necessary if you put the chip you are programming in an additional socket, and the programming socket is not used every day. If you cannot find or afford a ZIF socket, don't despair, you can locate good quality 18-pin IC socket in which to "program" the PIC chips. If you cannot locate this type of socket you could also use a good quality IC socket instead if you do not anticipate programming lots of chips.

An IC socket will typically have a small notch or cutout at one end of the plastic package. Just to the left

Serial Cable

Parts Bin

R1 2.2 k ohm, or 470 ohm, ¼-watt, 5% resistor

J1 4-pin RJ12 6P4C—PCB telco jack

P1 4-pin RJ12 6P4C—4-wire telco plug

P2 9-pin DB-9F serial connector

H1 9-pin serial connector hood

Misc 4-conductor telephone wire

The diagram in Figure 35-11 illustrates the wiring diagram for the serial cable which connects the Multi-Chip Programmer and your personal computer. The cable consists of a 4-pin telephone plug, a 9-pin DB-9F connector, a 2.2 k ohm resistor, and a length of 4-conductor telephone cable. You have to be careful when soldering the four wires to the pins of the 9-pin D-plug to make sure they are soldered to the correct places. It's very easy to make a mistake. Attach it to the board and bare the free end to see how the

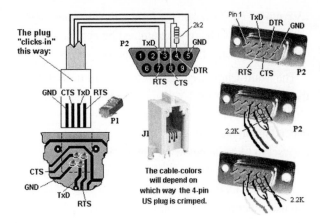

Figure 35-11 *Multi-Chip Programmer serial cable assembly*

4-core cable has been connected. From this information, you should determine how the leads will be connected to the 9-pin plug. Check the diagram for the correct pin-out and layout to see how to wire the conductors to the plug, including the 2.2 k resistor. This should be done neatly as it has to fit inside the connector hood. Screw the 9-pin D-plug together and you are ready to program a chip. Refer to Table 35-2, which displays the serial connector pin-outs.

When a chip is to be programmed for the first time, either the low voltage or high voltage method can be used. When a PIC16F628 is programmed in the "high voltage" mode, the chip can be re-programmed in the high voltage mode or you can set the LVP bit to "0" so that the chip can be re-programmed "in circuit" via the

Table 35-2
Serial Port Connector

9-Pin D-SUB MALE connector at computer DB-9M

PIN	NAME	I/O	Description
1	CD	I	Carrier Detect
2	RxD	I	Receive Data
3	TxD	O	Transmit Data
4	DTR	O	Data Terminal Ready
5	GND	–	System Ground
6	DSR	I	Data Set Ready
7	RTS	O	Request to Send
8	CTS	I	Clear to Send
9	RI	I	Ring Indicator

LVP mode. The Low Voltage Programming-mode allows the chip to be re-programmed by applying 5 V on pin 10 (instead of 12–14 V on pin 4).

The Multi-Chip Programmer "burns" a PIC16F628 in the "high voltage" ONLY. You can re-burn the chip "in-circuit" or in the Multi-Chip Programmer, depending on the setting of LVP. The chip comes with LVP set to "1." See below for details on this. When burning a chip for the first time, an instruction in your program sets LVP to "0" or "1." If it is set to "1" you can use either re-programming method, but you lose RB4 as an in-out-pin.

Programming the pic16F628

This chip has two programming modes:

1. **Normal Mode:** 12–14 V on pin 4
2. **Low Voltage Mode:** (LVP) 5 V on pin 10.

The PIC16F628 has a Low Voltage Programming-mode (LVP) for in-circuit programming. In this mode, the chip can be programmed with 5 V on the programming pin (pin 10) instead of 12–14 V on pin 4. Before deciding on the way you will program the chip, you need to know some of the differences and limitations. The PIC16F628 chip is supplied with the LVP bit as "1."

When the LVP bit is "1," RB4/PGM (pin 10) is dedicated to the programming function and is not available as in-out-pin RB4. The chip will enter programming mode when a HIGH (5 V) is placed on RB4/PGM (pin 10). This makes the chip "in-circuit" programmable and re-programmable "in-circuit." If you don't want the "in-circuit programmable" feature, LVP bit must be "0." To make LVP bit "0," the chip must be programmed via "Normal Mode," using 12–14 V on pin 4. The LVP bit cannot be changed when programming "in-circuit." When programming

via "Normal Mode," an instruction is available to change the value of LVP.

To recap: If you program via the "Normal Mode" (12–14 V to "activate" the chip to put it into "program mode"), you can use all the features of the chip. (Remember RA5 is input-only, so "Port A" is not a "complete port.") If you program via "Low Voltage Mode," output line RB4 (pin 10) is not available as you are reserving the pin for re-programming via LVP. This is very inconvenient as "Port B" is normally used as a complete 8-line output to drive displays, etc. To have one line missing from the port is like buying a book with 15 pages missing! Port A is already an incomplete Port, with RA5 as input-only. It would have been much more convenient to put LVP pin on port A and leave Port B complete! If you program a chip for the first time: "normally," you can re-program it "in-circuit" (via the 5 V feature) or re-program it via the "normal" method. If you program a chip for the first time: "in circuit," you can regain the RB4 as an in-out line by re-programming it "normally." You cannot regain RB4 as an in-out line by re-programming it "in-circuit."

The Multi-Chip Programmer is now ready to "burn" your game chips. Connect the serial cable between the Multi-Chip Programmer circuit and your computer. Turn on your PC, place the "blank" PIC chip into the programming socket. Download the *icprog105c-a.zip* from the http://www.books.mcgraw-hill.com/ authors/petruzzellis/elecgames directory and explode it into a new folder on your computer. Now run the *icprog105c-a.exe* program and select your particular "blank" chip, either a16F84 for the Tetris game or the 16F628 for the Simon or Tic-Tac-Toe game. Now load the **HEX** version of the program listing for the game program you want, within the *icprog105c-a.exe* program and select the "chip-parameters" as per the particular program and finally "burn" the chip. When the programming cycle is completed, then simply remove the PIC chip from the programmer and insert the "programmed" game chip into the game board PIC socket and apply power and your new game should spring-to-life!

Chapter 36

Simon Memory Game

Parts List

Parts Bin

R1   4.7 k ohm, ¼-watt, 5% resistor

R2, R3, R4, R5   220 ohm, ¼-watt, 5% resistor

R6, R7, R8, R9   10 k ohm, ¼-watt, 5% resistor

R10   3.9 k ohm, ¼-watt, 5% resistor

R11   560 ohm, ¼-watt, 5% resistor

R12   2.2 k ohm, ¼-watt, 5% resistor

C1   100 nF, 35-volt monoblock capacitor

C2   27 pF, 35-volt ceramic disk capacitor

L1   red LED

L2   green LED

L3   yellow LED

L4   orange LED

D1   1N 4148 silicon diode

Q1   BC 547 NPN transistor or NTE123AP

Q2   BC 338 NPN transistor or NTE123AP

U2   PIC16F628 microprocessor/Simon Game

T1   1-10 mH choke (for transformer core)

1-meter 0.5 mm dia winding wire

S1   SPST—on/off slide switch

S2, S3, S4, S5   Momentary pushbutton switches (PC mount)

SPK   piezo speaker (PCB mount)

B1, B2   "AA" batteries

Misc   PC board, IC socket, wire, "AA" battery holder, battery clip, hardware, etc.

The Simon Memory Game, highlighted in Figure 36-1, is a game many people know. The premise and the rules are easy to understand but the game can in fact be very challenging. The game requires careful attention and a good memory in order to be successful. The game has four pushbutton switches, four colored LED lamps and

Figure 36-1   *Simon Memory Game. Courtesy Velleman (VM)*

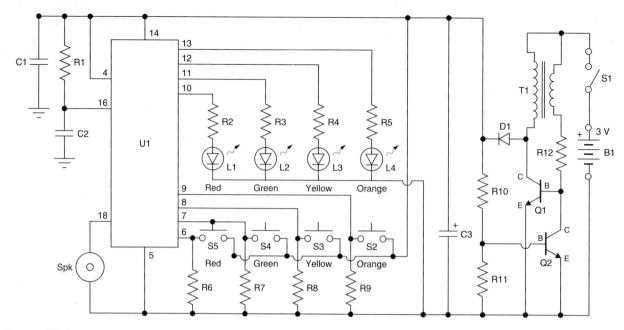

Figure 36-2 *Simon Memory Game. Courtesy Talking Electronics (TE)*

a speaker to produce a sequence of tones and flashes that has to be repeated. After each correct sequence, the computer adds another tone with its corresponding colored LED. The Simon Memory Game is our first more complex microprocessor game which uses the low cost PIC16F628 chip. The Simon game program is presented in both assembly language, in order to study how the program works, and in a hex format to make it easy to lead into the microprocessor. The project builder has only to load the hex file into the microprocessor using the multi-chip programmer, and the game will be ready for all to play.

How the Circuit Works

The heart of the Simon Memory Game is the PIC microprocessor, which is illustrated in the schematic diagram in Figure 36-2. The microprocessor chip has two ports, Port A and Port B, and any pin can be configured as an input or output. The microprocessor is like a "blank slate" and must be programmed with the Simon program, as mentioned. The assembly language program listing for the Simon program (simon.628.asm) can be downloaded from the McGraw-Hill website at http://www.books.mcgraw-hill.com/authors/petruzzellis/elecgames and it may help you to understand the

workings of the Simon Game. Once the program is loaded, the internal oscillator will read each instruction at a rate of 1 million lines (instructions) per second. You can load either the assembly ASM program or the HEX program into the microprocessor via the Multi-chip Programmer. Additional circuit hardware consists of four momentary pushbutton switches, which act as inputs to the microprocessor, and four colored LEDs, which act as output devices, along with the piezo speaker element which produces the musical notes.

The Simon Memory Game circuit consists of two main parts: the microprocessor and the power supply. This unique power supply was chosen because it is very efficient and consumes very little power, so the batteries will last quite some time. You could also choose to use a conventional LM7805 regulator, but battery life will not be as good as the circuit shown. The power supply circuit shown is a step-up voltage regulator and uses two "AA" or "AAA" cells to create a 3 V supply. The heart of the step-up regulator section is the "fly-back" transformer—actually a transformer in fly-back mode. The BC 338 is the driving transistor and is turned on via the 2.2 k ohm resistor. This produces magnetic flux in the primary winding that cuts the feedback winding and increases the voltage and current into the base to turn the transistor. This continues until the transistor is fully turned "on." At this point the magnetic flux in the primary winding is a maximum but it is not *expanding flux* and it ceases to produce a voltage in the feedback winding. The transistor is turned off slightly and the reduced current through the primary allows the flux to start to collapse and produce a voltage in the feedback winding that is in the opposite direction. This turns the transistor off more and finally the transistor is fully turned off. The transformer considers the transistor is omitted for the circuit and the collapsing magnetic flux produces a high voltage in the primary winding that is passed to the signal diode to charge the 100 μF.

The voltage across the 100 μF gradually increases and the two "detecting resistors," 3.9 k and 560 ohm, produce a voltage-divider network with the base of the BC 547 at the "detection-point." When the voltage on the base becomes 0.65 V, the transistor starts to turn on and effectively puts a resistance between the base and 0 V rail of the BC 338. This robs the BC 338 of "turn on" current and thus it does not turn on as hard.

The output energy from the secondary winding is reduced and the voltage across the 100 μF does not rise any higher. The "quiescent" or "idle" or "wasted" current has been kept to 5 mA by making the base resistor 2.2 k and the voltage-detection resistors 3.9 k and 560 ohm. When the micro turns on, the voltage on the 5 V rail drops slightly and this turns off the feedback transistor slightly. The BC 338 is driven a little harder and the voltage on the 5 V rail is maintained. Although this "switched" power supply is a brilliant design, it puts a lot of noise on the power-rail.

A 3½ inch × 5½ inch printed circuit board was used for this project and designed to accept the PIC16F628A microprocessor and all of the additional circuitry including the power supply components discussed.

Construction

Before we begin construction of the Simon Memory Game, you will need to locate a clean, well-lit workbench or table top in which to spread out all of the components for the project. You will also need to locate a 33 to 40-watt pencil tip soldering iron, a small spool of 60/40 rosin core solder, along with a wet sponge and a small jar of "Tip Tinner," a soldering iron tip cleaner/dresser available from your local Radio Shack

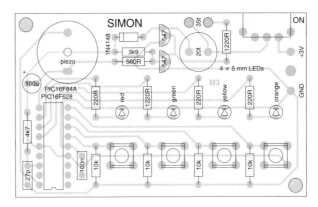

Figure 36-3 *Simon Memory Game pictorial diagram*

store. Also try to locate an anti-static wrist strap, also available from your local parts supply house. You will also need to assemble a few hand tools for construction. Find a pair of small needle-nose pliers, a pair of end-cutters, a magnifying glass, a pair of tweezers, and a small flat-blade and a Phillips screwdriver.

Once you have collected all the necessary tools and project parts, we can begin building the Simon Memory Game project. Next locate the Simon Memory Game pictorial diagram, depicted in Figure 36-3, which will help you identify parts placement on the PC board. Place all the project parts in front of you and heat up your soldering iron. Now refer to Table 36-1, which illustrates the resistor color chart. All resistors have color codes on the body of the resistor, which start at one end of the resistor body. The first color band represents the first digit of the resistor code. The second color band represents the second digit of the resistor's code and the third color band denotes the multiplier value used to compute the value. A fourth color band depicts the resistor's tolerance value. A silver band denotes a 10% tolerance value, while a gold band denotes a 5% tolerance. Absence of a fourth color band shows the resistor has a 20% tolerance. So look through the resistors for the project and try and see if you can locate a resistor with a 4.7 k ohm or 4700 ohm value. The first band will be yellow (4), the second band will be violet (7) and the third band will be (red), a multiplier of (00). Once you locate this resistor you can place it on the PC board at its proper location. Next you can solder the resistor to the circuit board, then follow-up by cutting the excess resistor lead flush to the edge of the circuit board with your end-cutters.

Table 36-1

Resistor Color Code Chart

Color Band	1st Digit	2nd Digit	Multiplier	Tolerance
Black	0	0	1	
Brown	1	1	10	1%
Red	2	2	100	2%
Orange	3	3	1000 (K)	3%
Yellow	4	4	10000	4%
Green	5	5	100000	
Blue	6	6	1000000 (M)	
Violet	7	7	10000000	
Gray	8	8	100000000	
White	9	9	1000000000	
Gold			0.1	5%
Silver			0.01	10%
No color				20%

Now locate the remaining resistors using the resistor color code chart. Install the remaining resistors, then go ahead and solder them in place on the PC board.

Next we will move on to installing the capacitors used in this project. Now refer to Table 36-2, which illustrates the capacitor code chart. Most larger capacitors, such as electrolytic types, will have their actual values printed on the body of the component. However, small capacitors do not have space on them for the values to be printed, so a three-digit code is used. Look through the capacitors in your parts pile for a capacitor marked with a code (104); this will denote a capacitor with a 100 nF or 0.1 µF value, and this capacitor is placed at C1. Install capacitor C1 at its proper location and then solder it in place. Remember to remove the excess component leads. Locate C2 and install it as well. The project also contains an electrolytic capacitor which is polarity sensitive. Electrolytic capacitors will have either a black or white color band along one side of the capacitor body. Near this color band, you will also see a plus (+) or minus (−) marking which denotes the polarity of that particular terminal. When installing electrolytic capacitors, make sure that you observe the polarity markings; for example, the plus (+) terminal of C3 connects to the plus terminal of U1 at pin 14. Solder the capacitor in place and then cut the excess component leads flush to the edge of the PC board.

The Simon Memory Game also contains a silicon diode at D1. Symbolically a diode has two terminals. One terminal is an arrow or triangle which points to the straight line, this is the anode lead of the diode. The straight line on the diode symbol is the cathode. Each diode will have either a black or white color band at one end of the diode body. This color band is the diode's cathode. Note that the cathode of D1 is connected to pin 14 of U1. Install the diode on the PC board and then solder it in place. Remember to cut the extra remaining lead lengths from the diode.

The Simon Memory Game has four colored LEDs, a red, yellow, green and orange one. Note that the red LED is connected to pin 10 of the IC, while the green LED is connected to pin 11 of the IC. The yellow LED is connected to pin 12 of the PIC16F82A, while the orange LED is connected to pin 13 of U1. The diagram in Figure 36-4 illustrates the LED pin-outs, which will be helpful when installing the four LEDs.

The circuit contains two NPN transistors at Q1 and Q2. Transistors have three leads, a Base lead, a

Table 36-2
Capacitance Codebreaker Information

This table is designed to provide the value of alphanumeric coded ceramic, mylar and mica capacitors in general. They come in many sizes, shapes, values and ratings; many different manufacturers worldwide produce them and not all play by the same rules. Most capacitors actually have the numeric values stamped on them; however, some are color coded and some have alphanumeric codes. The capacitor's first and second significant number IDs are the first and second values, followed by the multiplier number code, followed by the percentage tolerance letter code. Usually the first two digits of the code represent the significant part of the value, while the third digit, called the multiplier, corresponds to the number of zeros to be added to the first two digits.

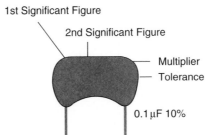

1st Significant Figure
2nd Significant Figure
Multiplier
Tolerance
0.1 µF 10%

Value	Type	Code	Value	Type	Code
1.5 pF	Ceramic		1000 pF/0.001 µF	Ceramic/Mylar	102
3.3 pF	Ceramic		1500 pF/0.0015 µF	Ceramic/Mylar	152
10 pF	Ceramic		2000 pF/0.002 µF	Ceramic/Mylar	202
15 pF	Ceramic		2200 pF/0.0022 µF	Ceramic/Mylar	222
20 pF	Ceramic		4700 pF/0.0047 µF	Ceramic/Mylar	472
30 pF	Ceramic		5000 pF/0.005 µF	Ceramic/Mylar	502
33 pF	Ceramic		5600 pF/0.0056 µF	Ceramic/Mylar	562
47 pF	Ceramic		6800 pF/0.0068 µF	Ceramic/Mylar	682
56 pF	Ceramic		0.01	Ceramic/Mylar	103
68 pF	Ceramic		0.015	Mylar	
75 pF	Ceramic		0.02	Mylar	203
82 pF	Ceramic		0.022	Mylar	223
91 pF	Ceramic		0.033	Mylar	333
100 pF	Ceramic	101	0.047	Mylar	473
120 pF	Ceramic	121	0.05	Mylar	503
130 pF	Ceramic	131	0.056	Mylar	563
150 pF	Ceramic	151	0.068	Mylar	683
180 pF	Ceramic	181	0.1	Mylar	104
220 pF	Ceramic	221	0.2	Mylar	204
330 pF	Ceramic	331	0.22	Mylar	224
470 pF	Ceramic	471	0.33	Mylar	334
560 pF	Ceramic	561	0.47	Mylar	474
680 pF	Ceramic	681	0.56	Mylar	564
750 pF	Ceramic	751	1	Mylar	105
820 pF	Ceramic	821	2	Mylar	205

(Continued)

Table 36-2 (*Continued*)

PicoFarad (pF)	NanoFarad (nF)	MicroFarad (mF, μF or mfd)	Capacitance Code
1000	1 or 1n	0.001	102
1500	1.5 or 1n5	0.0015	152
2200	2.2 or 2n2	0.0022	222
3300	3.3 or 3n3	0.0033	332
4700	4.7 or 4n7	0.0047	472
6800	6.8 or 6n8	0.0068	682
10000	10 or 10n	0.01	103
15000	15 or 15n	0.015	153
22000	22 or 22n	0.022	223
33000	33 or 33n	0.033	333
47000	47 or 47n	0.047	473
68000	68 or 68n	0.068	683
100000	100 or 100n	0.1	104
150000	150 or 150n	0.15	154
220000	220 or 220n	0.22	224
330000	330 or 330n	0.33	334
470000	470 or 470n	0.47	474

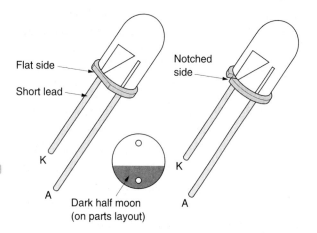

Flat side

Short lead

Notched side

K

A

K

A

Dark half moon (on parts layout)

Figure 36-4 *LED identification*

Collector lead, and an Emitter lead. The Base lead is the straight line symbol, while the Emitter lead has the small arrow pointing away from the transistor. The diagram in Figure 36-5, illustrates the pin-outs of the transistors and the PIC microprocessor chip. Install the transistors on the PC board at their respective locations and then solder them in place. Remove the excess transistor leads after soldering.

Finally, you have to construct the fly-back transformer. Carefully remove the winding on the choke by cutting the outer wire and unwinding the turns until only a few remain. Now put 20 turns (the feedback winding) onto the core and remove the enamel from the wire before winding it around the pin and soldering. Make sure the wire is not tight as the next winding is wound over this fine wire and you must be sure it will not stretch and break. The wire for the primary winding is included in the kit. Wind 35 turns over the feedback winding and twist the two ends together to keep it in position. Solder the two pins of the transformer to the board. Remove the enamel from the ends of the outer winding and solder them to the board.

This winding may be connected to the board around the wrong way, and will be corrected after the project is fully assembled.

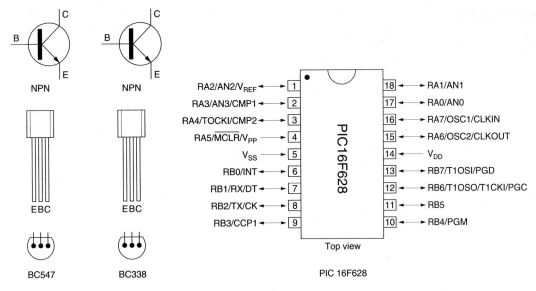

Figure 36-5 *Simon Memory Game semiconductors*

Do not install the PIC IC at this point. Now turn the project on very briefly and monitor the 5 V rail. If no voltage is detected, the connection of the primary winding is around the wrong way. Reverse the connections and measure again. The voltage should be very close to 5 V. Disconnect the power now and then install the integrated circuit into the socket, remember to line-up pin 1 of the chip with pin 1 of the socket.

Now go ahead and install the integrated circuit socket. Use of an integrated circuit socket is highly recommended when constructing the Simon Memory Game. On occasion an IC may fail, and in that situation, you can simply pull out the defective IC and replace it with a new one. Most electronic hobbyists cannot successfully unsolder a large integrated circuit without damaging the PC board, so an IC socket is cheap insurance.

Carefully handle the PIC IC package, with your anti-static wrist strap on and grounded, in order to avoid any electrostatic potential transfer to the integrated circuits. When installing the integrated circuits be extremely careful to avoid moving or walking on a carpet. Sit or stand in one place while installing these components, since moving generates high voltage electrostatic potential which may damage the IC.

An IC socket will typically have a small notch or cutout at one end of the plastic package. Just to the left of the notch or cutout you will find pin 1. Opposite pin 1 is pin 18, but note that pin 18 is not the plus (+)

power pin, as it is with most integrated circuits. Pin 16 in this project is connected to the clock RC circuit of R1 and C2 capacitor. Power is applied to the microprocessor on pin 14, while ground is on pin 5. When installing the PIC microprocessor chip into the IC socket, make sure that you line-up pin 1 on the socket with pin 1 of the integrated circuit. If you are having difficulties identifying the pin-outs of the integrated circuits, be sure to ask a knowledgeable electronics enthusiast for help. Finally go ahead and install the four PC board switches, and mini piezo sounder and the battery snap (or leads from the 3 V battery-box) and insert the cells.

Take a few moments to rest before inspecting the PC board for "cold" solder joints and possible "short" circuits. Turn the PC board over, so that the foil side is facing up toward you. Now look at all the solder joints and observe them carefully. All the solder joints should look clean, smooth and shiny. If any of the solder joints do not look good, you will have to remove the solder and re-solder the joint so that its looks clean, smooth and shiny with no big blobs of solder at the circuit pads. "Cold" solder joints will usually deteriorate in time and will cause the circuit to fail prematurely, so it's best to discover them early and to fix them quickly. Once you have inspected the board for "cold" solder joints, you can move on to inspect the board for possible "short" circuits. Take another look at the foil side of the circuit board; this time you are looking for "stray" component leads which may have attached themselves when the component leads were cut. Often these remaining "cut" leads can attach themselves to the underside of the board and just stick there, usually due to the residue from the rosin core solder. Applying power with these "short" can damage the circuit when power is first applied.

In order to use your new Simon Memory Game, you will have to "burn" the Simon.hex program into the PIC 16F628 microprocessor chip. If you are familiar with microprocessor programming, and are interested in how the program works, you will want to look over the Simon.asm program, so that you can follow the progression of the program from start to finish. If you are not interested in the programming aspect of the program then you can just load the Simon.hex program into the microprocessor chip directly. First, you will have to build the Multi-Chip Programmer project in this book or purchase a PIC microprocessor Programmer. You will have to connect the Multi-Chip Programmer to your computer via a cable described in the programmer article. Next you will have to insert the "blank" PIC chip into the socket in the programmer circuit. Since the Multi-Chip Programmer was designed for different chips, you will have to make sure that you insert your PIC chip correctly in the socket. Connect up the Multi-Chip programmer to your personal computer via the serial cable. Then, you will have to download the (*icprog105a-c.zip*) file from the http://www. books.mcgraw-hill.com/authors/ petruzzellis/ elecgames website into your computer and expand the zip file. Then run the *icprog105c-a.exe* program on your computer, and configure it in the following way:

Simon Program

Program for PIC16F628 micro (Simon.asm)

The following contains all the comments for the Simon program:

Note: __Config 3F18h covers the following:

_CP_off & _LVP_off & _PWRTE_off & Boden_off & _WDT_off & _IntRC_osc_I/O

where:

CP = code protection

LVP = low-voltage programming

PWRTE = power-up timer

Boden = Brown-out detector

WDT = watchdog timer

IntRC_osc_I/O = internal RC for 4MHz — all pins for in-out

You can simply write: _Config 3F18h

or:

__Config _CP_off & _LVP_off & _PWRTE_off & Boden_off & _WDT_off & _IntRC_osc_I/O

Note, you can also select the boxes in *icprog105a-c.exe* program at the time of burning the chip. In all cases the chip will get burnt with the correct configuration value and when it is copied, the copy will automatically have the correct configuration value.

Next, you will have to download the *Simon.hex* program. Once configured, run the *Simon.hex* file within the *icprog105c-a.exe* program. You simply have to connect the serial cable between the programmer and your computer, and the computer will "burn" the PIC chip with the Simon.hex program. Once the program has been "burned" into the PIC chip, remove the PIC chip from the programmer circuit and insert it into the IC socket on the Simon Memory Game circuit board.

Now apply power to the Simon Memory Game board. You should hear the "start" tune begin playing. The object of the Simon Memory Game is to have the player repeat the sequence of flashes and beeps, quickly and accurately, by pressing the corresponding switch associated with the LED which has just flashed. As the speed gets faster and faster between the flashes it gets difficult to follow quickly without making mistakes. The game is really simple but you must pay strict attention which makes the game a real challenge if you want to win the game. Now you are ready to play the Simon Memory Game, and challenge your friends and family to a match of wits. Have fun!

Chapter 37

Tetris Game Project

Parts List

R1, R2, R3, R4, R5
100 k ohm, 5%, ¼-watt
resistor (SIL 5-pack
resistor)

R6, R7, R8, R9, R10 1k
ohm, 5%, ¼-watt
resistor

R11, R12, R13, R14, R15
100 k ohm, 5%, ¼-watt
resistor (SIL 5-pack
resistor)

R16, R17, R18, R19, R20
1k ohm, 5%, ¼-watt
resistor

R21 220 ohm, 5%,
¼-watt resistor

R22 180 ohm, 5%,
¼-watt resistor

R23 470 ohm, 5%,
¼-watt resistor

R24 1 k ohm, 5%,
¼-watt resistor

R25 180 ohm, 5%,
¼-watt resistor

R26 10k ohm, 5%,
¼-watt resistor

C1 100 µF, 35-volt
electrolytic capacitor

C2, C3 3.3 µF, 35-volt
electrolytic capacitor

D1 1N4148 silicon diode

U1 LM7805 5-volt
regulator

U2 PIC16F84 PIC
microprocessor

XTL 12 mHz crystal

J1, J2 RCA jacks

P1, P2 9-pin RS-232
male connector

S1 momentary pushbutton
switch

Misc PC board, wire,
solder, battery
holder, battery clip,
plastic box

Tetris is an old Russian computer game where you should try to fit in blocks into a play field. It is a simple game but really a lot of fun and really addicting. The name Tetris is derived from the ancient Greek word for four: "tetra." There are seven combinations of the four bricks as seen in Figures 37-1 and 37-2. This version of the Tetris Game utilizes a PIC16F84 running @ 12 MHz, generating a video signal in software.

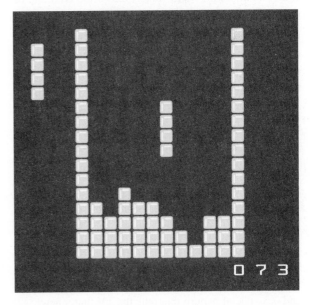

Figure 37-1 *Tetris Game Display*

Figure 37-2 *Tetris Game blocks*

Software

The only hardware used for the video generation is two resistors forming a 2-bit DA converter. Usually the video signal as generated in video games is created with dedicated video chips, reading the image data from a graphics memory. In this project the video signal is calculated in real time by the microprocessor as the electron beam sweeps over the screen.

With a processor performing 3 MIPS, it is not easy to make a video signal in software. Each instruction performed takes ⅓-μs. Each scan-line on the screen takes 64 μs, where 52 μs are visible, so it gives 52*3=156 visible clock cycles per line. Maximum resolution that can be obtained is 156 pixels in x-axis if the software is setting one pixel per clock (using for example only bcf and bsf), but more is needed to make a game, like loops and such. A loop quantifies the time to 3-clock pieces, giving a resolution of 52 pixels. (One could obtain a kind of 156 pixels resolution with one or two offset nops, but the code to select this would eat too many clock cycles to do any good.) However, Tetris is quite simple, the resolution is quite low, and there is no motion, the blocks of pixels are just turned on and off. The most demanding part of the game is to show the score at the bottom of the screen. It obtains higher resolution by loading the PORTB with the bitmap for the number and shifts it out one pixel per clock cycle.

So far we've only talked about the graphic generation. But there is more to it to get a video signal. All scan-lines first have a 4 μs-sync pulse, then black for 8 μs, then the 52 μs graphic comes. These horizontal sync-pulses makes the TV understand when a scan-line starts, but there is a need to send information about when a new picture starts too; it is called vertical sync, and is a special pattern that tells the TV that a new image is coming. There are two kinds of vertical sync,

because the image is divided into two part images, showing even and odd lines, to get less flickering. In Tetris, the two images are identical, so the game is not using the full y-resolution possible, but it doesn't matter because it is way better than the x-resolution anyway, making the x-resolution the biggest problem.

The game field is kept in memory as a 32 byte array, 16×16 bits, where one bit is one pixel-block on the screen. The area to the upper left is for showing the next block, and by making it a part of the game-field it is possible to use the same block-drawing routines as for the game, and thereby saving memory. Each frame, the falling block is first removed from the game-field, and then tests are performed if the block can move, as the player wants it to. Then the block is drawn back to the screen at the new position. When a block is to be tested, put or removed, it first must be generated. To generate a block means compressing it from the compressed data, rotating it and then storing the relative coordinates of the block in the block array. The block data is compressed in relative coordinates. In compressed format, each coordinate is stored in two bits for both x and y, where the two bits can represent the numbers — 1, 0, 1, 2. These values need to be uncompressed to 4*2 byte sized values representing the coordinates in two's complement format. Depending on the angle the block should have, the coordinates might need to be mirrored or/and swapped. When the block has been created it can easily be put, removed or tested. The test routine checks if there are any pixels set on the block positions where the block should be put. If pixels are set, then the block can't be put there. New blocks are selected at random, where the random number is a counter that increases for every frame, making the random number dependent on how long it takes for the player to place the block, and making a quite good random number.

The game stuff, like checking joystick and moving stuff around, is taken care of in the first scan-lines, when no graphics are drawn. During the time before the play-field is shown, there is a little bit of free time to play the music, but there is not time to play it on all lines, and that makes the music sound distorted. The music is stored in the data EEPROM, in a compressed one byte format, where one byte contains length and note. The note's frequency is looked up in a table, and so is the length too. (The frequencies are based on the line frequency so they are not exactly the correct

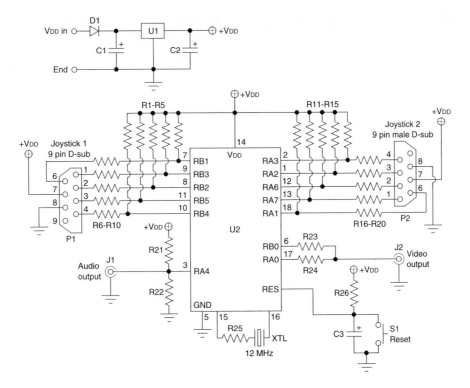

Figure 37-3   *Tetris Game. Courtesy R. Gunee*

frequencies.) The speed of the game is increasing constantly and music-speed increases as the game-speed increases.

Developing this kind of software is mostly a clock-cycle-counting project, all timings are quite critical, so whatever paths the execution-flow of the program takes, it must take the same number of clock cycles. This is quite hard, and I've not managed to do this on all lines, so the image is a little bit bent in some places. (Most analog TV-sets fix this, but on some digital projectors it is more visible.) The Tetris Assembly language program (tetris.asm) can be downloaded from the McGraw-Hill website at http://www.books.mcgraw-hill.com/authors/petruzzellis/elecgames and it will help those who are interested in following the program operation.

The hardware

The circuit for the Tetris Game, as shown in Figure 37-3, is really quite straightforward, since much of the work is done in software. Two resistors, forming a DA converter together with the input impedance of the TV, generate the video signal. This can generate the levels

0 V (sync), 0.3 V (black), 0.7 V (gray), and 1.0 V (white). To be able to handle the variation of input resistance of different audio equipment two resistors are used to make a 1-bit DA to generate the audio. When generating the video, the PORTB is used as a shift register to get one pixel per instruction when high-resolution text is shown on the screen. Shifting a port requires the port to be set as output if a whole byte is to be shifted out. First, this seems like a problem, the whole port can't be used for anything else than video generation, but that is not quite correct. A port can be used as an input when not used as a shift register, so in Tetris PORTB it is used for joystick input when not used as a shift register. The digital joystick is a switch to ground, so all that is needed to connect it to the PIC is a couple of pull-up resistors, and that is available inside the PIC. Unfortunately it is not that simple; if a pin on a port is grounded when used as an output, the output buffer of the PIC would burn up, so this is solved by adding one extra 1 k resistor on each pin to limit the current. What about those pull-up resistors? There are 10 k pull-up resistors built into the PIC that can be switched on and off. However, using them would be a too strong pull-up, so the 1 k current limiting resistor (plus bad switches in the joystick) can't pull the input low enough. Therefore

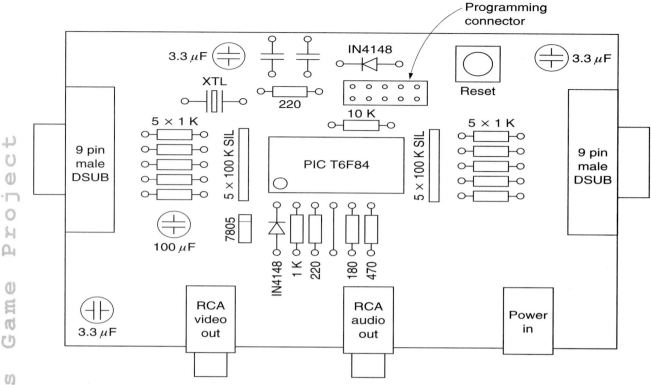

Figure 37-4   *Tetris Game pictorial diagram*

Construction

an external 100 k resistor pull-up network is added. The power supply part of the circuit is quite simple, it uses a standard 7805 to get a 5 V supply. The input can be 8–18 volt, DC or AC. (Thanks to the diode at the power input.)

Construction

The Tetris Game was constructed on a 3¼ inch × 4¾ inch printed circuit-board. The parts layout is shown on the pictorial diagram Figure 37-4. Notice that the LED display is at the right side of the circuit board. Before we begin building the Tetris Game, you will need to secure a clean, well-lighted worktable, so that you can place all the diagrams, the printed circuit and all the parts in front of you. Once you have secured a good spot, you will want to grab a 30–37-watt pencil tip soldering iron, some 60/40 lead/tin solder, a wet sponge and some "Tip Tinnerr," a soldering iron tip dresser, available at your local Radio Shack store. First heat up your soldering iron and clean off the tip once it's hot and then wipe the heated iron tip across the "tip dressing." Now you are about ready to begin. Open the

parts bag. With a large project, you may want to place the components in small trays or egg cartons to hold all the small parts from flying off the worktable. Locate the resistors from the parts and now refer to Table 37-1, which lists the resistor color code information. Resistors will generally have at least three or four color bands which help identify the resistor values. The color band will start from one edge of the resistor body, this is the first color code which represents the first digit, the second color band depicts the second digit, while the third band is the resistor multiplier value. The fourth color band represents the resistor tolerance value. If there is no fourth band then the resistor has a 20% tolerance, if the fourth band is silver, then the resistor has a 10% tolerance value, and if the fourth band is gold then the resistor has a 5% tolerance value. Try to locate a resistor with the first band having a brown color (1), the second band with a black color (0) and a third band with a yellow color. This first resistor will be R1, with a value of 100 K ohm. Place R1 on the circuit board in its proper location and repeat the procedure for each of the resistors until all have been installed. Then solder all the resistors in their proper locations and clip the excess component leads. Note, you could also use a SIL 5-pack

Table 37-1
Resistor Color Code Chart

Color Band	1st Digit	2nd Digit	Multiplier	Tolerance
Black	0	0	1	
Brown	1	1	10	1%
Red	2	2	100	2%
Orange	3	3	1000 (K)	3%
Yellow	4	4	10000	4%
Green	5	5	100000	
Blue	6	6	1000000 (M)	
Violet	7	7	10000000	
Gray	8	8	100000000	
White	9	9	1000000000	
Gold			0.1	5%
Silver			0.01	10%
No color				20%

resistor for R1 through R5 and for R11 through R15 if desired.

Now let's move to installing the three capacitors. Note that all three capacitors are electrolytic types. The electrolytic capacitors will have their values printed on the capacitor body. You will also note that there will be either a white or black band on the electrolytic capacitors. This band will denote the minus (–) or plus (+) capacitor lead. Refer to the schematic to determine the polarity of the capacitor in the circuit, before inserting the electrolytic capacitors into the circuit board. The polarity of the capacitor must match its notation on the schematic and circuit board. Capacitors installed backwards are often the reason circuits do not work upon power-up and can sometimes cause damage to the circuit, so pay close attention to the correct orientation of the capacitors when installing them. Install all the capacitors on the circuit board, and solder them to the circuit board, remembering to cut the excess capacitor leads flush to the edge of the circuit board.

The Tetris Game has a single silicon diode at D1. Diodes are two-lead devices with a designated polarity. Schematically the anode lead looks much like an arrow, pointing toward a thin line, which is the cathode. The cathode on the diode's body will appear asa black or white line at one end of the body of the diode.

The heart of the Tetris Game is the PIC16F84 microprocessor, which is shown in the pin-out diagram depicted in Figure 37-5. The 18 pin PIC microprocessor is static sensitive and should be handled using a grounded anti-static wrist strap. Moving up and down in your fabric covered chair or moving around your worktable can generate large static charges which could damage the chip. It is highly recommended that you install an 18 pin integrated circuit socket prior to installing the PIC microprocessor chip on the circuit board. In the event of a circuit failure it will be a simple matter to just unplug the chip and replace it with a new chip. Since most electronic hobbyists cannot unsolder a large IC without damaging it, it is recommended that you install an IC socket first before installing the PIC chip into the circuit. Integrated circuit sockets are cheap insurance, in the event that the PIC chip becomes defective. Most IC sockets will have a notch or cutout at one end of the plastic package. Note that pin 1 will be just to the left of this notch or cutout. When installing the PIC16F84 chip into the socket, just make sure that you line-up pin 1 of the IC socket with pin of the actual PIC16F84 chip. Each PIC chip will have a small cutout, notch or indented circle designated for pin 1 of the chip. Note that pin 1 will be just to the left of the notch or indented circle.

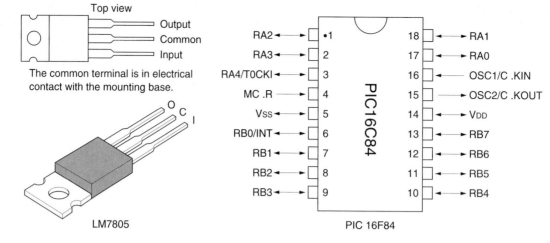

Figure 37-5 *Tetris Game semiconductor pin-out diagram*

Next, you will need to locate and install the 12 mHz crystal resonator in series with a 180 ohm resistor between pins 15 and 16 of the PIC16F84 chip. The momentary pushbutton Reset switch is wired between pin 4 of U2 and ground in parallel with a 3.3 µF capacitor. Solder in the switch and the crystal and then cut the excess component leads flush to the edge of the circuit board.

Now locate the LM7805, 5-volt regulator and install it onto the circuit board; note the 100 µF capacitor at C1 is connected to the input of the regulator and the 3.3 µF electrolytic capacitor is connected to the output side of the regulator. The PIC16F84 chip requires a 5-volt power source, which the regulator provides. A 9 to 12-volt power supply or battery could be used to power the Tetris Game circuit. You could elect to install an on/off power switch in series with the Vdd input line at diode D1 if desired.

Go ahead and install the two RCA jacks. Jack J1 is used to port the audio from the game to your TV set, while jack J2 is used to port the video signal from the Tetris Game to video input jack on your TV set. Solder these two jacks onto the circuit board.

Finally locate and install the two 9-pin RS232 male plugs onto the PC board and solder them in place. These jacks are used for connecting two joysticks to your game circuit in order to move the Tetris blocks around while playing the game.

The Tetris Game is now completed, but before we test it out, you should take a short break. After the break we will inspect the circuit board for any "cold" solder joints and any "short" circuits. Pick up the circuit board and place the circuit board with the copper foil side up facing toward you. Inspect the solder joints on your PC board, you will want to make sure that all the solder joints look clean, smooth and shiny. If any of the solder joints look dull or "blobby" then you will need to remove solder from that particular joint with a solder-sucker or brain material and then re-solder the joint so that it looks clean, smooth and shiny. "Cold" solder joints will cause the circuit to fail prematurely, so it is best to attend to these problems quickly. Next, you will want to inspect the PC board for any possible "short" circuits which can be caused from any "stray" cut component leads which may have attached themselves to the underside of the board when the component leads were initially cut. "Short" circuit can damage the circuit when power is first applied to the circuit. So you will want to find any "short" circuits before connecting up the battery.

Chip "Burning"

In order to use your new Tetris Game, you will have to "burn" the *tetris.hex* program into the PIC 16F84 microprocessor chip. First, you may want to study the *tetris.asm* program in more detail if you are interested in how the program works. If you are familiar with microprocessor programming you will want to look over the *tetris.asm* program where you can follow the progression of the program from start to finish. If you are not interested in the programming aspect of the program then you can just load the *tetris.hex* program into the microprocessor chip directly.

First, you will have to build the Multi-Chip Programmer project in this book or purchase a PIC microprocessor programmer. Next, you will have to connect the Multi-Chip programmer to your computer's serial port via a cable described in the programmer article. Now you will have to insert the "blank" PIC16F628 chip into the socket in the programmer circuit. Since the Multi-Chip Programmer was designed for different chips, you will have to make sure that you insert your PIC chip correctly in the socket.

Go to http://www.books.mcgraw-hill. com/authors/ petruzzellis/elecgames website and download the (*icprog105-a.zip*) program into your computer. Unzip the file and open up the *icprog105c-a.exe* program. Locate the *tetris.hex* program from the website and open it up within the *icprog105c-a.exe* program and configure it in the following way:

p=16F84,r=hex __config 0x3FFA

You want to make sure that the *icprog105a-c.exe* program recognizes that you will be burning a PIC 16F84 microprocessor chip. Run the *tetris.hex* program file within the *icprog105c-a.exe* program and "burn" **tetris.hex** file into the microprocessor. Once the PIC chip has been programmed, simply remove the PIC chip from the Multi-Chip programmer. Now insert the 16F84 "programmed" or "burned" chip into the Tetris Game circuit board.

Now for the "moment of truth"! Find a 9 to 12-volt battery or AC to DC power supply and connect it up to the Tetris Game circuit. Connect an RCA to RCA patch cord between the Tetris Game board video output and your TV set. Now, connect a second RCA to RCA patch cord from the audio output on the Tetris Game board to the audio input on your TV set or to an audio amplifier. Connect up the joystick to the joystick port. If you installed the optional power switch, turn it to the "on" position. With power applied the Tetris Game should start up immediately.

If all went well, you can start playing your new Tetris Game. In the event that your Tetris Game does not work, you will have to go back and re-check a few things. First, remove the battery! Carefully read the color codes on each resistor, and make sure that you have placed the correct resistor in the correct PC board hole. Next take a look at the two transistors and make sure that you have installed them correctly. Carefully compare the transistors with the pin-out diagram and make sure that each of the transistor leads goes to the correct place on the PC board. Now make sure that the polarity sensitive capacitors are installed with respect to the polarity marking band at one end of the capacitor body, and finally make sure the you have installed the LEDs correctly. If in doubt, check the LEDs mounting diagram to be able to identify the flat edge of the LED and its negative or ground lead.

When the power is turned on the game starts! The score is shown in the bottom right corner, and the next block to come is shown in the upper left corner of the screen. See picture of the game in action here to the left. As the blocks fall down, they can be moved sideways by using the joystick (left gameport on hardware); the fall speed can temporarily be increased by moving the joystick down. The fire-button is used to rotate the blocks. When one horizontal line is full, then it is removed. You get points for full lines and placed blocks. As you get more points the difficult level is increased by increased block falling speed. The music's speed is increased as the game speed increases. You get game over when the playing field level has reached to the top and there is not room for more blocks.

How to play the game

The first screen is where you select how you want to play by moving the joystick: DOWN: Human vs. Human (H-H), LEFT: Human vs. Computer (H-C) or RIGHT: Computer vs. Computer (C-C). Start with FIRE. Unfortunately it is impossible to beat the computer, since there was not enough room to make the computer beatable. That makes the computer vs. computer game to play forever until someone resets the game using the reset switch. You start serving by pressing FIRE; it is also possible to change direction and speed of the ball using FIRE. The player who has the serve will get points. If the player with the serve misses the ball, then the serve goes over to the other player. When someone wins a game over a picture will show and tell who won. Go ahead and challenge your friends to a game of Tetris and have some fun!

Chapter 38

Tic-Tac-Toe Game

Parts List

Parts Bin

R1, R2 10 k ohm, ¼-watt, 5% resistor

R3, R4, R5, R6, R7 330 ohm, ¼-watt, 5% resistor

R8, R9, R10 330 ohm, ¼-watt, 5% resistor

R11, R12, R13 2.2 k ohm, ¼-watt, 5% resistor

R14 47 k ohm, ¼-watt, 5% resistor

C1 100 nF, 35-volt disk capacitor (104)

C2 100 µF, 35-volt electrolytic capacitor

D1 1N 4004 power silicon diode

Q1, Q2, Q3 BC 547 NPN transistors 2N3904 or NTE123AP

U1 PIC16F628 IC w/Tic Tac Toe program

L1, L2, L3, L4, bi-colored red/green LEDs

L10 green LED

L11 red LED

S1, S2 momentary pushbutton switches (PC mount)

S3 SPST on/off slide switch

B1, B2, B3, B4 "AA" cells

Misc PC board, IC socket, battery clip, battery holder, wire, etc.

Everyone knows the game of Tic-Tac-Toe! This electronic version of the Tic-Tac-Toe game introduces two features, i.e. pushbuttons, bi-colored LEDs, and the microprocessor, see Figure 38-1. You can use the project to learn the skills of creating the Tic-Tac-Toe program.

Tic-Tac-Toe is one of the simplest yet most-challenging games to be invented. With just a choice of nine locations, two players can pit their wits and very quickly work out who is superior. Even though the game is very simple, it has an enormous fascination. The game of Tic-Tac-Toe provides an interesting challenge for creating an electronic game. While it is possible to produce a computer program capable of playing with consistent number of "computer wins," we have created a game which does not use high level strategy, but it plays a very interesting game. The program has been kept as simple

Figure 38-1 *Tic-Tac-Toe Game*

as possible to show how to produce routines that carry out a function.

Many of the sub-routines are algorithms; these are software routines that solve a specific problem. The routine looks to see if a certain condition is present and produces an answer. This is the basis of "Computer Intelligence" or "Artificial Intelligence" (AI). When a number of these routines are combined and a result is obtained in a very short period of time, the computer appears to be "clever." There is nothing more rewarding than producing a program that delivers "feedback" of this type. It's the programmers' highlight of the day.

The microprocessor used for this project is a PIC16F628. This is one of the latest low-cost microprocessors on the market and is an ideal starting-point for beginners to the "art of programming." The micro is easy to use and has a re-programmable feature that allows it to be programmed almost any number of times. A game is the ideal place to begin programming, as you know how it is played or how it is used and it's just a matter of seeing how the routines are created—so you can copy them or use them in other programs.

There is one other advantage of using a game. It introduces strategy. If the game is played against the "computer" and the computer has a chance of winning, it appears to have "intelligence." The program for this game can be developed in two different ways—as an algorithm or in "linear-mode." An algorithm is essentially a routine consisting of instructions that come up with a definite answer. The program could consist of a single, extremely complex algorithmic sub-routine. It would take hours to explain the thinking behind the structure of the routine and beginners would be left floundering. The solution is to produce a program with very simple sub-routines.

A linear-mode layout makes the micro run through the program and find a set of instructions that applies to the particular condition. This type of layout requires more instructions but it is much easier to follow. The operation of the program can best be seen from the following *Flow Chart,* shown in Table 38-1. The Tic-Tac-Toe assembly language program (tictac.asm) can be downloaded from the McGraw-Hill website at http://www.books.mcgraw-hill.com/authors/petruzzellis/elecgames, it will help the builder understand how the program functions.

Game Rules

Everyone knows the rules of Tic-Tac-Toe, so we don't need to describe them here. But there is one point we should explain. There are various levels of play to the game. The simplest is to make a series of moves in the hope of achieving a win. The other is to "set up" the board to create a win in two directions. In this ploy there are two approaches. One is an obvious "split" by taking opposite corners, the other is more sneaky. It involves placing pieces on either side of your opponent and thereby creating a non-threatening situation. The next move will place a key piece to link-up the two directions and the game is won. This can be called "levels of play" and the computer has been programmed to recognize the first two but not the third.

Tic-Tac-Toe Program

A replica of the board is created in memory via Pre-load. It clears nine locations, from 31h to 39h. These correspond to the nine squares on the board. The cursor uses one location before and after these so it can "hide." The cursor starts in location 30h and moves across the board to location 3A. The program takes the cursor from location 3A and puts it in location 30h. Each time "Move" button is pushed, the cursor moves across the display. This creates an orange flashing LED on the display. When PLAY button is pressed, the flashing cursor is converted to red to represent the player piece and a number of sub-routines are executed to produce the best outcome for the computer.

The program consists of a "Main" routine that is constantly looping. It scans the display and illuminates the LEDs according to values in memory locations 31h to 39h. If a location has 01, a red LED on the display is illuminated. If the value is 04, a green LED is illuminated. If the file holds 05, both red and green are illuminated to produce an *orange* color. The cursor is "flashing orange" and to make it flash, the display is scanned a number of times (determined by the number of loops in file 22h) and then the cursor is "hidden." It is hidden by changing the value from 05 to 08. The display is looped a number of times and the cursor is placed back on the display. The

Table 38-1
Tic-Tac-Toe Flow Chart

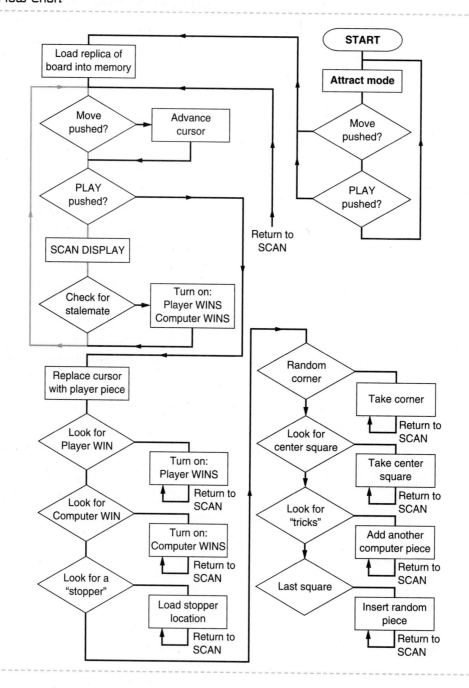

display is scanned by outputting a "High" to LEDs on six lines of Port B. RB0, RB1, RB2, RB3, RB4 and RB5. These six LEDs represent the red and green elements of the top three bi-colored LEDs. A single sinking transistor takes these LEDs to the 0V rail. The transistor is turned off and the information for the middle three bi-colored LEDs is outputted on the six lines of Port B. The middle transistor is then turned on and the elements are illuminated. This is continued for the lower three LEDs. By repeating the process very rapidly, the whole display is illuminated; see the *tictac.asm* program listing. The assembly program listing can be very helpful in understanding how the Tic-Tac-Toe Game works.

Attract Mode

When the project is turned on, the screen flashes and shows the effectiveness and capability of the bi-colored LEDs. This is called the "*attract mode*" and is sometimes used in amusement machines to attract players. The LEDs flash from red to orange to green and then a single LED in the center of the display gradually changes from red to green. There are 256 steps in the change. To produce this gradual effect, it takes about 20 instructions and the program shows how this is done. The routines for the "*attract mode*" are separate from the Tic-Tac-Toe game and the buttons are monitored during the "*attract mode*" to allow the player to go to the game. When either button is pressed, the micro goes to Tic-Tac-Toe.

Game Circuit

The heart of the game is the PIC16F628, an 18 pin chip with 16 in/out lines. Not all the lines are both input and output and the pin-out of the chip is shown in Figure 38-2. The program is loaded into a microprocessor and runs the program that controls the nine bi-colored LED display.

The first step in getting the prototype Tic-Tac-Toe project to work was to figure out how the microprocessor would drive the display. The display consists of 9 bi-colored LEDs, each LED contains two elements—a red and green emitter. The output is orange when both emitting chips are driven, and when either chip is driven the output is red or green. Other LEDs are also available with three chips inside to produce red, green and blue. When all chips are driven, the output is white light. Bi-colored LEDs have three leads, see Figure 38-3. The two light-emitting parts of the LED have the cathodes connected together and this is taken to the 0 V rail. In our project, we have connected the LED so that the lead to the red chip is on the left and the green is on the right. This means the display is very similar to driving 18 separate LEDs. This allows a scanning routine with a "run-of-three." Since we do not have enough outputs on the micro to drive all the LEDs at the same time, we need to introduce a drive method called "multiplexing." This involves turning on one row at a time and repeating the process at a rate that cannot be detected by the eye. The effect is to produce a display that can be fully illuminated with any LED producing any of the three colors. As each row is turned on, six outputs from the micro take the six elements of the three LEDs to the positive rail, via current-limiting resistors. Each line for the micro can only supply 25 mA and if each LED is illuminated for 33% of the time, the average current will be about 8 mA.

The brightness of a LED can be produced in two different ways. It can be constantly illuminated with a

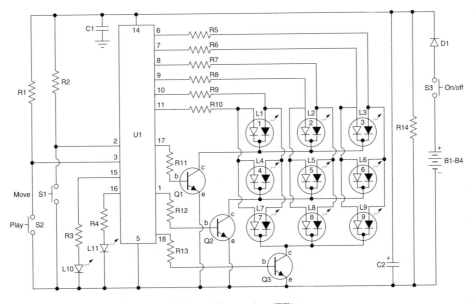

Figure 38-2   *Tic-Tac-Toe Game. Courtesy Talking Electronics (TE)*

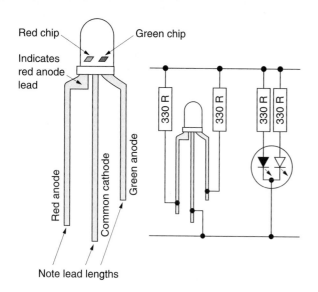

Red chip   Green chip

Indicates
red anode
lead

Red anode   Common cathode   Green anode

Note lead lengths

330 R   330 R   330 R   330 R

Figure 38-3 *LED identification*

current or pulsed with a higher current for a short period of time. In our case, the pulse of 25 mA for a duty cycle of 33% is equivalent to a constant current of not 8 mA but about 12 mA and this is sufficient to give a very good output brightness. Thus we will have no problem with the illumination of the display.

Once we have the scan routine worked out, we can decide on the best pins of the chip for each row of LEDs—mainly to simplify the track layout on the PC board. The current capability of each drive-line is approx. 25 mA and since the duty-cycle for each LED is 33%, the average current for a LED is a maximum of 8 mA. We have already explained one of the unusual characteristics of a LED. If it is driven by a high current for a short period of time, the brightness is better than a lower, constant current. In our case, peaks of 25 mA, for a 33% duty cycle, produce a very bright output from the LEDs and the drive-current needed to be reduced. This was done by increasing the current-limiting resistors to 330 ohm. It is important that the current limiting resistors are placed on each of the drive-lines instead of the sinking line, as the sinking transistor will be carrying a varying current, depending on how many elements are illuminated during each part of the scanning cycle. With a varying current, any dropper resistor in the transistor-line will drop a different voltage and this will produce a varying brightness.

The Tic-Tac-Toe game was powered by a 6-volt DC source made up of four "AA" cells wired in series with a silicon diode to drop the voltage down to 5.2 volts, through power switch S3.

Construction

All the parts for Tic-Tac-Toe can be purchased in a kit. Before we begin construction of the game, you will need to locate a clean, well-lit workbench or table top in which to spread out all the components for the project. You will also need to locate a 33 to 40-watt pencil tip soldering iron, a small spool of 60/40 rosin core solder, along with a wet sponge and a small jar of "Tip Tinner," a soldering iron tip cleaner/dresser available from your local Radio Shack store. Also try to locate an anti-static wrist strap, also available from your local parts supply house. You will also need to assemble a few hand tools for construction. Find a small pair of needle-nose pliers, a pair of end-cutters, a magnifying glass, a pair of tweezers, and a small flat-blade and a Phillips screwdriver.

Once you have collected all the necessary tools and project parts, we can begin building our project. Locate the schematic diagram and the Tic-Tac-Toe pictorial parts placement diagram shown in Figure 38-4, along with all the project parts in front of you while your soldering iron heats up. Now refer to Table 38-2, which illustrates the resistor color chart. All resistors have color codes on the body of the resistor, which start at one end of the resistor body. The first color band represents the first digit of the resistor code. The second color band represents the second digit of the resistor's code and the third color band denotes the multiplier value used to compute the value. A fourth color band depicts the resistor's tolerance value. A silver band denotes a 10% tolerance value, while a gold band denotes a 5% tolerance. Absence of a fourth color band shows the resistor has a 20% tolerance. So look through the resistors for the project and try to see if you can locate a resistor with a 10 k ohm or 10,000 ohm value. The first band will be brown (1), the second band will be black (0), and the third band will be (orange), a multiplier of (000). Once you locate this resistor you can place it on the PC board at its proper location. Next you can solder the resistor to the circuit board, then follow-up by cutting the excess resistor lead, flush to the edge of the circuit board with your end-cutters. Now

Table 38-2
Resistor Color Code Chart

Color Band	1st Digit	2nd Digit	Multiplier	Tolerance
Black	0	0	1	
Brown	1	1	10	1%
Red	2	2	100	2%
Orange	3	3	1000 (K)	3%
Yellow	4	4	10000	4%
Green	5	5	100000	
Blue	6	6	1000000 (M)	
Violet	7	7	10000000	
Gray	8	8	100000000	
White	9	9	1000000000	
Gold			0.1	5%
Silver			0.01	10%
No color				20%

locate the remaining resistors using the resistor color code chart. Install the remaining resistors, then go ahead and solder them in place on the PC board.

Next we will move on to installing the capacitors used in this project. Now refer to Table 38-3, which illustrates the capacitor code chart. Most larger capacitors, such as electrolytic types, will have their actual values printed on the body of the component; however small capacitors do not have space on them for the values to be printed, so a three-digit code is used. Look through the capacitors in your parts pile for a capacitor marked with a code (104); this will denote a capacitor with a 100 nF or 0.1 μF value. This capacitor is placed at C1. Install capacitor C1 at its proper location and then solder it in place. Remember to remove the excess component leads. Locate C2 and install it as well. The project also contains an electrolytic capacitor which is polarity sensitive. Electrolytic capacitors will have either a black or white color band along one side of the capacitor body. Near this color band you will also see a plus (+) or minus (−) marking which denotes the polarity of that particular terminal. When installing electrolytic capacitors, make sure that you observe the polarity markings, for example the plus (+) terminal of C3 connects to the plus terminal of U1 at pin 14. Solder the capacitor in place and then cut the excess component leads flush to the edge of the PC board.

The game also contains a silicon diode at D1. Symbolically a diode has two terminals. One terminal is an arrow or triangle which points to the straight line, this is the anode lead of the diode. The straight line on the diode symbol is the cathode. Each diode will have either a black or white color band at one end of the diode body. This color band is the diode's cathode. Note

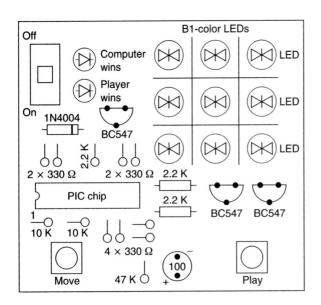

Figure 38-4   *Tic-Tac-Toe pictorial diagram*

Table 38-2

Capacitance Codebreaker Information

This table is designed to provide the value of alphanumeric coded ceramic, mylar and mica capacitors in general. They come in many sizes, shapes, values and ratings; many different manufacturers worldwide produce them and not all play by the same rules. Most capacitors actually have the numeric values stamped on them; however, some are color coded and some have alphanumeric codes. The capacitor's first and second significant number IDs are the first and second values, followed by the multiplier number code, followed by the percentage tolerance letter code. Usually the first two digits of the code represent the significant part of the value, while the third digit, called the multiplier, corresponds to the number of zeros to be added to the first two digits.

1st Significant Figure
2nd Significant Figure
Multiplier
Tolerance
0.1 µF 10%

Value	Type	Code	Value	Type	Code
1.5 pF	Ceramic		1000 pF/0.001 µF	Ceramic/Mylar	102
3.3 pF	Ceramic		1500 pF/0.0015 µF	Ceramic/Mylar	152
10 pF	Ceramic		2000 pF/0.002 µF	Ceramic/Mylar	202
15 pF	Ceramic		2200 pF/0.0022 µF	Ceramic/Mylar	222
20 pF	Ceramic		4700 pF/0.0047 µF	Ceramic/Mylar	472
30 pF	Ceramic		5000 pF/0.005 µF	Ceramic/Mylar	502
33	Ceramic		5600 pF/0.0056 µF	Ceramic/Mylar	562
	Ceramic		6800 pF/0.0068 µF	Ceramic/Mylar	682
	Ceramic		0.01	Ceramic/Mylar	103
68 pF	Ceramic		0.015	Mylar	
75 pF	Ceramic		0.02	Mylar	203
82 pF	Ceramic		0.022	Mylar	223
91 pF	Ceramic		0.033	Mylar	333
100 pF	Ceramic	101	0.047	Mylar	473
120 pF	Ceramic	121	0.05	Mylar	503
130 pF	Ceramic	131	0.056	Mylar	563
150 pF	Ceramic	151	0.068	Mylar	683
180 pF	Ceramic	181	0.1	Mylar	104
220 pF	Ceramic	221	0.2	Mylar	204
330 pF	Ceramic	331	0.22	Mylar	224
470 pF	Ceramic	471	0.33	Mylar	334
560 pF	Ceramic	561	0.47	Mylar	474
680 pF	Ceramic	681	0.56	Mylar	564
750 pF	Ceramic	751	1	Mylar	105
820 pF	Ceramic	821	2	Mylar	205

Table 38-3 (*Continued*)

PicoFarad (pF)	NanoFarad (nF)	MicroFarad (mF, μF or mfd)	Capacitance Code
1000	1 or 1n	0.001	102
1500	1.5 or 1n5	0.0015	152
2200	2.2 or 2n2	0.0022	222
3300	3.3 or 3n3	0.0033	332
4700	4.7 or 4n7	0.0047	472
6800	6.8 or 6n8	0.0068	682
10000	10 or 10n	0.01	103
15000	15 or 15n	0.015	153
22000	22 or 22n	0.022	223
33000	33 or 33n	0.033	333
47000	47 or 47n	0.047	473
68000	68 or 68n	0.068	683
100000	100 or 100n	0.1	104
150000	150 or 150n	0.15	154
220000	220 or 220n	0.22	224
330000	330 or 330n	0.33	334
470000	470 or 470n	0.47	474

that the cathode of D1 is connected to pin 14 of U1. Install the diode on the PC board and then solder it in place. Remember to cut the extra remaining lead lengths from the diode.

The circuit contains three NPN transistors at Q1, Q2 and Q3. Transistors have three leads, a Base lead, a Collector lead, and an Emitter lead. The Base lead is the straight line symbol, while the Emitter lead has the small arrow pointing away from the transistor. The diagram in Figure 38-5, illustrates the pin-outs of the transistors, and the PIC microprocessor. Install the transistors on the PC board at their respective locations and then solder them in place. Remove the excess transistor leads after soldering.

Using an integrated circuit socket for a microprocessor project is highly recommended, in the event that you might want to update the program in the microprocessor. On occasion an IC may fail; in that situation, you can simply pull out the defective IC and replace it with a new one. An IC socket will typically have a small notch or cutout at one end of the plastic package. Just to the left of the notch or cutout you will find pin 1. In this circuit pin 1 is the (RA_2) transistor driver at Q1, while pin 18, which is opposite pin 1, will be connected to the "Computer Wins" (RA_7)LED.

The microprocessor IC will also have a small notch or cutout or indented circle at one end of the IC package. Again pin 1 will be just to the left of the notch or cutout. Be sure to line-up pin 1 of the IC with pin 1 of the IC socket when inserting the microprocessor IC into the socket. When handling the microprocessor IC be sure to wear your grounded anti-static wrist strap to avoid damaging the IC. High voltage static charges can easily build up from walking around your workbench and from getting up and down from your chair. A little insurance can go a long way in protecting your investment.

The only components requiring careful identification are the LEDs. The schematic shows how they are installed on to the board. There are nine bi-color LEDs arranged in a square, and two additional LEDs. The "Player" LED is red and the "Computer" LED is green. Push all the LEDs until they touch the board and make

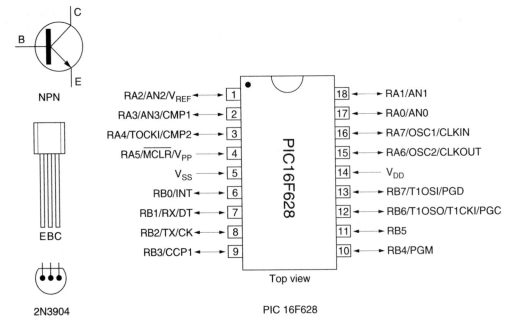

Figure 38-5 *Tic-Tac-Toe semiconductor pin-out diagram*

sure they are upright. Solder the leads very quickly to prevent damage to the light-emitting crystals inside the body of the LED. The 100 nf across the chip (to prevent spikes getting into the chip) has been mounted on the underside of the board. This component is a surface-mount capacitor (commonly called a "chip" capacitor) and will need a little care when soldering—mainly to prevent losing it on the floor!

All the other components are easy to fit and the transistors should be pushed up until they are slightly less than ¼ in (3 mm) from the board. The switch mounts on its side and don't forget the IC socket. We are one of the first suppliers to include an IC socket in all kits.

The tactile switches can only be fitted with the leads on the side. They can be fitted around either way as they are effectively double-contact devices. When all the components are soldered to the board, the battery box fits on the underside with a small piece of double-sided tape.

Once construction of the printed circuit has been completed, you can take a short break and when we return we will inspect the PC board for possible "cold" solder joints and "short" circuits. First, we will inspect the PC board for "cold" solder joints. Pick up the circuit board with the foil side of the board facing upwards toward you. Take a look at all of the solder joints, they should all appear to look clean, smooth and shiny. If you find that any of the solder joints look dull or "blobby"

then simply unsolder the joint, remove the solder and then re-solder the joint so that it looks clean, smooth and shiny. Next, we will inspect the circuit board for possible "short" circuits. Often "cut" or "stray" component leads will attach themselves to the underside of the board, and get "stuck" due to the sticky nature of the rosin core solder. Inspect the board and remove any component leads or wires that may have attached to the underside of the board during construction.

Install the battery into the battery holder and attach the battery clip and switch the project to the "on" position. The chip is pre-programmed and the *Attract Mode* will appear on the display.

Don't forget the 100 nF and jumpers: the 100 nF capacitor is a surface-mount type. It is soldered under the PC board in the kit version of this project.

Two jumpers are needed. They are different to a normal through-hole jumper. They are surface-mount jumpers.

The micro is constantly scanning the display and when Move or Play is pressed, it goes through a number of sub-routines that fixes the player piece on the board (changes the cursor into a red piece) and then places a computer piece (green) on the board. These operations are carried out so quickly that any interruption in the scanning of the display is not noticed.

How to Play

When the project is switched "on," the "*attract mode*" is displayed. Press pushbutton "A" to cancel the effect and introduce an orange flashing cursor. By pushing button "A," the cursor will move across the display. Press button "B" to change the cursor into a red LED to indicate your move. The computer will then make its move and show a green LED on the display. Press button "A" again to bring the cursor onto the display. Press "A" again to increment the position for your next "X." The computer will then make its move. Continue in this way to complete the game. The "Player Wins" or "Computer Wins" LED will turn on to indicate a winner. If both LEDs turn on the game is a draw.

Improving the Program

You will also learn a lot by reading through the program. The program has plenty of room for improvement. It has not been optimized nor has every strategy of the game been covered. These are things you can investigate. The whole object of this project is to teach the concepts of programming and games are one of the best ways to develop impressive sub-routines. Each sub-routine takes a number of possibilities and produces a result. The result may be a "nil result" and the micro goes to the next sub-routine. Even so, the generation of a "nil result" requires a functional sub-routine. Not only is the operation of each sub-routine important, but the order in which they are placed in the program is very important.

Chip "Burning"

In order to use your new Tic-Tic-Toe Game, you will have to "burn" the *tictac.hex* program into the PIC microprocessor chip. First, you may want to study the program in more detail if you are interested in how the program works. If you are familiar with microprocessor programming you will want to look over the *tictac.asm* program where you can follow the progression of the program from start to finish. If you are not interested in the programming aspect of the program then you can just load the *tictac.hex* program into the microprocessor chip directly. First, you will have to build the Multi-Chip Programmer project in this book or purchase a PIC microprocessor programmer. Next, you will have to connect the Multi-Chip programmer to your computer via a cable described in the programmer article. Now you will have to insert the "blank" PIC16F628 chip into socket in the programmer circuit. Since the Multi-Chip Programmer was designed for different chips, you will have to make sure that you insert your PIC chip correctly in the socket. Connect up the Multi-Chip programmer circuit to the serial port of your computer. Go to the http://www.books.mcgraw-hill.com/authors/petruzzellis/elecgames website and download the (*icprog105c-a.zip*) program into your computer. Unzip the file and open up the *icprog105c-a.exe* program. Locate the *tictac.hex* program from the website and open it up within the *icprog105c-a.exe* program and configure it in the following way:

```
;TicTacToe.asm
;Project: TIC TAC TOE
; use internal 4 MHz oscillator
LIST P=16F628 ;f=inhx8m
#include "P16F628.inc"

__CONFIG _CP_OFF &_BODEN_OFF
&_PWRTE_OFF &_WDT_OFF &_LVP_OFF
&_MCLRE_OFF &_INTRC_OSC_NOCLKOUT
```

You can select the boxes in IC Prog at the time of burning the chip. In all cases the chip will get burnt with the correct configuration value, and when it is copied the copy will automatically have the correct configuration value.

Once configured, run the *tictac.hex* file within the **IC-PROG** program, and the computer will "burn" the PIC chip with the Tic-Tac-Toe program. Once the program has been "burned" into the PIC chip, remove the PIC 16F628 chip and insert it into the IC socket on the Tic-Tac-Toe Game PC board and you are ready to play the Tic-Tac-Toe Game. Connect up the battery power supply, then turn "on" the Tic-Tac-Toe game. The game will be in the Attract Mode and you will have to switch to the "run" mode to begin playing the game. Have fun!

Electronic Kits Suppliers

Chaney Electronics (CH)

PO Box 4116

Scottsdale, Arizona 85261

1-800-227-7312

http://www.chaneyelectronics.com

Manufacturer of kits, sells wholesale

large quantity to schools

The Electronic Goldmine

PO Box 5408

Scottsdale, AZ 85261

http://www.goldmine-elec-products.com

Sells kits and electronic components

Distributor of Chaney Electronics kits

Ramsey Kits

590 Fishers Station Drive

Victor, N.Y. 14564

1-800446-2295

http://www.ramseykits.com

Retail sales of electronic kits

Jaycar Electronics Group (JC)

100 Silverwater Road

Silverwater NSW 2128

Australia

International: +61 2 9741 8555

http://www.jaycar.com.au/

Large selection of kits, many auto kits

Future Kits (FK)

Hobby World Center Co., Ltd

708 Happyland Soi 1, Ladprow Road Klongchun, Bangkapi, Bangkok 10240 Thailand

Phone: (662) 375 7740 / 1, Fax: (662) 734-3005, Mobile: (666) 342-9326

http://futurekit.com

Manufacturer of many kits

BAKATRONICS

PMB 328 994 North Colony Road, Wallingford, CT 06492 USA Tel: 203-213-4498 Fax: 203-237-7197, 860-291-8270

www.bakatronics.com

Distributor of Future Kits

Qkits

Quality Kits

49 McMichael Street

Kingston K7M 1M8

Canada

1-888-464-5487

Distributor of Future Kits

Talking Electronics (TE)

35 Rosewarne Ave., Cheltenham, 3192

Victoria, Australia.

Tel: +61 3 9584 2386 Fax: +61 3 9583 1854

email: talking@tpg.com.au

Elenco Electronics (EE)

150W. Carpenter Ave

Wheeling, IL 60090

1-847-541-3800

info@elenco.com

Test equipment and game and educational kits

Velleman-USA, Inc. (VM)

7354 Tower Street

Fort Worth, Texas 76118

817 284 7785

817 284 7712

US site : www.vellemanusa.com

Main site : www.velleman.be

Many electronics games and kits

Electronics Component Suppliers

DigiKey Electronics

701 Brooks Ave South

Thief River Falls, MN 56701

1-888-344-4539

http://www.digikey.com

Electronics Parts Supplier

Mouser Electronics

Mouser Electronics® | 1000 North Main Street | Mansfield, TX 76063-1514

(800)346-6873

http://www.mouser.com

Electronics parts supplier, no minimum order

Newark Electronics

Newark InOne

4801 N. Ravenswood

Chicago, IL 60640-4496

T: (773) 784-5100

F: (888) 551-4801

www.newarkinone.com

(1.800.463.9275)

Electronics parts supplier

NTE Electronics

44 Farrand Street

Bloomfield, NJ 07003

973-748-5089

Replacement semiconductors

Index

Index